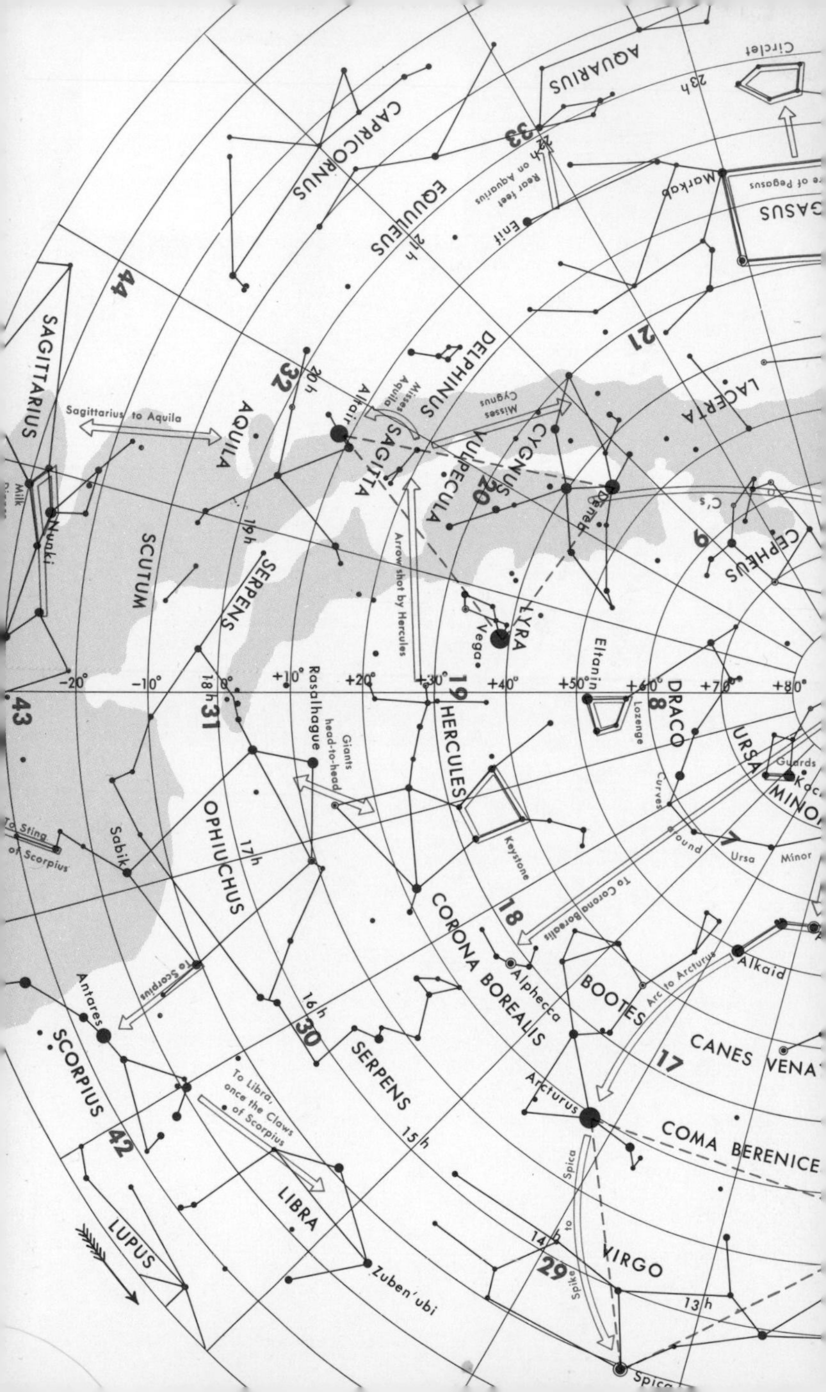

AQUARIUS
Circlet
23h
CAPRICORNUS
33
22h
Rear feet on Aquarius
Markab
of Pegasus
PEGASUS
EQUULEUS
21h
Enif
44
SAGITTARIUS
DELPHINUS
21
LACERTA
32
20h
Altair
Misses Aquila
Misses Cygnus
Sagittarius to Aquila
AQUILA
SAGITTA
VULPECULA
CYGNUS
20
Deneb
C's
9
CEPHEUS
Milk
Nunki
SCUTUM
19h
SERPENS
Arrow shot by Hercules
LYRA
Vega
Eltanin
-20°
-10°
0°
+10°
+20°
+30°
+40°
+50°
+60°
+70°
+80°
18h
31
43
19
Rasalhague
HERCULES
DRACO
8
Lozenge
URSA MINOR
Giants head-to-head
Curves
Guards
Kochab
Round
7
Ursa Minor
To Sting of Scorpius
Sabik
OPHIUCHUS
17h
Keystone
To Corona Borealis
18
CORONA BOREALIS
Alphecca
Alkaid
To Scorpius
Antares
BOOTES
Arc to Arcturus
16h
30
CANES VENATICI
SCORPIUS
17
SERPENS
To Libra, once the Claws of Scorpius
Arcturus
COMA BERENICES
15h
42
Spica
LIBRA
LUPUS
14h
29
Spike to
VIRGO
Zuben'ubi
13h
Spica

A Field Guide
to the
Stars and Planets

THE PETERSON FIELD GUIDE SERIES

EDITED BY ROGER TORY PETERSON

1. A Field Guide to the Birds by Roger Tory Peterson
2. A Field Guide to Western Birds by Roger Tory Peterson
3. A Field Guide to Shells of the Atlantic and Gulf Coasts and the West Indies by Percy A. Morris
4. A Field Guide to the Butterflies by Alexander B. Klots
5. A Field Guide to the Mammals by William H. Burt and Richard P. Grossenheider
6. A Field Guide to Pacific Coast Shells (including shells of Hawaii and the Gulf of California) by Percy A. Morris
7. A Field Guide to Rocks and Minerals by Frederick H. Pough
8. A Field Guide to the Birds of Britain and Europe by Roger Tory Peterson, Guy Mountfort, and P. A. D. Hollom
9. A Field Guide to Animal Tracks by Olaus J. Murie
10. A Field Guide to the Ferns and Their Related Families of Northeastern and Central North America by Boughton Cobb
11. A Field Guide to Trees and Shrubs (Northeastern and Central North America) by George A. Petrides
12. A Field Guide to Reptiles and Amphibians of Eastern and Central North America by Roger Conant
13. A Field Guide to the Birds of Texas and Adjacent States by Roger Tory Peterson
14. A Field Guide to Rocky Mountain Wildflowers by John J. Craighead, Frank C. Craighead, Jr., and Ray J. Davis
15. A Field Guide to the Stars and Planets by Donald H. Menzel
16. A Field Guide to Western Reptiles and Amphibians by Robert C. Stebbins
17. A Field Guide to Wildflowers of Northeastern and North-central North America by Roger Tory Peterson and Margaret McKenny
18. A Field Guide to the Mammals of Britain and Europe by F. H. van den Brink
19. A Field Guide to the Insects of America North of Mexico by Donald J. Borror and Richard E. White
20. A Field Guide to Mexican Birds by Roger Tory Peterson and Edward L. Chalif
21. A Field Guide to Birds' Nests (found east of Mississippi River) by Hal H. Harrison

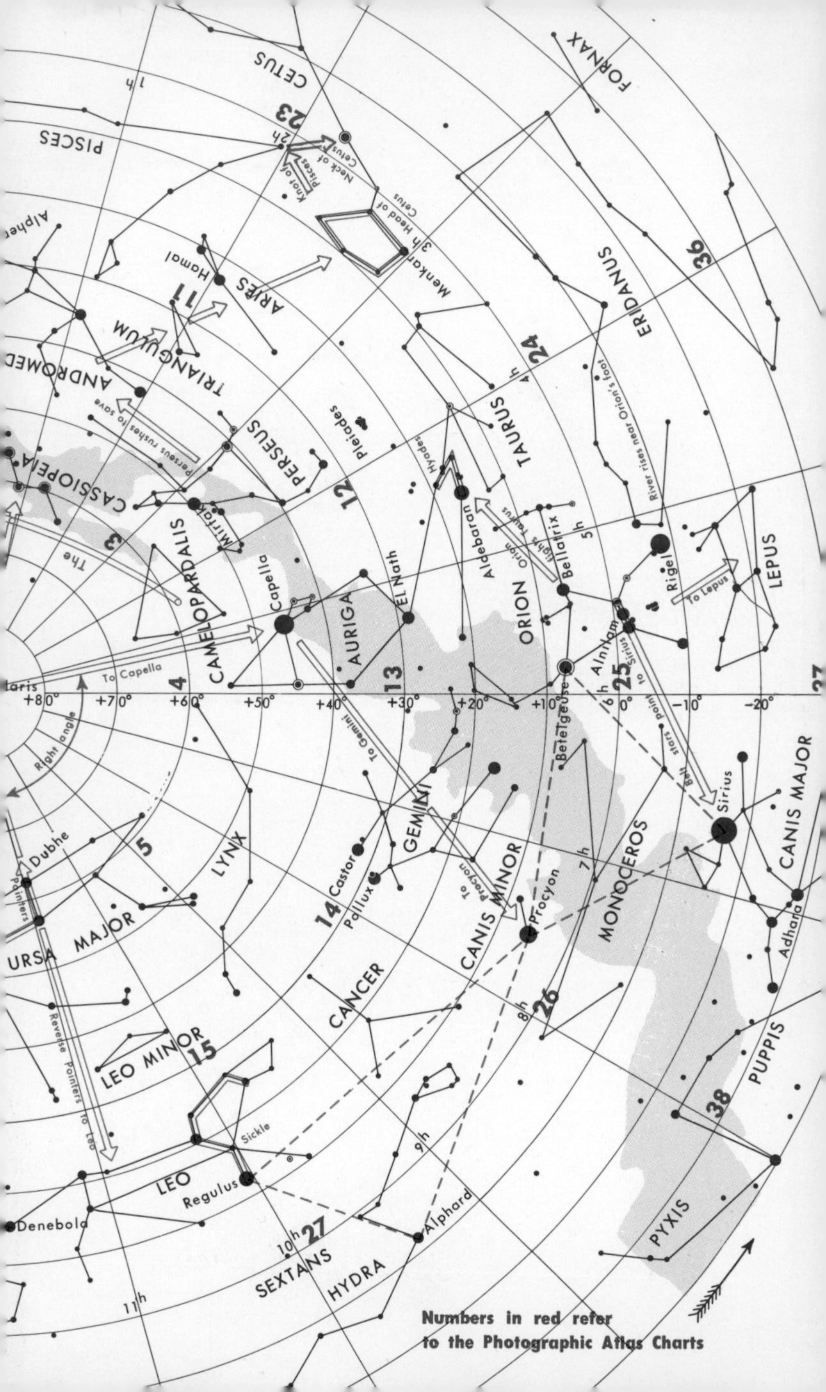

Numbers in red refer to the Photographic Atlas Charts

THE PETERSON FIELD GUIDE SERIES

A Field Guide to the Stars and Planets

Including the moon, satellites, comets, and other features of the universe

BY DONALD H. MENZEL

Former Director, Harvard College Observatory

With photographs, and with sky maps and other illustrations by Ching Sung Yü Professor of Astronomy, Hood College

HOUGHTON MIFFLIN COMPANY BOSTON

ISBN 0-395-07998-5 hardbound
ISBN 0-395-19422-9 paperbound
Library of Congress Catalog Card Number: 63-7017

Printed in the United States of America

V 16 15 14 13 12 11

Editor's Note

ALTHOUGH we expect never to publish a guide to the creatures of outer space, it is inevitable that the Field Guide Series should include this volume on recognition of the stars and planets.

Among all the inhabitants of the minor planet earth, man alone has systematically considered the heavenly bodies. He has given names to the constellations, has charted their relative positions and movements. Until recently he believed that he alone was able to navigate by celestial means, but now we learn through the experiments of E. G. Franz Sauer that nocturnal bird migrants apparently take their direction by means of an innate ability to read the night sky.

In recent years, and particularly since the development of rocket-launched satellites, more people than ever before have become aware of outer space and want to know what is out there. The tiny dots of light in the dome of the night sky may be obscured by city fog, but on clear nights they cannot fail to stir the inquiring mind.

The first step in astronomy, as in zoology, is to put names to things, to identify them. This *Field Guide* will facilitate the process and is equally usable for the observer depending on his naked eye, his binocular, or a small telescope. Unlike most of the books in the Field Guide Series, which tend to be regional in scope or at least confined to a single continent, it may be used at any point on the earth's surface and on any day of the year. In line with the general policy of the other *Field Guides*, emphasis has been put on new and simplified techniques of recognition. So that identification is made doubly certain, each map of the sky is presented twice — first with the recognition lines based on the imaginative system of H. A. Rey and then as things actually appear in the sky. Two unique features are (1) the complete photographic atlas of the sky made from plates taken at the Harvard College Observatory, over which the author of this book, Donald H. Menzel, presides; and (2) the most detailed atlas of the moon available in any portable guide.

The book should present no problems to the beginner interested in finding his way around in the heavens, but at the same time the completeness of its charts and tables should make it a useful tool for the serious amateur and even for the professional.

It is a joy to thumb through the book while relaxing in an armchair, but inasmuch as it is basically a field guide, put it to practical use. Use it on clear nights to interpret the free show put on by the heavens.

Roger Tory Peterson

Acknowledgments

I HAVE WRITTEN this book for people who wish to know more of the sky and its wonders. The lore of the stars and constellations goes back into antiquity. My efforts to reconstruct the origin of various star groups led me to write an article, "The Heavens Above," published in the *National Geographic Magazine* for July 1943. Since part of the present text is based on the material there presented, I wish to thank the editors of the *National Geographic Magazine* for permission to use this material. That article also contained a set of sky maps and charts of the major constellations. I originally intended to use these in the current book, but later decided that a somewhat different projection would give a better representation, over a wider range of latitude.

The constellation forms herein presented derive, for the most part, from the ingenious figures constructed by H. A. Rey for his books *The Stars* and *Find the Constellations*. I also acknowledge helpful suggestions concerning the history and forms of constellations from the late Mr. George Davis and Mr. William Barnholth. Mr. J. F. Chappell prepared the unusual Moon Maps from Lick Observatory photographs. The photographs of celestial objects are chiefly from Lick, Mount Wilson–Palomar, Harvard, Boyden, and Sacramento Peak Observatories. Other specific acknowledgments appear in the list of illustrations. I am indebted to Dr. George van Biesbroeck, of Yerkes Observatory, for advice in the selection of interesting double stars. Mrs. Lyle Boyd and Mrs. Margoret Smith gave valuable editorial assistance. Mrs. Menzel helped me especially in the collection and interpretation of data concerning the early history of the constellations. Mrs. Margaret Mayall furnished the basic data for the variable star charts, Figures 4–9, from the files of the American Association of Variable Star Observers. Mrs. Elinor West drew the imaginative Figures 21–23. Above all, I express my appreciation to Dr. Ching Sung Yü, Professor of Astronomy at Hood College, for his superb and painstaking art work.

DONALD H. MENZEL

Contents

	Editor's Note	v
	Acknowledgments	vii
I	Introduction: How to Use This Book	1
II	The Monthly Sky Maps	3
III	Pathways in the Sky	104
IV	Order and System of the Constellations	107
V	The Nature of the Stars and Nebulae	116
VI	The Photographic Atlas Charts	138
VII	The Moon	250
VIII	The Sun	285
IX	The Planets and Their Positions	292
X	Other Bodies of the Solar System	304
XI	The Telescope and How to Use It	309
XII	Photography in Astronomy	317
XIII	Time	322
	Appendixes	
	I Glossary	335
	II Bibliography	340
	III Tables	346
	Index	389

56. Photograph of Big Dipper Taken with Polaroid Land Camera 319
57. Northern Star Clock 327
58. Southern Star Clock 328
59. Precession 330
60. Precession and Nutation 330
61. Aberration of Light 331

Sky Maps
Northern Horizon, Numbers 1–24 8
Southern Horizon, Numbers 25–48

Photographic Atlas Charts Numbers 1–54 140

Moon Maps A and Numbers 1–12 256

Ecliptic Star Maps 302

Endpapers
Front: North Polar Region
Back: South Polar Region

List of Tables

(*see Appendix III*)

1. Curves Defining Mask for Different Latitudes 346
2. The Greek Alphabet 346
3. Stars Fainter than Magnitude 4.55 Included in the Sky Maps 347
4. Sky Map Numbers, Standard Times 347
5. The Constellations 348
6. Asterisms 351
7. The Brightest Stars 352
8. Spectral Classes and Star Colors 355
9. Variable Stars with Maxima Brighter than Magnitude 6.0 355
10. Double Stars (A) between Declinations +90° and −30° 360
11. Double Stars (B) between Declinations −30° and −90° 365
12. Open Star Clusters (oc) 367
13. Globular Star Clusters (gc) 370
14. Diffuse Galactic Nebulae (gn) 373
15. Planetary Nebulae (pn) 375
16. Galaxies (eg) 376
17. Harvard College Observatory Plates Used for Photographic Atlas Charts 379
18. Planetary Data 381
19. Latitudes of Mercury, Venus, and Mars at Unit Distance from the Earth 382
20. Latitudes of Jupiter and Saturn as Seen from the Sun 383
21. Meteor Showers 384
22. Recommended Telescopes 386
23. Recommended Cameras 386
24. Almanac Data, 1959 387
25. Code for Use with Table 24 388

A Field Guide to the Stars and Planets

I

Introduction: How to Use This Book

A Field Guide to the Stars and Planets is designed for the novice as well as for the advanced amateur. Since satellites have made man more conscious of outer space, people are turning as never before to astronomy to satisfy their curiosity about the stars, the sun, the moon, and the planets.

The key to understanding the mechanical universe lies in knowledge of the constellations, the star groups of antiquity which rapidly become old friends as you learn to recognize them. The monthly Sky Maps will facilitate the learning process and guide you through the sky. Read carefully the instructions for their use in Chapter II. Then follow the hints of Chapters III and IV for associating one star group with another in an orderly and consistent fashion.

When you have learned some of the major constellations, you may wish to probe the stars more deeply, with the aid of field glasses or telescope. The Photographic Atlas Charts of Chapter VI, with their annotations, will guide the student toward many of the interesting classes of celestial objects discussed in Chapter V and listed in the tables referred to in that chapter. The Atlas Charts also furnish a ready index to the identification of stars by their Greek letters. The guide lines on these photographs correspond to those used on the monthly Sky Maps. The tables discussed in Chapter V list the positions of the various objects in terms of right ascension and declination, a system of celestial coordinates analogous to latitude and longitude on earth. For details, see page 117.

Chapter VII contains an extensive annotated photographic atlas of the moon which will help the student become an expert selenographer, as the lunar equivalent of geographer is sometimes termed. Chapters VIII, IX, and X are, in essence, guidebooks to the solar system: the sun, the planets and their satellites, the asteroids, meteors, comets, and artificial satellites of the earth.

Chapter IX contains, as a special feature, new and simple charts for predicting the apparent positions of Mercury, Venus, Mars, Jupiter, and Saturn for centuries into the past or future.

Chapters XI and XII give directions for using a telescope and for taking photographs of the sky with ordinary cameras.

Finally, Chapter XIII gives a full explanation, for the more advanced amateur, of the intricacies of time. The development of the standardized year and calendar is first taken up. This is followed by the numerous varieties of solar time in common use, then sidereal time or star time, and full directions for calculating the sidereal time for a specified location. Such information enables the astronomer to point his equatorially mounted telescope skyward and find the sought-for object within or near the field of view.

At the back of this *Field Guide* will be found certain information to which the user of the book will constantly turn. Appendix I is a glossary of terms and Appendix II a selective bibliography for those who wish to increase further their knowledge of the heavens. The bibliography also gives the basic references used for the compilation of Tables 9–16. Works cited in the text are fully documented in the bibliography. Appendix III contains the 25 tables to which frequent reference is made throughout the book. It concentrates in one easily found place all the essential technical information for study of the sky. Tables of star positions for purposes other than navigation are generally referred to such equinoxes as 1850, 1900, 1950, and so on. This book has employed 1950 in the text and tables as the basic date unless otherwise stated.

This *Field Guide* could also serve as a textbook for constellation study in elementary courses of astronomy. And even the professional astronomer may find the carefully selected list of celestial objects useful in his work. The book need not be read consecutively except perhaps by those to whom astronomy is a new interest. Each reader will make use of the book according to his own purposes and degree of familiarity with the subject.

II

The Monthly Sky Maps

THE MONTHLY Sky Maps here presented are of special design to permit their use anywhere over the greater portion of the earth's inhabited surface. Since each map is constructed for a range of latitude, before you try to relate it to your sky you must determine the limits of the particular horizon visible to you. By way of analogy, imagine that you are on an airplane flying south. As you proceed, the northern stars will sink below the horizon behind you and southern stars will rise over the curved surface of the earth in front. The areas of northern and southern skies you can view at a given moment will depend on the position of your plane, and will change as the position of your plane changes.

Basically we have two sets of maps, one for observers in the Northern Hemisphere (northern observers) and one for observers in the Southern Hemisphere (southern observers). Also, for a given month and time of day we have two maps, one for the observer to use when he faces north and one when he faces south.

Each map appears twice, on opposite pages. The right-hand map represents the sky as it appears to the eye, without lines or designations, and indicates corresponding stellar magnitudes for the dots on the map. The left-hand map shows constellation and star names, has lines drawn on it to outline the forms of various star groups, and gives other significant reference marks.

The heavens present the appearance of a globe or vault over our heads. Mapping the stars, therefore, is like mapping the curved surface of the earth. It is impossible to chart without distortion the surface of a sphere on a flat map. Several methods of mapping exist which minimize the effect. The one used here — stereographic projection — preserves the shape of small areas, of the constellations themselves, for example; but away from the central point the areas tend to increase, and to look relatively larger than they actually are.

Only two points on each map remain fixed, whatever the latitude of the observer may be: the east and west points of the horizon, designated E and W. Each map represents one half of the visible hemisphere, from either the north or south horizons up to the circle drawn from east to west directly through the zenith, a line called the *prime vertical.*

To draw in his horizon, a northern observer located at latitude

+35° merely connects the points marked E and W by a straight line. The prime vertical, then, is just a semicircle erected on this line, with E and W at opposite ends of the diameter. The zenith lies on the circumference, halfway between E and W. The center of the straight line represents either the north or south points of the horizon, according to the direction one faces.

Some observers will wish to draw light pencil lines to mark the horizon and prime vertical. Others may find it useful to cut a semicircular window, with a radius of 2¾ inches, in a black piece of paper and use it as a mask to hide those stars lying outside the specified region; but most people will find the continuity beyond this boundary helpful, since certain of the constellations would otherwise be cut in two.

The mask is useful, however, in another way. There are only 12 basic maps, one for each month; and since there are 24 hours in a day, you will probably want to look at the stars at some time other than the one specified for the map. The arrows along the curved upper boundary indicate the direction in which the stars appear to move. You can simulate this motion by sliding the mask in the opposite direction. On the left-hand map of each pair you will find a circle labeled *equator*. The index marks along it, including the points E and W, represent the amount of motion in an hour. For one hour earlier than the given time, slide the mask *westward along the equator* until the curved corners touch the hour index. For intermediate times move the mask proportionally.

A northern observer at any latitude other than 35° will need two such masks, one for the northern and one for the southern view. As mentioned above, the position of the horizon and prime vertical depends on the latitude of the observer. Figure 1 displays the nature of the mask for different latitudes. Superpose a piece of semitransparent paper and carefully outline the two masks appropriate for your latitude. The lines A, B, through I represent the horizons and the arcs *a*, *b*, through *i* the corresponding prime verticals for the different latitudes. Portions of the vertical circle *i* lie beyond the boundaries of the map. The limits of the mask for various latitudes appear in Table 1.

Southern observers follow the same procedure as the northern, except that the straight horizon lines occur for latitude 25° instead of 35°. This difference arises because the maps allow for the fact that land masses of the Southern Hemisphere tend to lie nearer the equator than do those of the Northern Hemisphere.

The Sky Maps have been carefully constructed from astronomical catalogs. The size of each circle indicates approximately the brightness of the star, as shown by the scales on the right-hand maps. Stars of 0 or 1st magnitude are among the brightest that we can see. A star of 6th magnitude is barely visible to the naked eye under the best conditions. The Sky Maps are complete down to magnitude 4.55. A few stars fainter than this limit have been

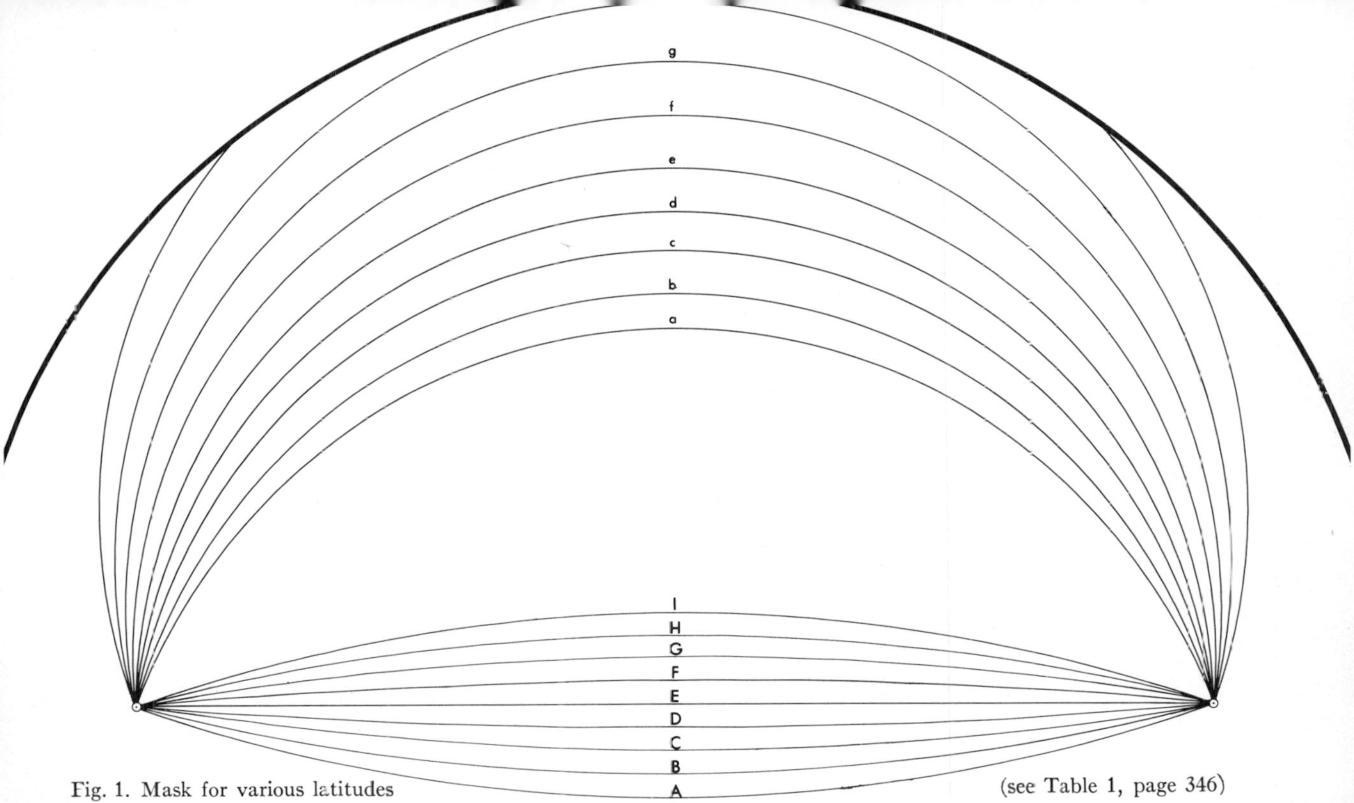

Fig. 1. Mask for various latitudes

(see Table 1, page 346)

arbitrarily included when they are vital for recognition of the constellation figure. These stars are listed in Table 3.

Many of the stars in the sky, for one reason or another, are not constant in brightness. Such stars are indicated by an outer ring of diameter appropriate for maximum magnitude. So great is the range of variation at times that failure to indicate such a star might cause serious confusion. Some become completely invisible to the naked eye. The observer may find it interesting to follow the fluctuations in brightness of many of these peculiar objects. For details see Table 9, which lists all variable stars down to 6.0 magnitude.

The lines connecting the individual stars of the constellations have been put in to help you in the recognition and memorization of the outstanding patterns.

Before you begin, you must have a rough idea of direction. A pocket compass is useful, but remember that magnetic north will not generally coincide with true north. The sun rises exactly in the east and sets exactly in the west on or about March 21 and September 21. During the (northern) summer it rises somewhat north of true east, and during the winter somewhat south. But even an approximate orientation will suffice. For then you will shortly locate the Big Dipper and the North Star. When you face east, north will lie to your left. In the Southern Hemisphere you will need other reference points, because there is no South Star.

To use the Sky Maps at night you will need a flashlight, and to minimize the glare, which will reduce the sensitivity of your eyes for the perception of faint starlight, cover the lens with a layer or two of dark red cellophane, held on with a rubber band.

First of all, you must select the *pair* of maps appropriate for your location and time of day. Sky Maps 1–24 are for observers in the Northern Hemisphere and 25–48 for observers in the Southern Hemisphere. Therefore, if you live north of the equator you may disregard the second set, at least until your knowledge of northern constellations piques your curiosity about the star groups visible only from the Southern Hemisphere. The observer will use the odd-numbered maps when he faces north and the even-numbered ones when he faces south. Table 4 lists the appropriate pair of charts to be used for various months at different times of day. As the earth moves around its orbit, the sun drifts slowly eastward across the sky. Consequently, the stars rise about 4 minutes earlier each day. This interval adds up to 1 hour for 15 days, to 2 hours for a month, and to an entire day in the course of a year.

The effect is clearly shown by the diagonal shift of map numbers across Table 4. For example, Sky Maps 7 and 8, designed for 11:30 P.M. on April 1, also apply for 1:30 A.M. on March 1 or 9:30 P.M. on May 1, etc. Using Table 4, allow 4 minutes per day for intermediate days. Thus for April 11 we subtract $4 \times 11 = 44$

minutes from the tabulated time and get 10:46 P.M. as the specific time to use Sky Maps 7 and 8. The figures represent standard, not daylight, time. If you live in a region operating on daylight time, subtract 1 hour from your clock time to get standard time. In the strictest sense we should reduce the Standard Mean Time (SMT) to Local Mean Time (LMT) by correcting for the difference between the longitude of the observer and that of the standard meridian (see p. 324). Standard time, however, is usually amply accurate for the general location and identification of the constellations.

Many of the brighter stars have names: Vega, Polaris, Sirius, etc. These stars as well as the fainter ones have other names derived from constellations in which they lie. Within each group astronomers have assigned the Greek letters α, β, γ, δ to the stars. The letter is followed by the *genitive* of the Latin name. Thus γ Ursae Majoris means "gamma of Ursa Major," and so on. Table 2 gives the Greek alphabet and Table 5 the Latin names and the standard three-letter abbreviations, the genitives of these names, the English equivalent, and the page on which the description occurs. Fainter stars are known by numbers or Latin letters. A given star may thus have a number of designations and the choice of which one to use is arbitrary. To avoid complexities, monthly Sky Maps use only the star's assigned name. The Atlas Charts, however, give first priority to the Greek letters, second priority to the lower-case Latin letters assigned by Bayer, next priority to the Flamsteed numbers, and lowest priority to upper-case Latin letters, except those assigned to variable stars.

Although to some extent astronomers have assigned the earlier Greek letters to the stars roughly in order of their brightness, there are many exceptions. For example, the seven Dipper stars carry, in order, the letters α–η around the outline. Also, when the ancient constellation Argo Navis the Ship was cut up to make the four modern groups (Puppis the Stern, Carina the Keel, Vela the Sails, and Pyxis the Compass), no reassignment of letters occurred.

Many groups of stars or geometric patterns are of interest and beauty even though they have not been raised to the dignity of a constellation. The most familiar examples of these "asterisms" include: the Big Dipper, the Pleiades, and the Hyades. Indeed, Ptolemy regarded the second of these as an independent constellation. Table 6 lists the better-known asterisms.

Do not try to learn too many constellations in one night. Review your knowledge frequently. Find the star groups by means of the pathways described in Chapter III and relate them with the aid of myths and other associations given in Chapter IV. The diagrams appearing as endpapers of this book should be a helpful guide. The endpapers also can serve as monthly maps for observers near the equator. The horizon line is a diameter of the circle.

SKY MAP 1

Northern Hemisphere
North Horizon

TIME
January 1 at 11:30 P.M.
January 15 at 10:30 P.M.
January 30 at 9:30 P.M.
etc.

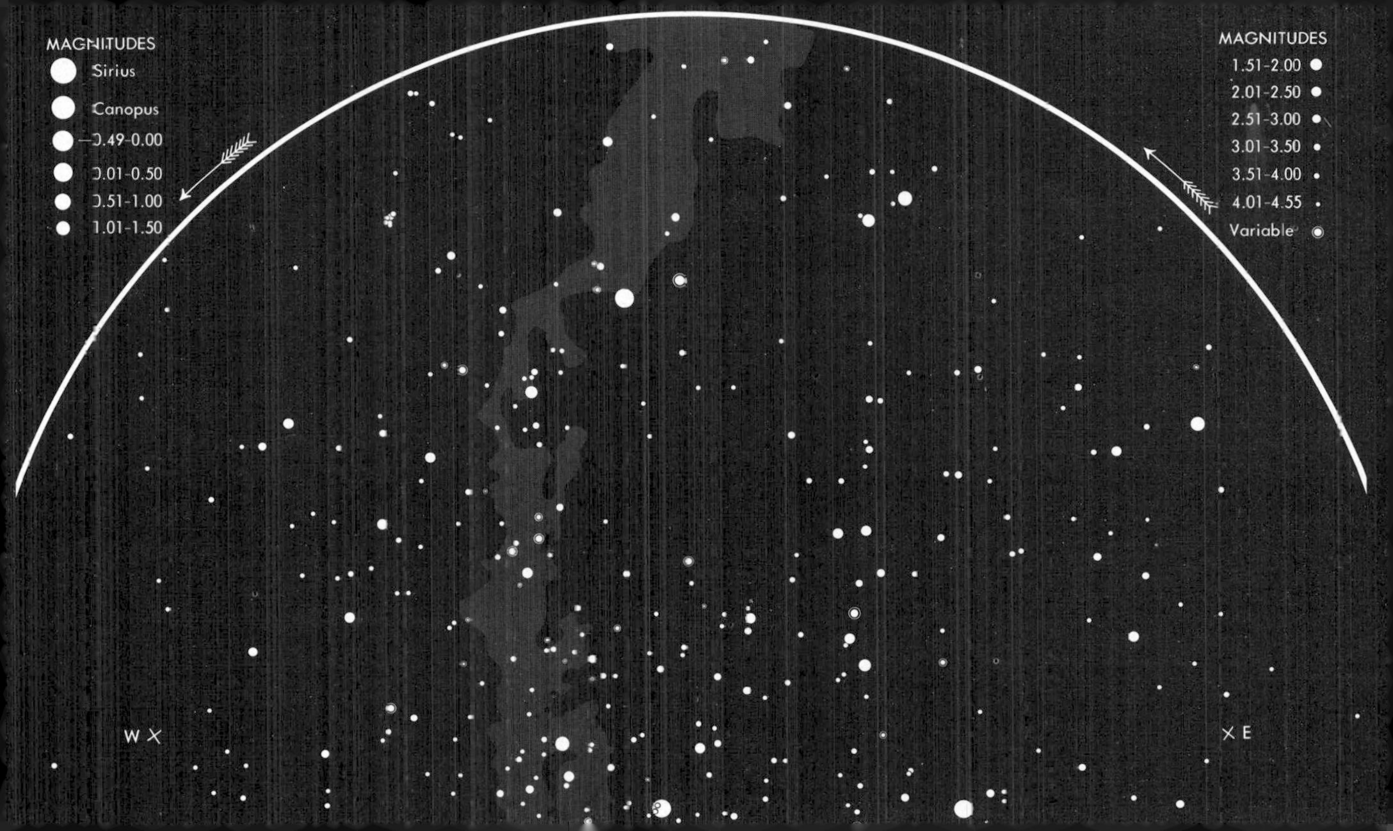
MAGNITUDES
Sirius
Canopus
−0.49–0.00
0.01–0.50
0.51–1.00
1.01–1.50
MAGNITUDES
1.51–2.00
2.01–2.50
2.51–3.00
3.01–3.50
3.51–4.00
4.01–4.55
Variable
W
E

SKY MAP 2

Northern Hemisphere
South Horizon

TIME

January 1 at 11:30 P.M.
January 15 at 10:30 P.M.
January 30 at 9:30 P.M.
etc.

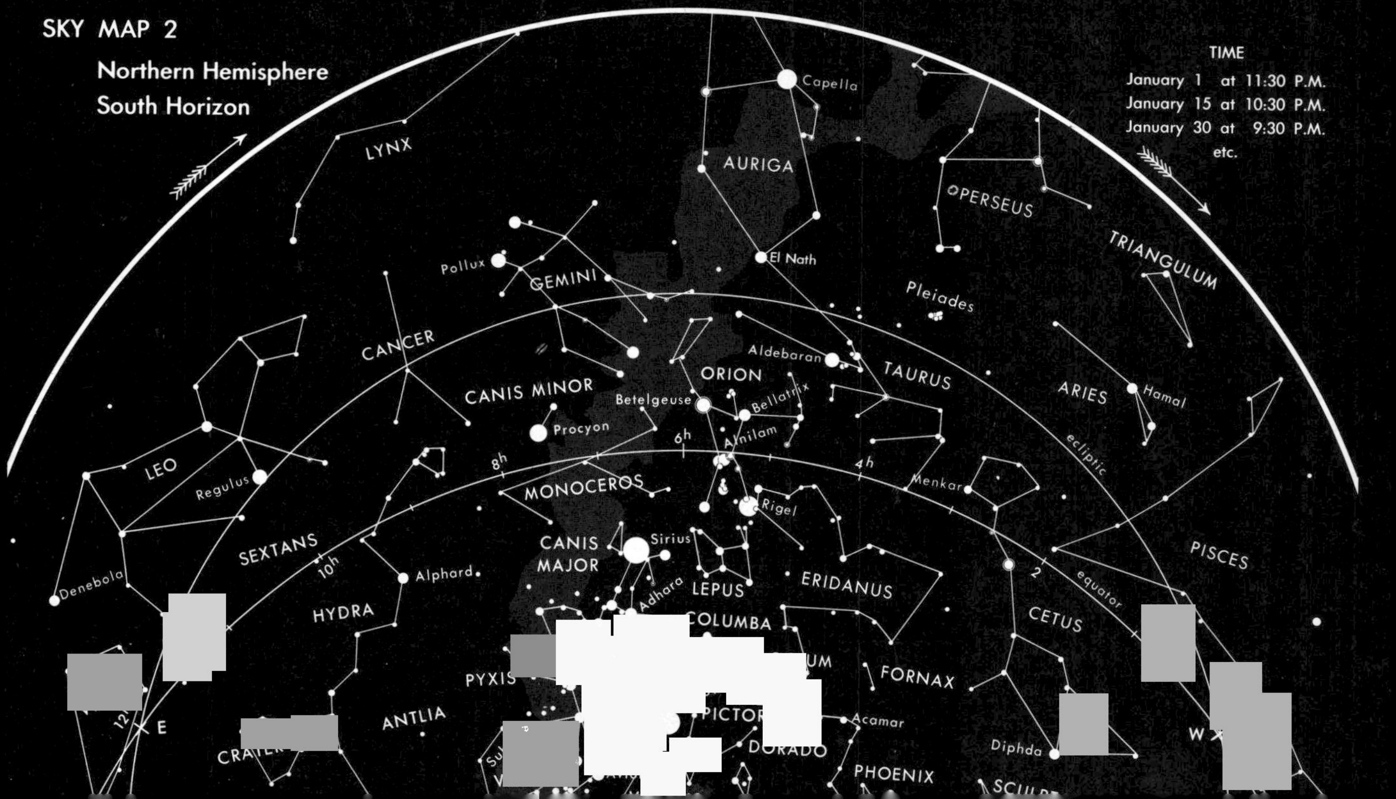

MAGNITUDES
Sirius
Canopus
—0.49-0.00
0.01-0.50
0.51-1.00
1.01-1.50
MAGNITUDES
1.51-2.00
2.01-2.50
2.51-3.00
3.01-3.50
3.51-4.00
4.01-4.55
Variable
E
W

SKY MAP 3

Northern Hemisphere
North Horizon

TIME
February 1 at 11:30 P.M.
February 15 at 10:30 P.M.
March 2 at 9:30 P.M.
etc.

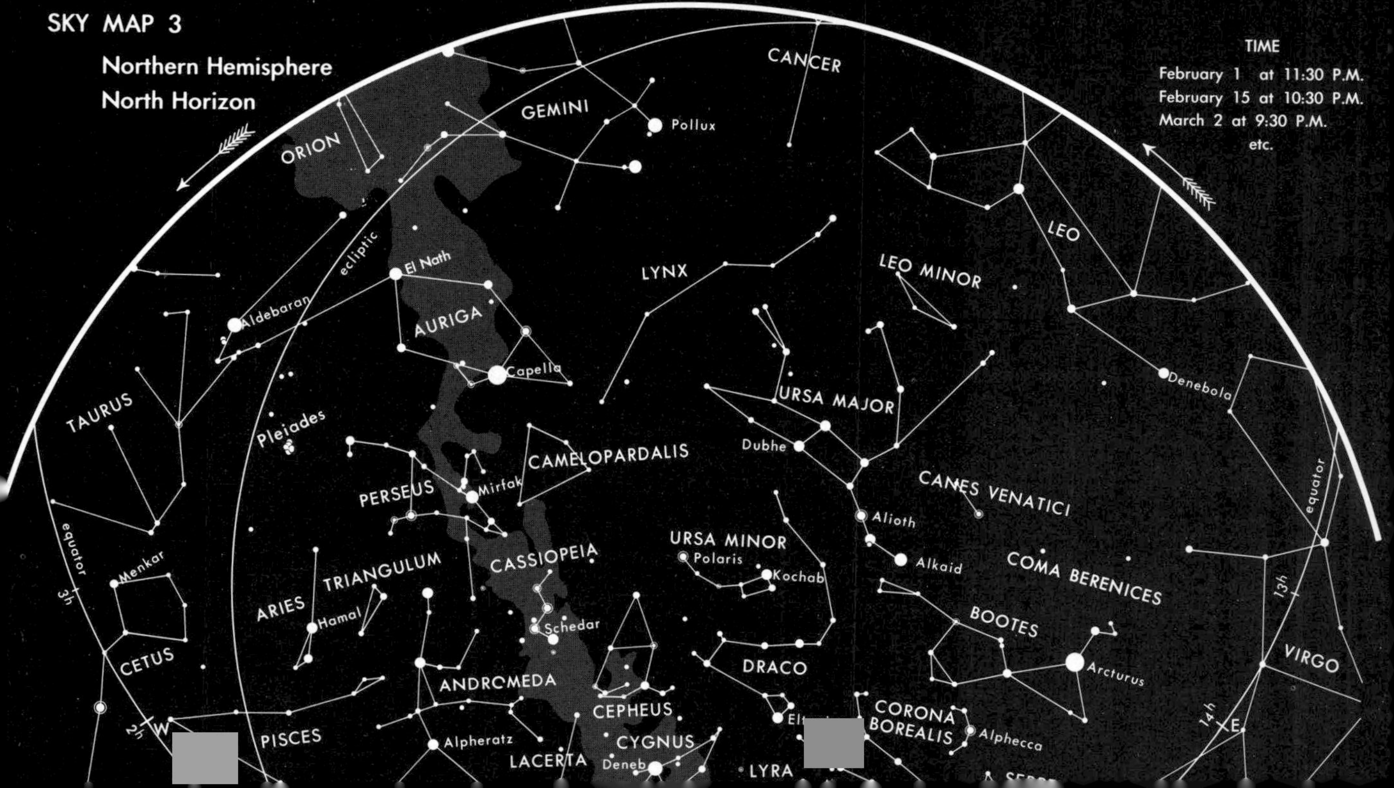

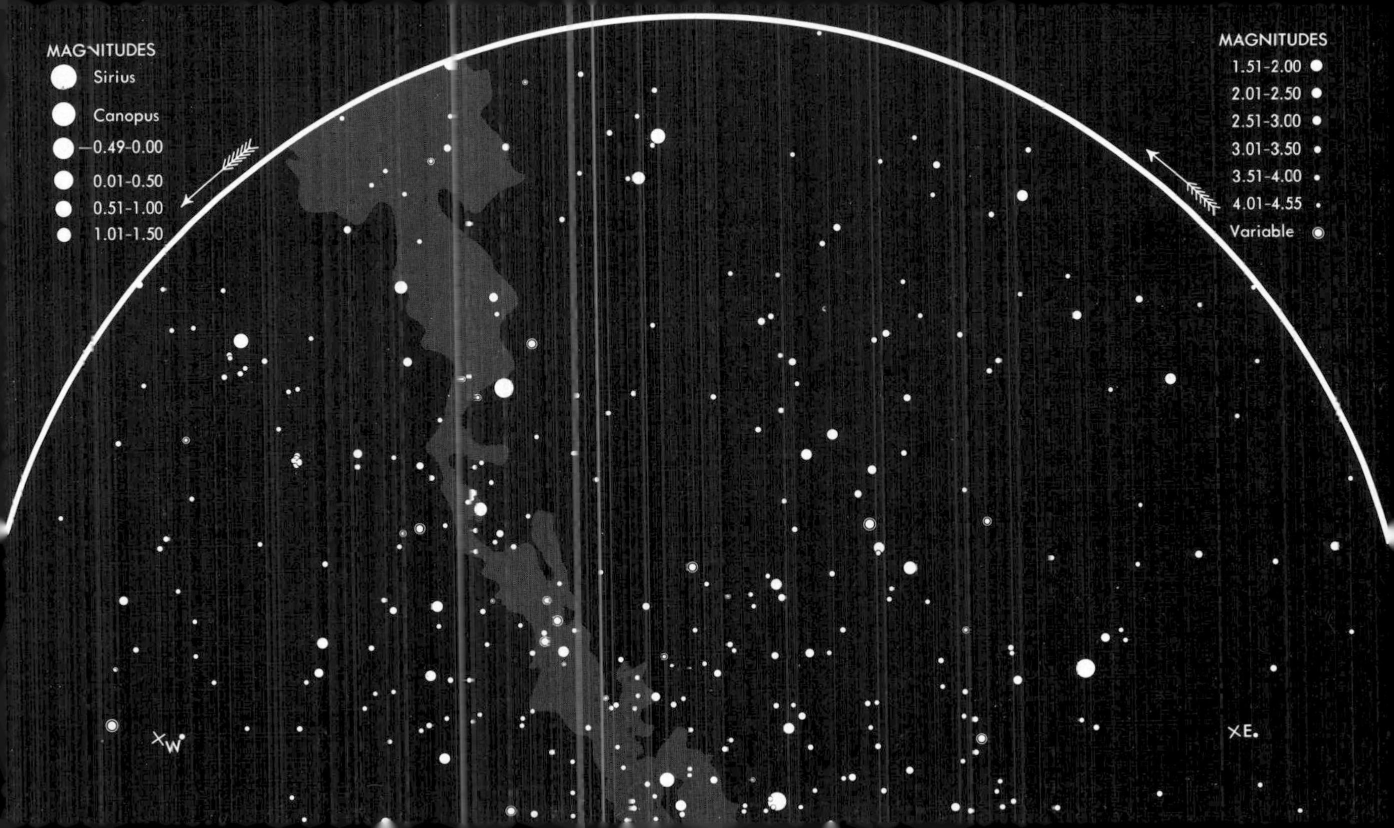
MAGNITUDES
Sirius
Canopus
—0.49-0.00
0.01-0.50
0.51-1.00
1.01-1.50
MAGNITUDES
1.51-2.00
2.01-2.50
2.51-3.00
3.01-3.50
3.51-4.00
4.01-4.55
Variable
W
E.

SKY MAP 4

Northern Hemisphere
South Horizon

TIME
February 1 at 11:30 P.M.
February 15 at 10:30 P.M.
March 2 at 9:30 P.M.
etc.

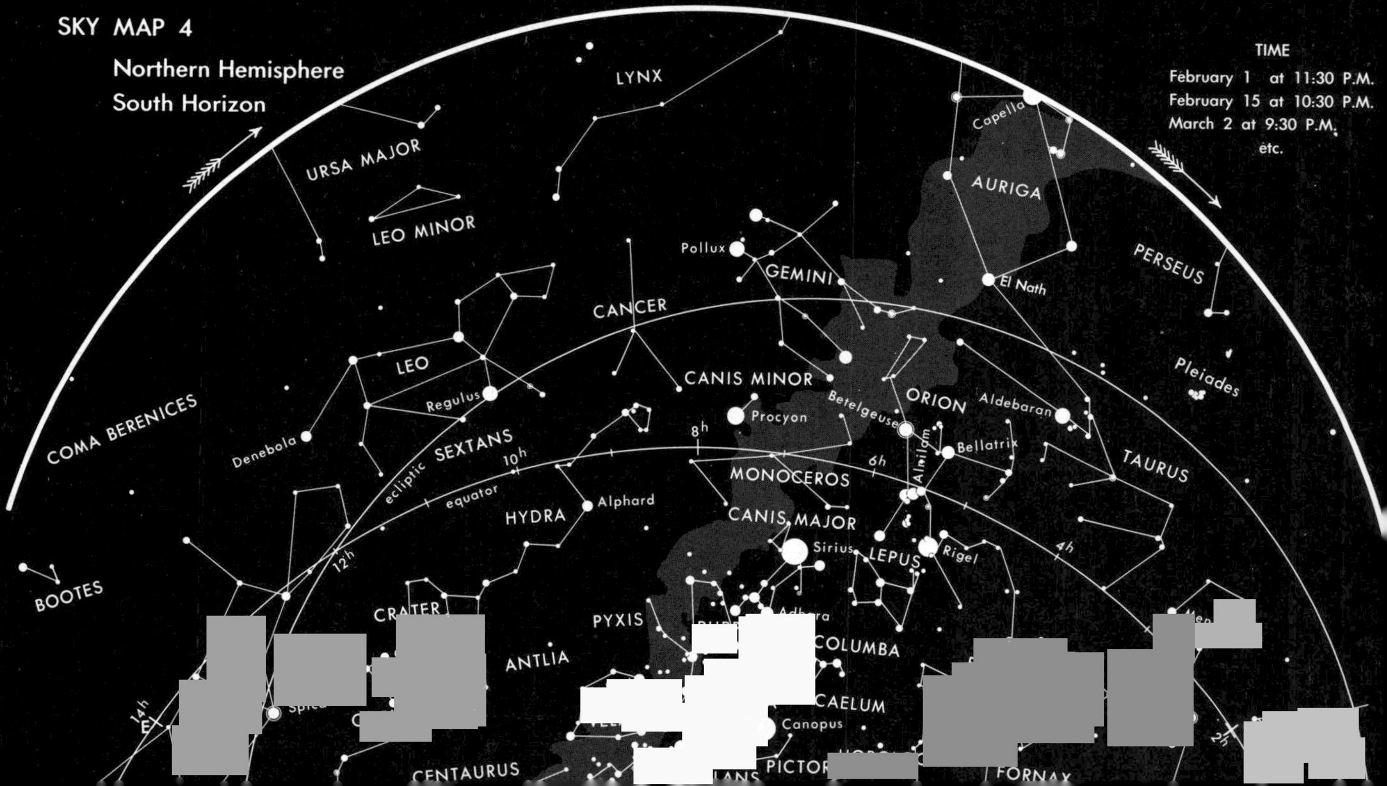

MAGNITUDES
Sirius
Canopus
—0.49-0.00
0.01-0.50
0.51-1.00
1.01-1.50
MAGNITUDES
1.51-2.00
2.01-2.50
2.51-3.00
3.01-3.50
3.51-4.00
4.01-4.55
Variable
E
W

SKY MAP 5

Northern Hemisphere

North Horizon

TIME

March 1 at 11:30 P.M.

March 15 at 10:30 P.M.

March 30 at 9:30 P.M.

etc.

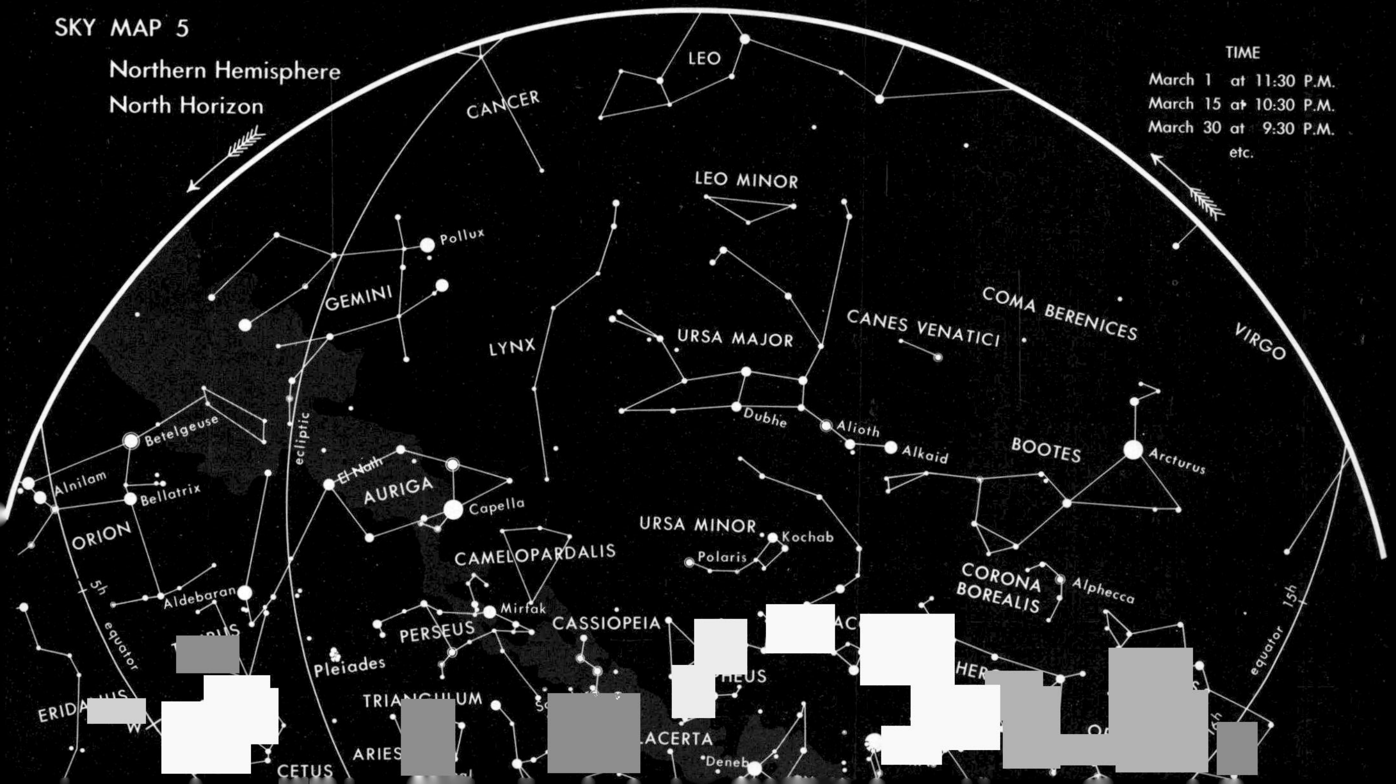

MAGNITUDES
Sirius
Canopus
—0.49-0.00
0.01-0.50
0.51-1.00
1.01-1.50
MAGNITUDES
1.51-2.00
2.01-2.50
2.51-3.00
3.01-3.50
3.51-4.00
4.01-4.55
Variable
W
E

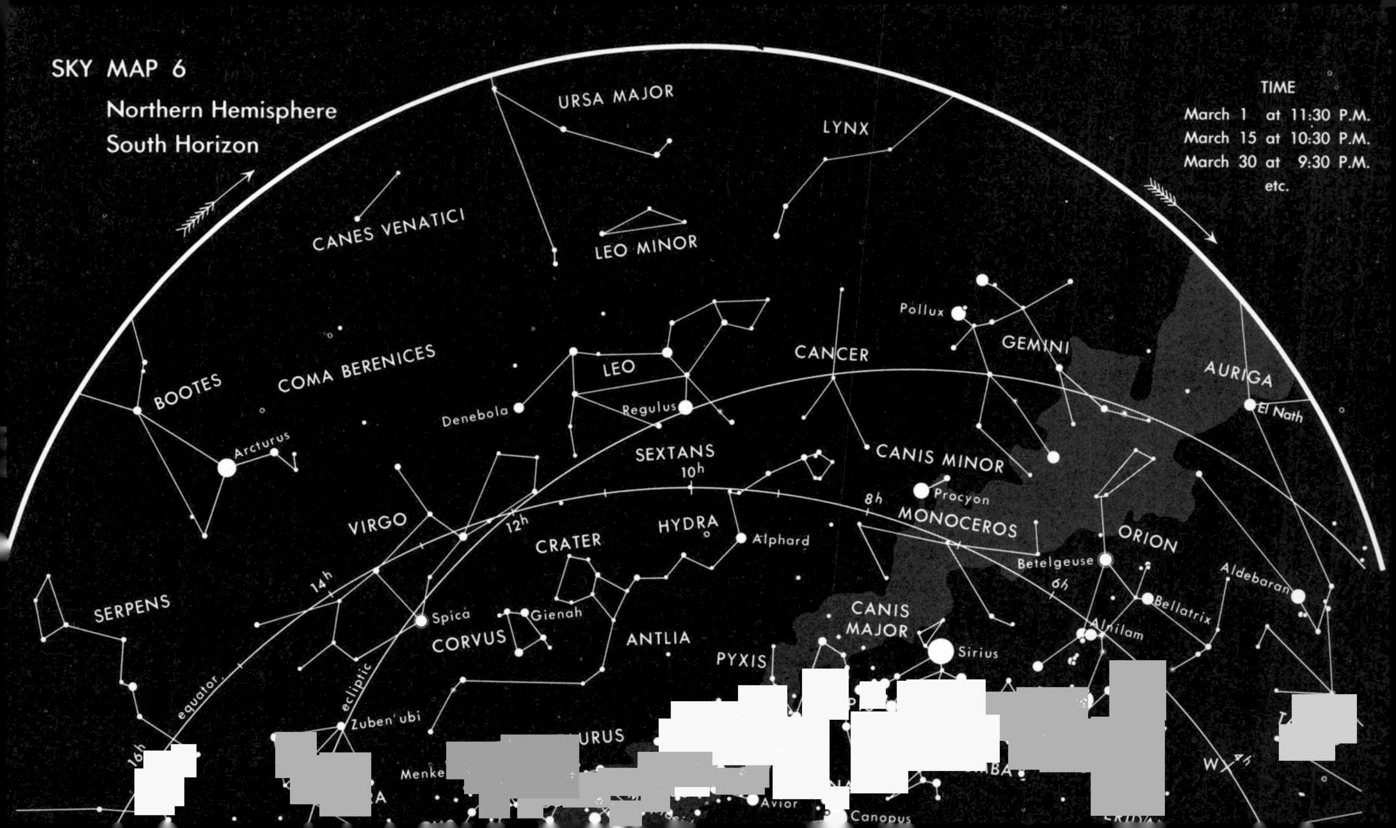
SKY MAP 6
Northern Hemisphere
South Horizon
TIME
March 1 at 11:30 P.M.
March 15 at 10:30 P.M.
March 30 at 9:30 P.M.
etc.
URSA MAJOR
LYNX
CANES VENATICI
LEO MINOR
BOOTES
COMA BERENICES
LEO
CANCER
GEMINI
AURIGA
Pollux
El Nath
Denebola
Regulus
Arcturus
SEXTANS
10h
CANIS MINOR
Procyon
8h
MONOCEROS
ORION
Betelgeuse
Aldebaran
Bellatrix
Alnilam
6h
VIRGO
12h
CRATER
HYDRA
Alphard
14h
SERPENS
Spica
Gienah
CORVUS
ANTLIA
PYXIS
CANIS MAJOR
Sirius
equator
ecliptic
Zuben'ubi
16h
Avior
Canopus

MAGNITUDES
Sirius
Canopus
−0.49–0.00
0.01–0.50
0.51–1.00
1.01–1.50
MAGNITUDES
1.51–2.00
2.01–2.50
2.51–3.00
3.01–3.50
3.51–4.00
4.01–4.55
Variable
E
W

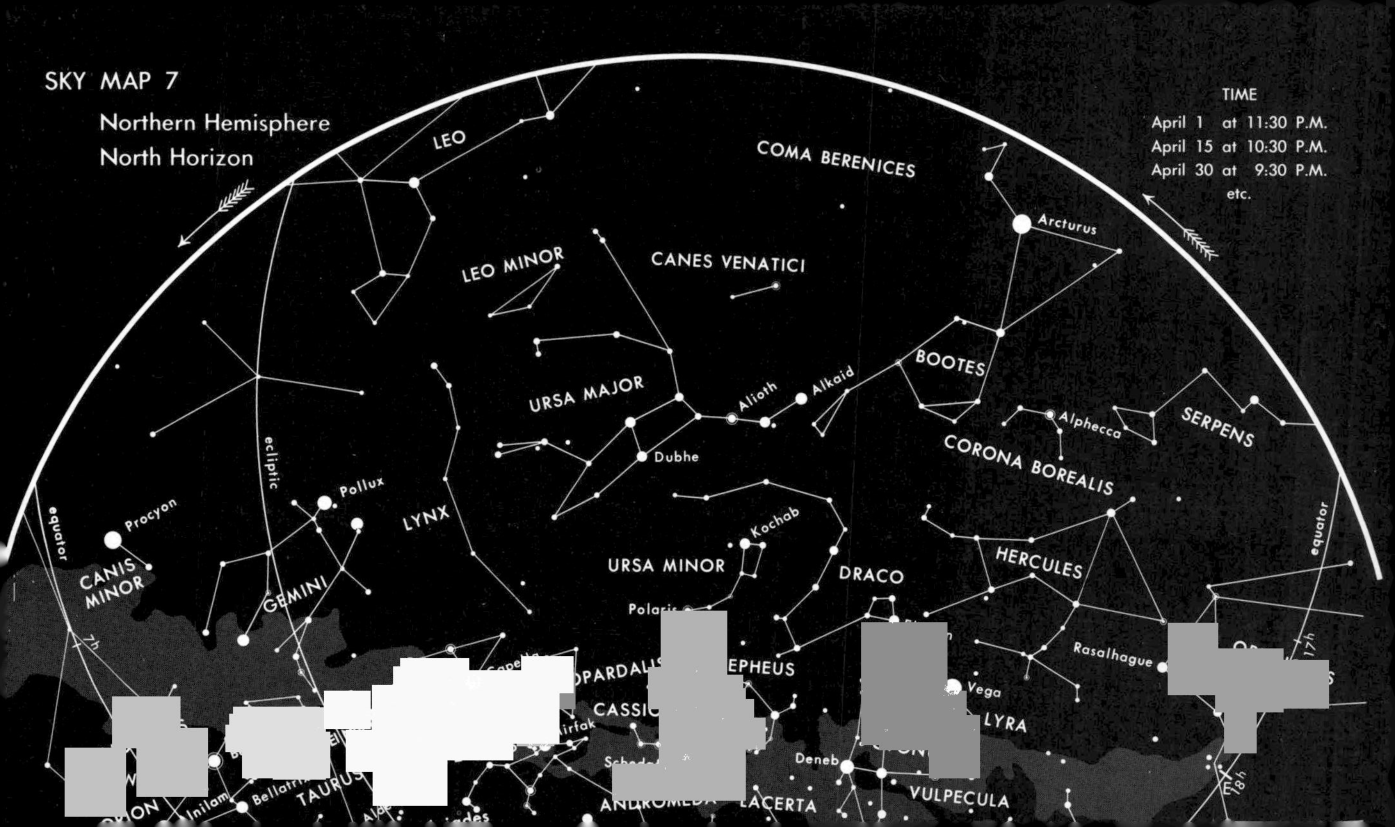
SKY MAP 7
Northern Hemisphere
North Horizon
TIME
April 1 at 11:30 P.M.
April 15 at 10:30 P.M.
April 30 at 9:30 P.M.
etc.
LEO
COMA BERENICES
Arcturus
LEO MINOR
CANES VENATICI
BOOTES
URSA MAJOR
Alioth
Alkaid
Alphecca
SERPENS
CORONA BOREALIS
Dubhe
ecliptic
Pollux
LYNX
Procyon
equator
CANIS MINOR
GEMINI
Kochab
URSA MINOR
DRACO
HERCULES
Polaris
Rasalhague
Vega
LYRA
Deneb
VULPECULA
LACERTA
TAURUS
Bellatrix
7h
17h
18h

MAGNITUDES
Sirius
Canopus
−0.49–0.00
0.01–0.50
0.51–1.00
1.01–1.50
MAGNITUDES
1.51–2.00
2.01–2.50
2.51–3.00
3.01–3.50
3.51–4.00
4.01–4.55
Variable
W
E

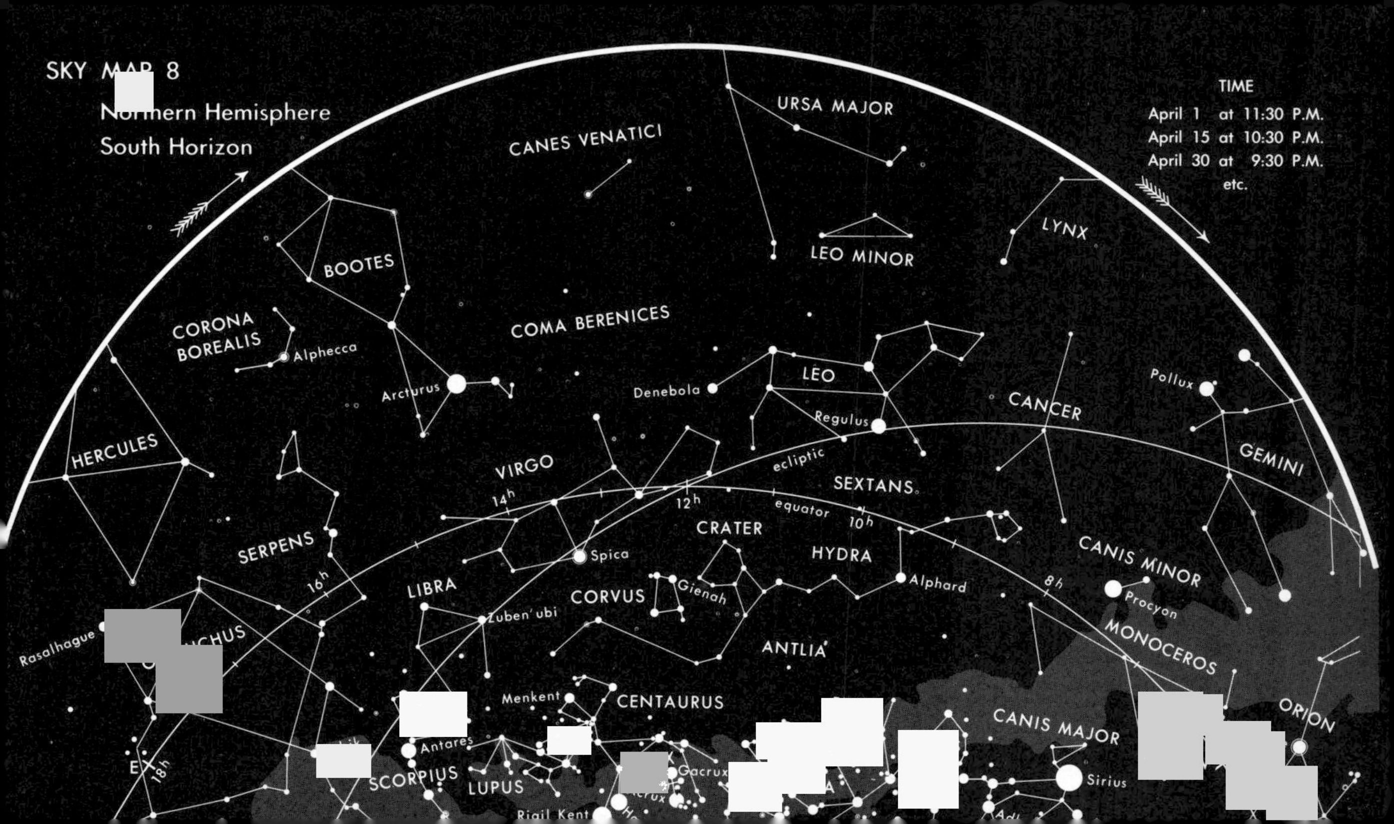

SKY MA[redacted] 8
N[redacted]rthern Hemisphere
South Horizon
TIME
April 1 at 11:30 P.M.
April 15 at 10:30 P.M.
April 30 at 9:30 P.M.
etc.
URSA MAJOR
CANES VENATICI
LEO MINOR
LYNX
BOOTES
CORONA BOREALIS
Alphecca
COMA BERENICES
Arcturus
LEO
Denebola
Regulus
CANCER
Pollux
GEMINI
HERCULES
VIRGO
ecliptic
SEXTANS
14h
12h
equator
10h
CRATER
HYDRA
SERPENS
Spica
CANIS MINOR
16h
LIBRA
CORVUS
Gienah
Alphard
8h
Procyon
Zuben'ubi
MONOCEROS
Rasalhague
[redacted]CHUS
ANTLIA
Menkent
CENTAURUS
CANIS MAJOR
ORION
Antares
SCORPIUS
LUPUS
Gacrux
Sirius
18h
Rigil Kent

MAGNITUDES
Sirius
Canopus
—0.49-0.00
0.01-0.50
0.51-1.00
1.01-1.50
MAGNITUDES
1.51-2.00
2.01-2.50
2.51-3.00
3.01-3.50
3.51-4.00
4.01-4.55
Variable
E
W

SKY MAP 9

Northern Hemisphere
North Horizon

TIME	
May 1	at 11:30 P.M.
May 15	at 10:30 P.M.
May 30	at 9:30 P.M.
etc.	

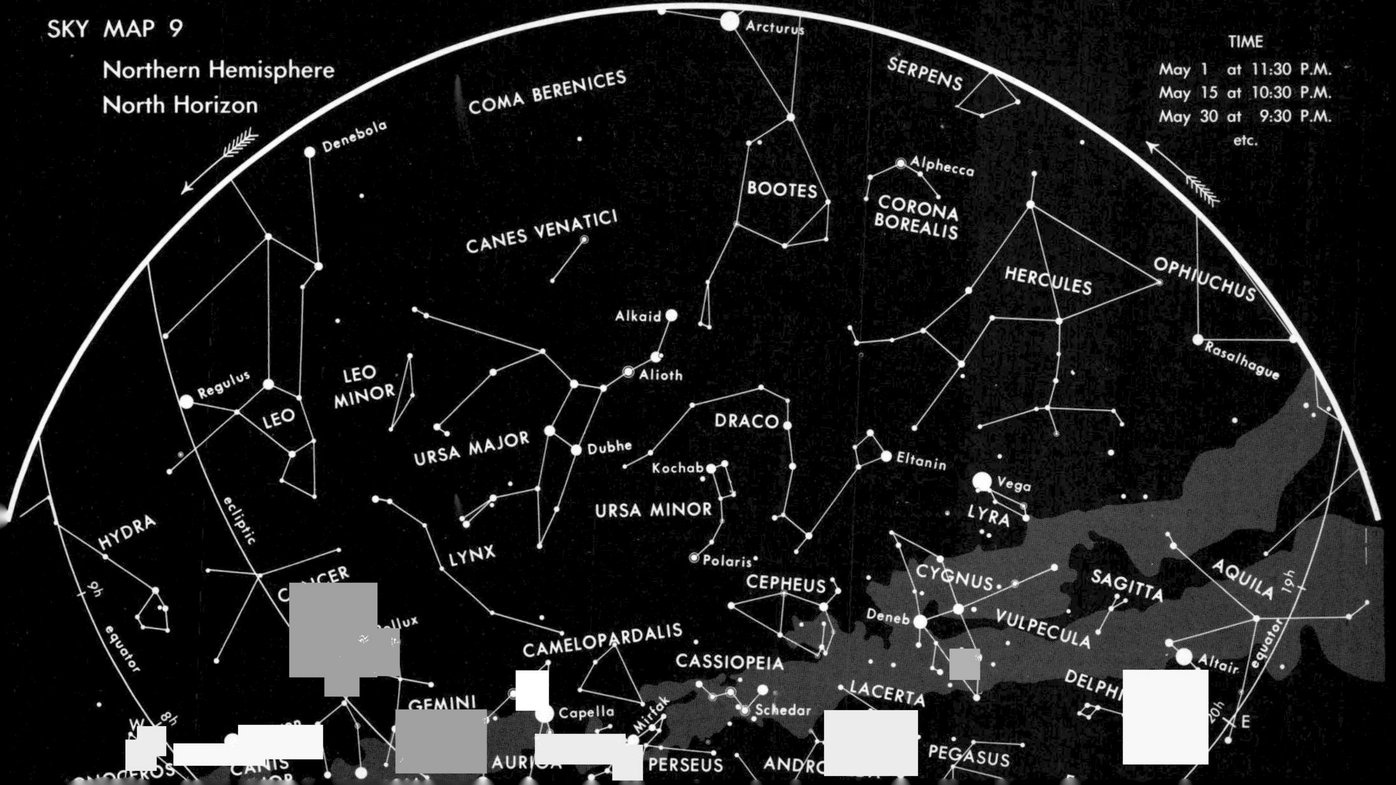

MAGNITUDES
Sirius
Canopus
−0.49-0.00
0.01-0.50
0.51-1.00
1.01-1.50
MAGNITUDES
1.51-2.00
2.01-2.50
2.51-3.00
3.01-3.50
3.51-4.00
4.01-4.55
Variable
W
E

SKY MAP 10

Northern Hemisphere
South Horizon

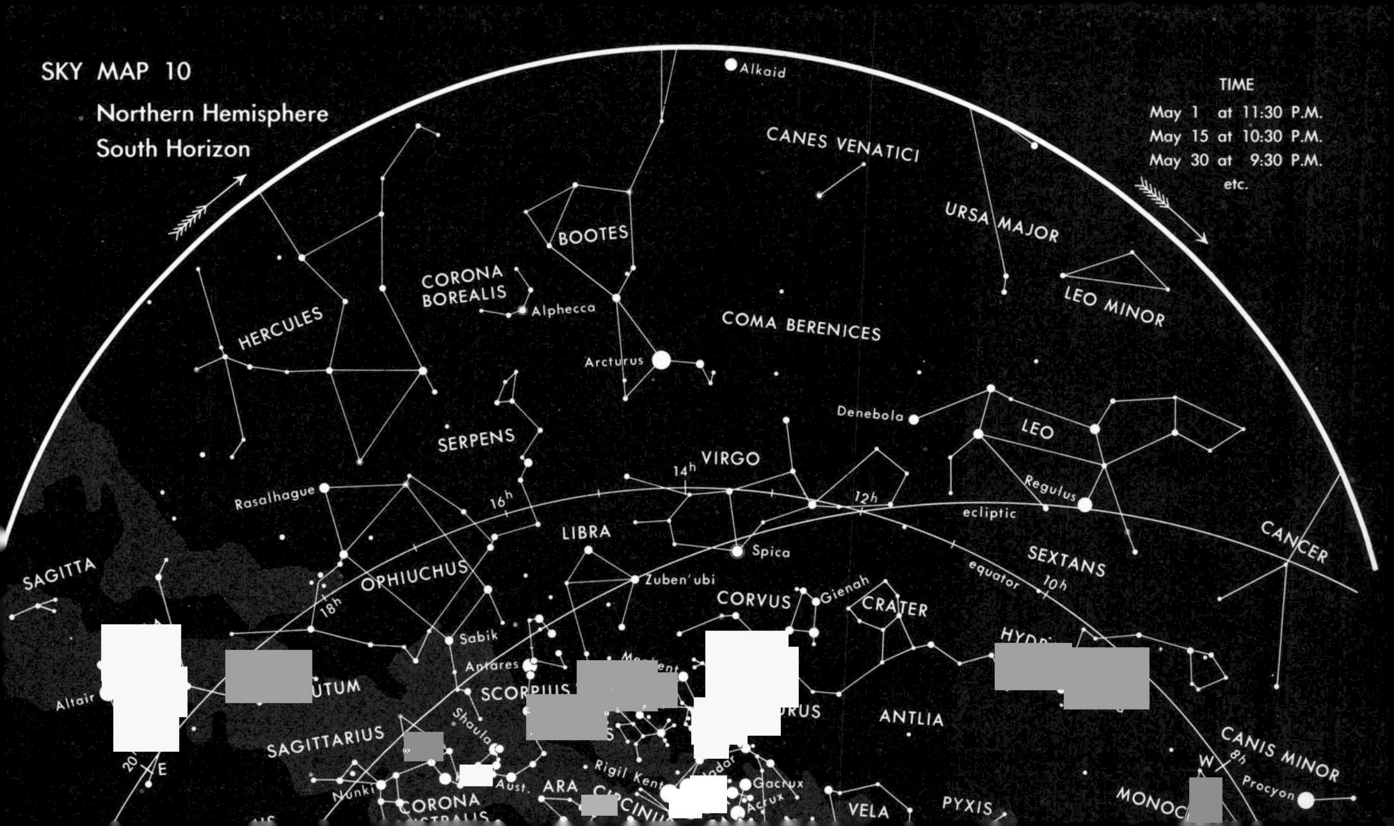

MAGNITUDES
Sirius
Canopus
−0.49-0.00
0.01-0.50
0.51-1.00
1.01-1.50
MAGNITUDES
1.51-2.00
2.01-2.50
2.51-3.00
3.01-3.50
3.51-4.00
4.01-4.55
Variable
E
W

SKY MAP 11

Northern Hemisphere
North Horizon

	TIME
June 1	at 11:30 P.M.
June 15	at 10:30 P.M.
June 30	at 9:30 P.M.
	etc.

MAGNITUDES
Sirius
Canopus
−0.49–0.00
0.01–0.50
0.51–1.00
1.01–1.50
MAGNITUDES
1.51–2.00
2.01–2.50
2.51–3.00
3.01–3.50
3.51–4.00
4.01–4.55
Variable
W
E

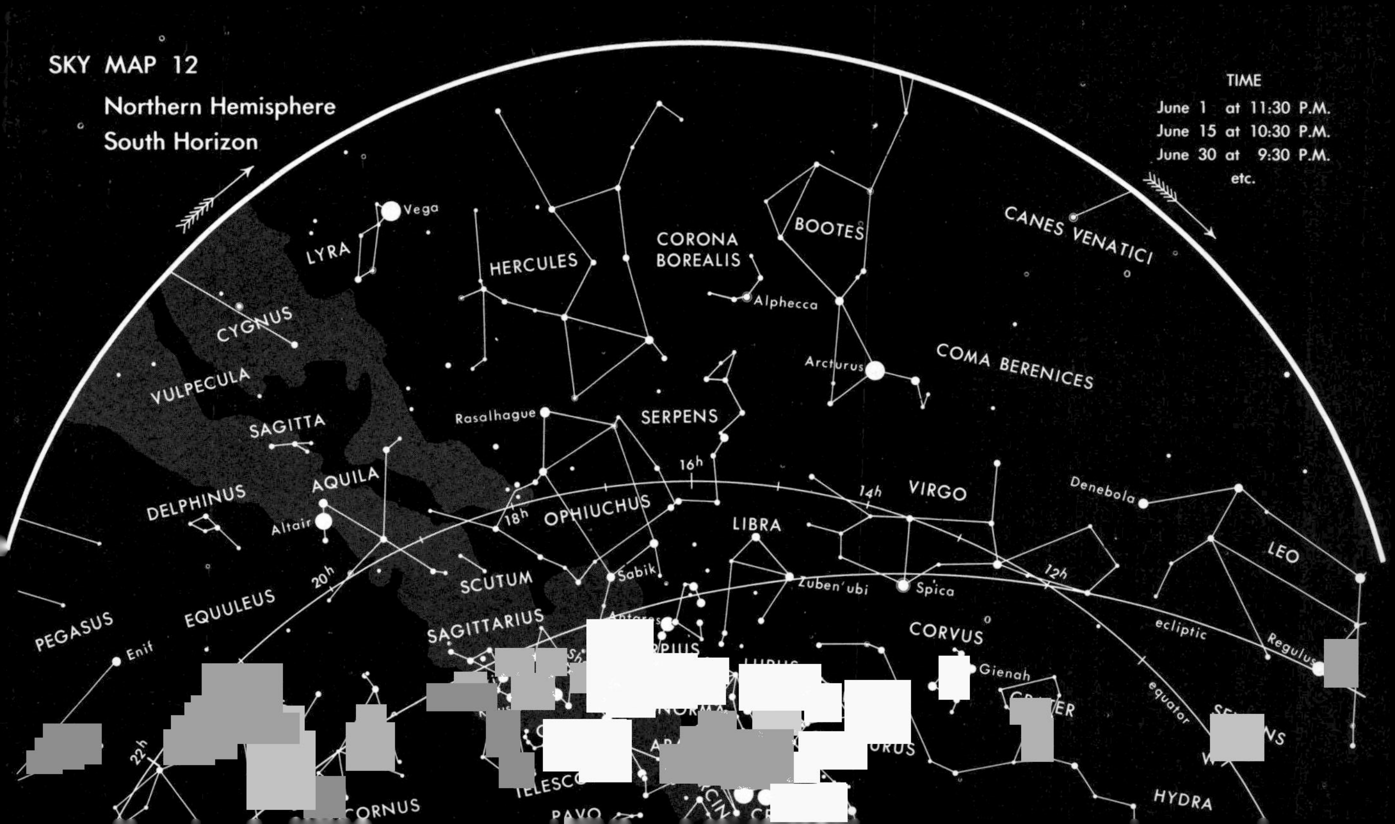
SKY MAP 12
Northern Hemisphere
South Horizon
TIME
June 1 at 11:30 P.M.
June 15 at 10:30 P.M.
June 30 at 9:30 P.M.
etc.
LYRA
Vega
HERCULES
CORONA BOREALIS
Alphecca
BOOTES
Arcturus
CANES VENATICI
COMA BERENICES
CYGNUS
VULPECULA
SAGITTA
Rasalhague
SERPENS
16h
14h
VIRGO
Denebola
LEO
DELPHINUS
AQUILA
Altair
18h
OPHIUCHUS
LIBRA
12h
Sabik
Zuben'ubi
Spica
SCUTUM
20h
EQUULEUS
PEGASUS
Enif
SAGITTARIUS
CORVUS
ecliptic
Regulus
Gienah
equator
HYDRA
22h
PAVO

MAGNITUDES
Sirius
Canopus
−0.49-0.00
0.01-0.50
0.51-1.00
1.01-1.50
MAGNITUDES
1.51-2.00
2.01-2.50
2.51-3.00
3.01-3.50
3.51-4.00
4.01-4.55
Variable
E
W

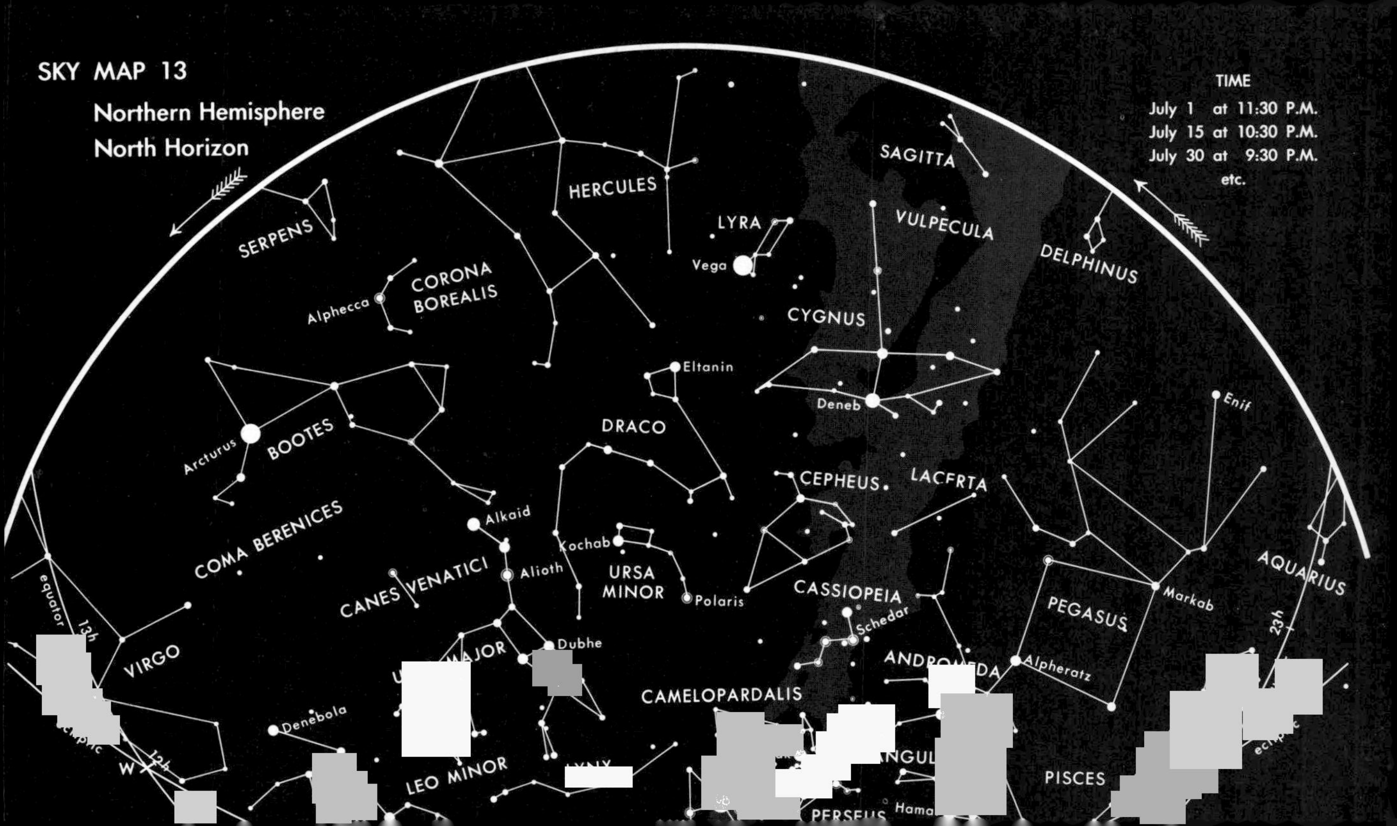
SKY MAP 13
Northern Hemisphere
North Horizon
TIME
July 1 at 11:30 P.M.
July 15 at 10:30 P.M.
July 30 at 9:30 P.M.
etc.
SERPENS
CORONA BOREALIS
Alphecca
HERCULES
LYRA
Vega
SAGITTA
VULPECULA
DELPHINUS
CYGNUS
Deneb
Eltanin
BOOTES
Arcturus
DRACO
Enif
CEPHEUS
LACERTA
AQUARIUS
COMA BERENICES
Alkaid
Kochab
URSA MINOR
Alioth
CANES VENATICI
Polaris
CASSIOPEIA
Markab
PEGASUS
Schedar
Dubhe
VIRGO
equator
13h
23h
ANDROMEDA
Alpheratz
CAMELOPARDALIS
Denebola
LEO MINOR
PISCES
PERSEUS
Hamal

MAGNITUDES
Sirius
Canopus
—0.49–0.00
0.01–0.50
0.51–1.00
1.01–1.50
MAGNITUDES
1.51–2.00
2.01–2.50
2.51–3.00
3.01–3.50
3.51–4.00
4.01–4.55
Variable
W
E

SKY MAP 14

Northern Hemisphere
South Horizon

TIME
July 1 at 11:30 P.M.
July 15 at 10:30 P.M.
July 30 at 9:30 P.M.
etc.

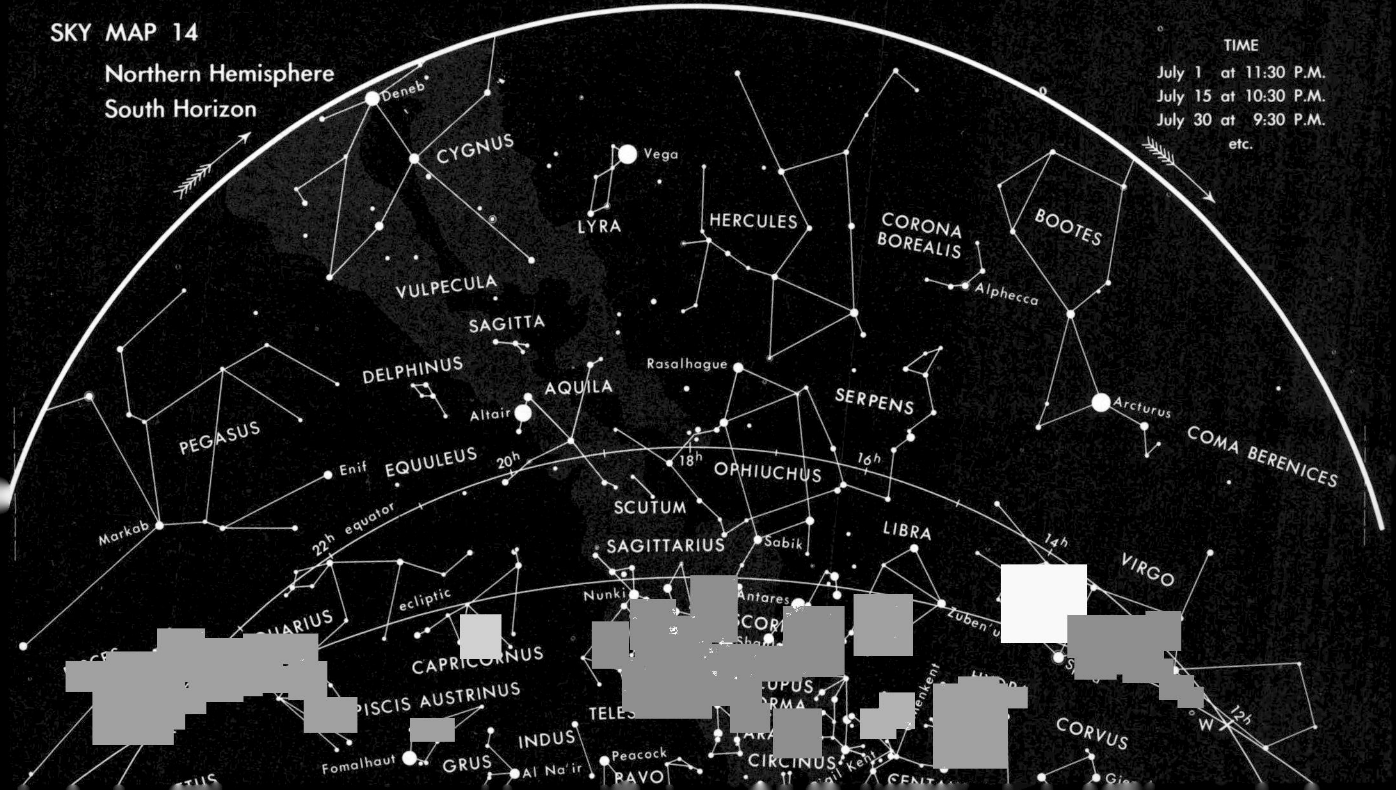

MAGNITUDES
Sirius
Canopus
−0.49-0.00
0.01-0.50
0.51-1.00
1.01-1.50
MAGNITUDES
1.51-2.00
2.01-2.50
2.51-3.00
3.01-3.50
3.51-4.00
4.01-4.55
Variable
E
W

SKY MAP 15

Northern Hemisphere
North Horizon

TIME
August 1 at 11:30 P.M.
August 15 at 10:30 P.M.
August 30 at 9:30 P.M.
etc.

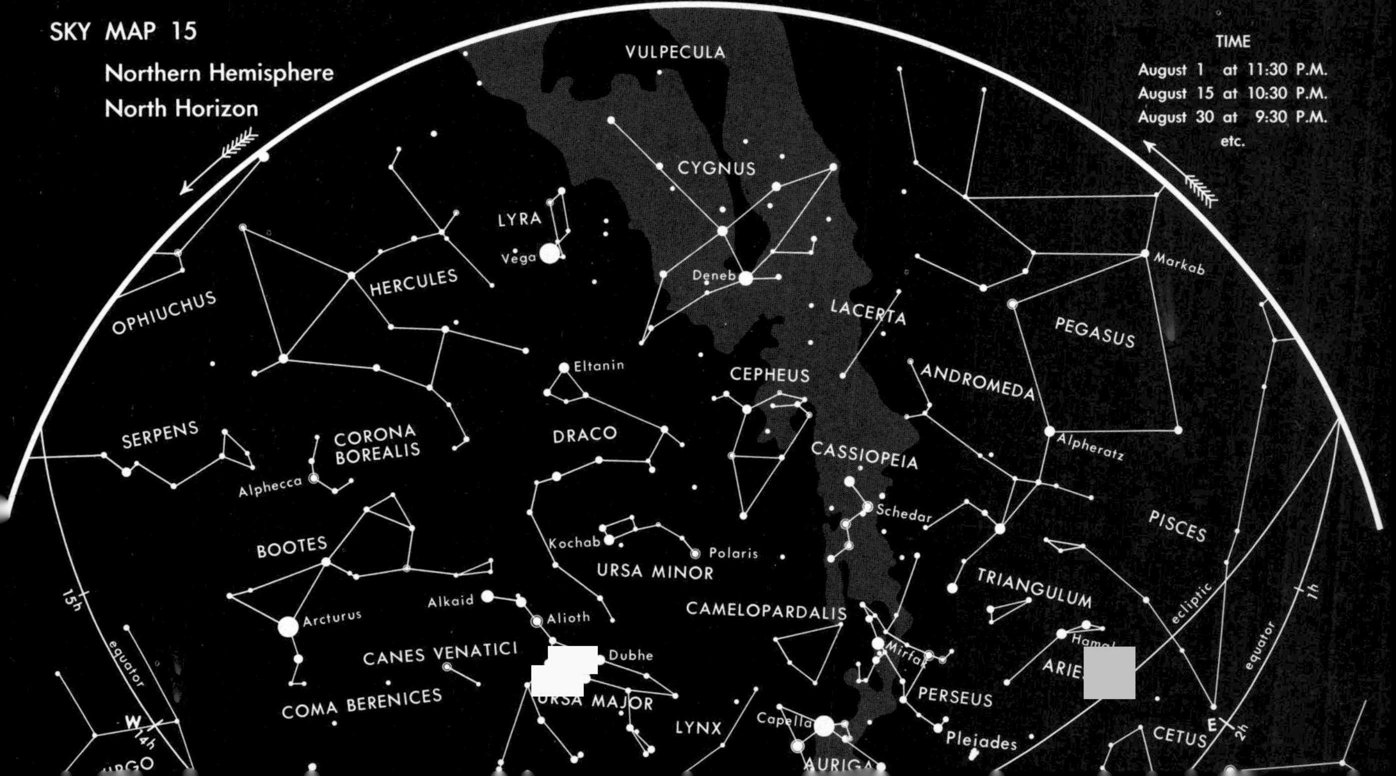

MAGNITUDES
Sirius
Canopus
−0.49-0.00
0.01-0.50
0.51-1.00
1.01-1.50
MAGNITUDES
1.51-2.00
2.01-2.50
2.51-3.00
3.01-3.50
3.51-4.00
4.01-4.55
Variable
W
E

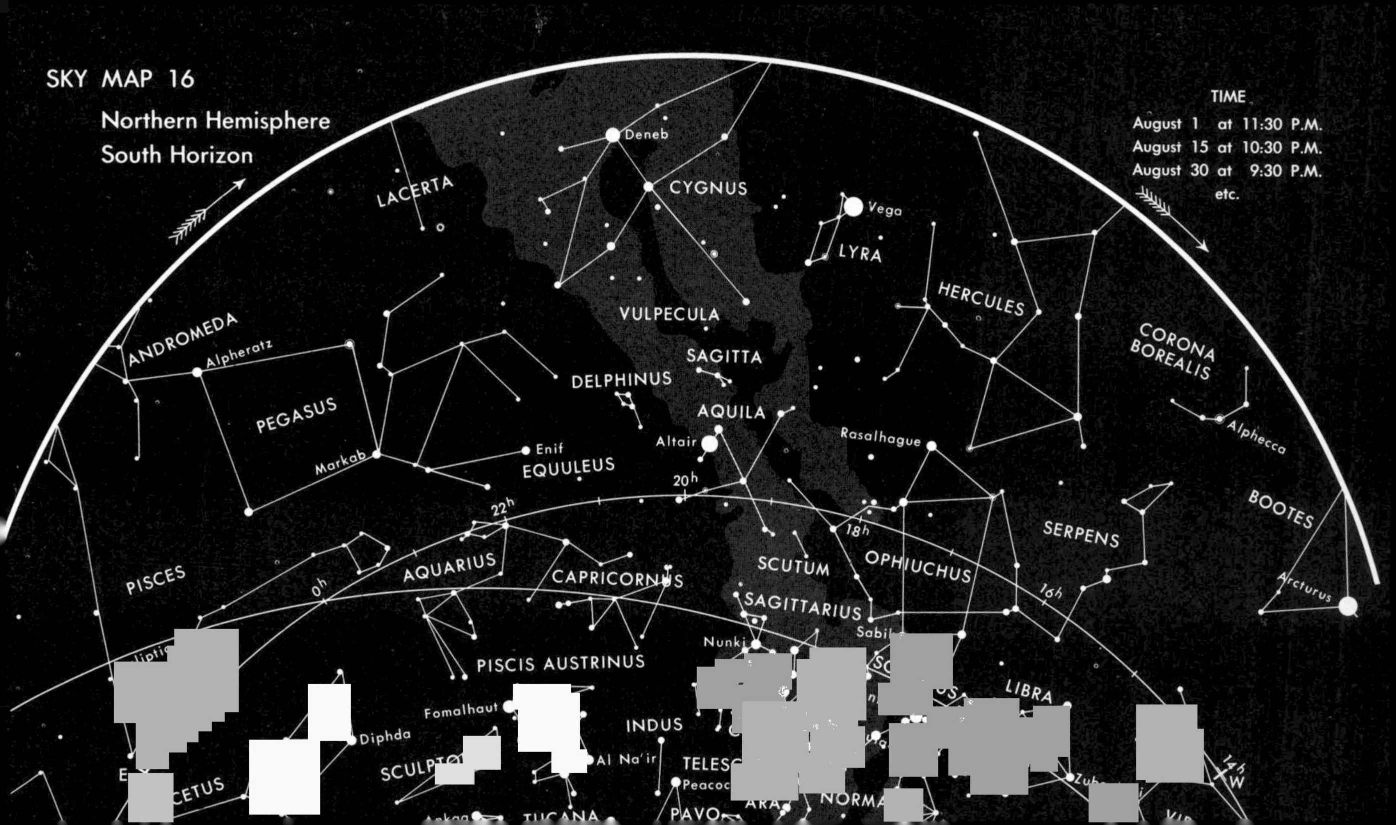
SKY MAP 16
Northern Hemisphere
South Horizon
TIME
August 1 at 11:30 P.M.
August 15 at 10:30 P.M.
August 30 at 9:30 P.M.
etc.
LACERTA
Deneb
CYGNUS
Vega
LYRA
HERCULES
CORONA BOREALIS
Alphecca
BOOTES
Arcturus
ANDROMEDA
Alpheratz
PEGASUS
Markab
VULPECULA
SAGITTA
DELPHINUS
AQUILA
Altair
Rasalhague
Enif
EQUULEUS
20h
22h
18h
16h
SERPENS
OPHIUCHUS
SCUTUM
SAGITTARIUS
Nunki
PISCES
0h
AQUARIUS
CAPRICORNUS
PISCIS AUSTRINUS
LIBRA
Fomalhaut
Diphda
INDUS
Al Na'ir
CETUS
14h
W

MAGNITUDES
Sirius
Canopus
—0.49-0.00
0.01-0.50
0.51-1.00
1.01-1.50
MAGNITUDES
1.51-2.00
2.01-2.50
2.51-3.00
3.01-3.50
3.51-4.00
4.01-4.55
Variable
E
W

SKY MAP 17

Northern Hemisphere
North Horizon

TIME
September 1 at 11:30 P.M.
September 15 at 10:30 P.M.
September 30 at 9:30 P.M.
etc.

MAGNITUDES
Sirius
Canopus
−0.49–0.00
0.01–0.50
0.51–1.00
1.01–1.50
MAGNITUDES
1.51–2.00
2.01–2.50
2.51–3.00
3.01–3.50
3.51–4.00
4.01–4.55
Variable
W
E

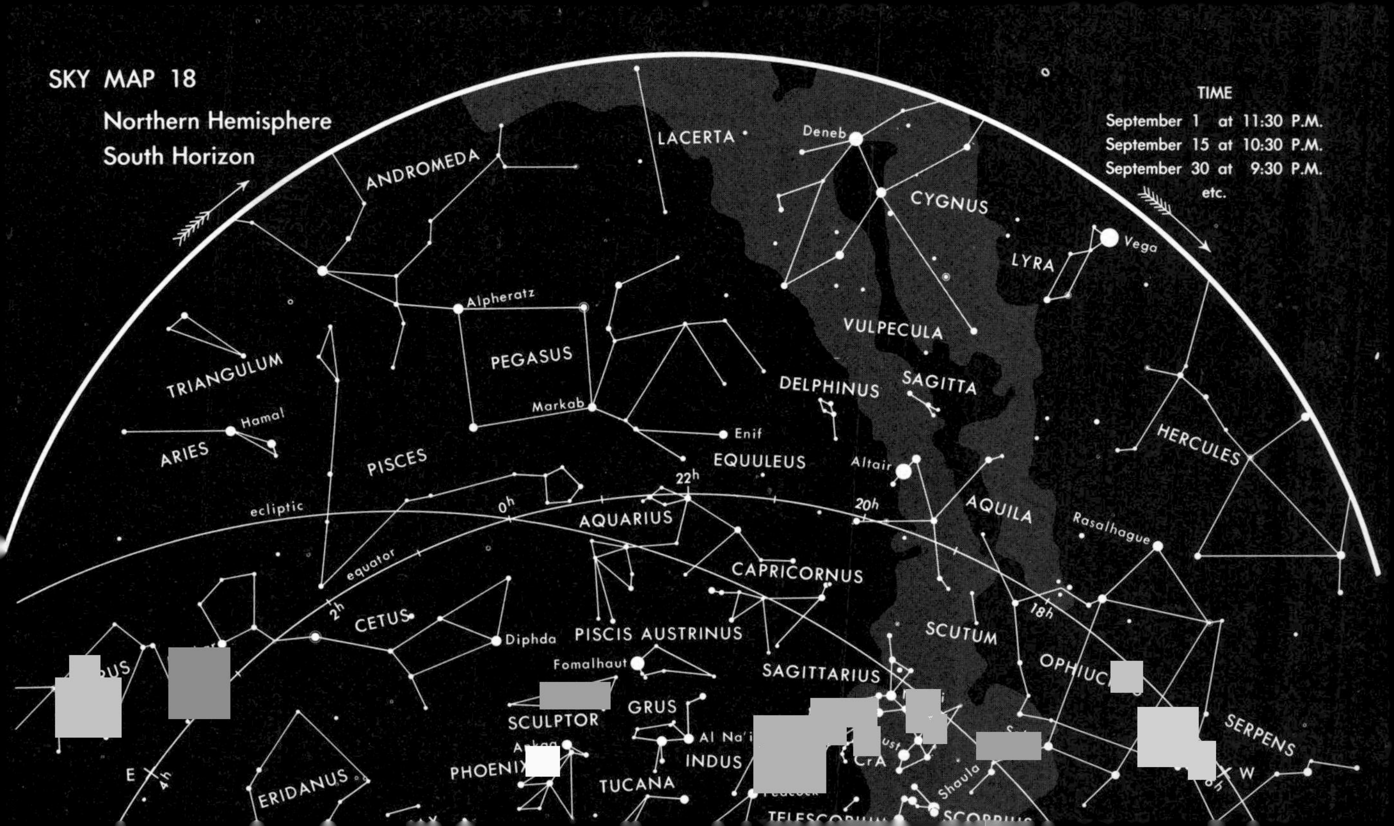
SKY MAP 18
Northern Hemisphere
South Horizon
TIME
September 1 at 11:30 P.M.
September 15 at 10:30 P.M.
September 30 at 9:30 P.M.
etc.
LACERTA
Deneb
CYGNUS
LYRA
Vega
ANDROMEDA
Alpheratz
PEGASUS
Markab
TRIANGULUM
Hamal
ARIES
PISCES
VULPECULA
SAGITTA
DELPHINUS
Enif
EQUULEUS
Altair
AQUILA
HERCULES
Rasalhague
ecliptic
equator
0h
22h
20h
18h
2h
AQUARIUS
CAPRICORNUS
CETUS
Diphda
PISCIS AUSTRINUS
Fomalhaut
SCUTUM
SAGITTARIUS
SERPENS
GRUS
SCULPTOR
Al Na'ir
INDUS
PHOENIX
TUCANA
ERIDANUS
Shaula
E
W

MAGNITUDES
Sirius
Canopus
−0.49–0.00
0.01–0.50
0.51–1.00
1.01–1.50
MAGNITUDES
1.51–2.00
2.01–2.50
2.51–3.00
3.01–3.50
3.51–4.00
4.01–4.55
Variable
E
W

SKY MAP 19

Northern Hemisphere

North Horizon

TIME

October 1 at 11:30 P.M.
October 15 at 10:30 P.M.
October 30 at 9:30 P.M.
etc.

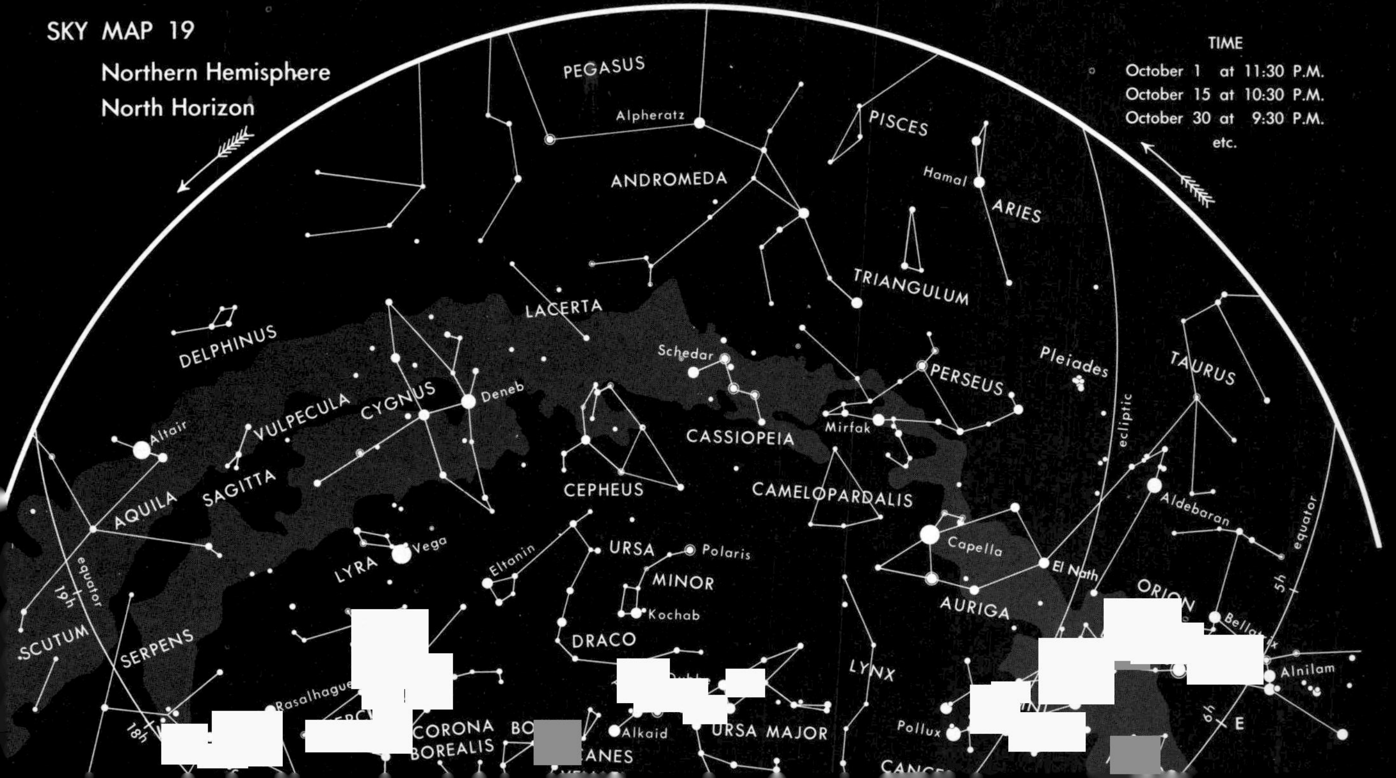

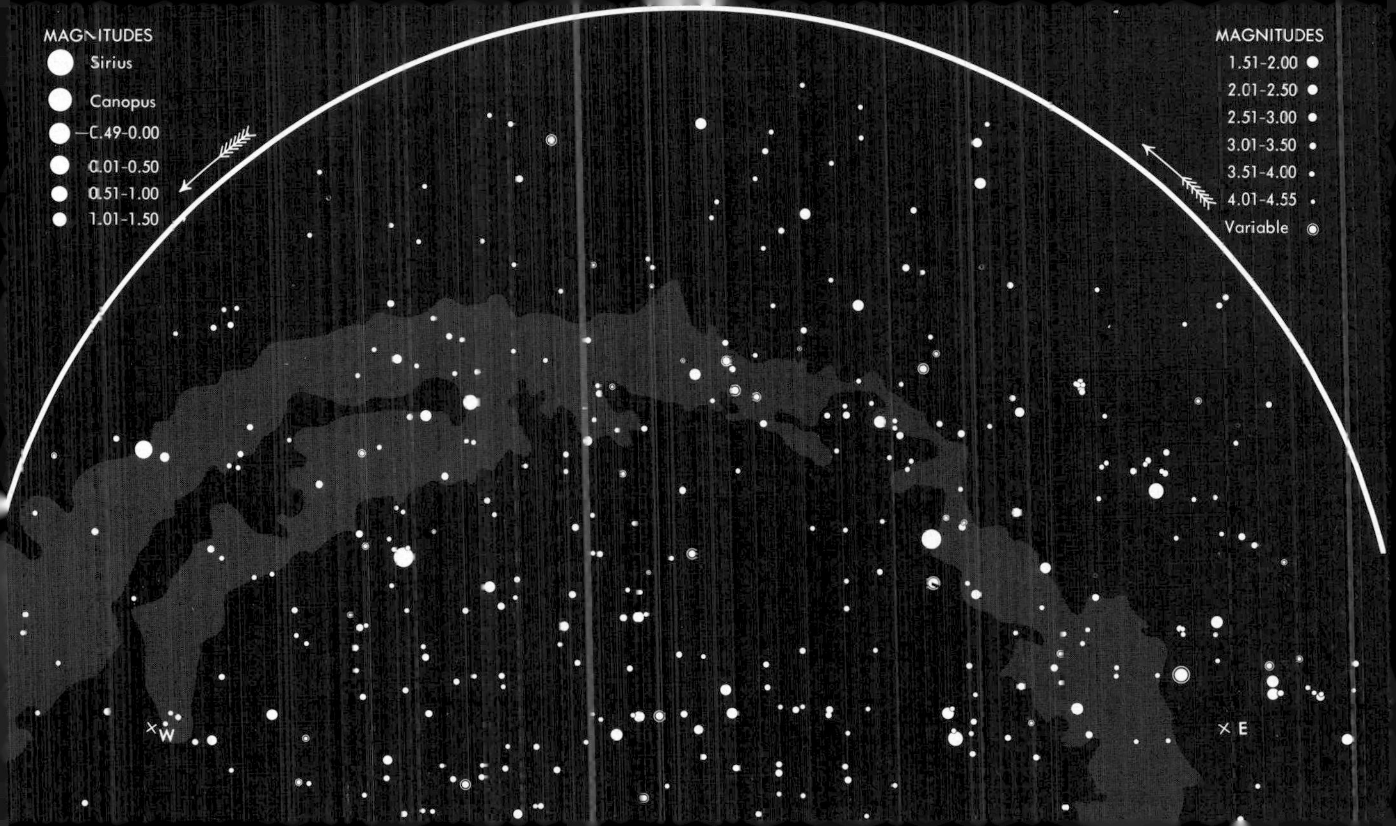
MAGNITUDES
Sirius
Canopus
—0.49-0.00
0.01-0.50
0.51-1.00
1.01-1.50
MAGNITUDES
1.51-2.00
2.01-2.50
2.51-3.00
3.01-3.50
3.51-4.00
4.01-4.55
Variable
W
E

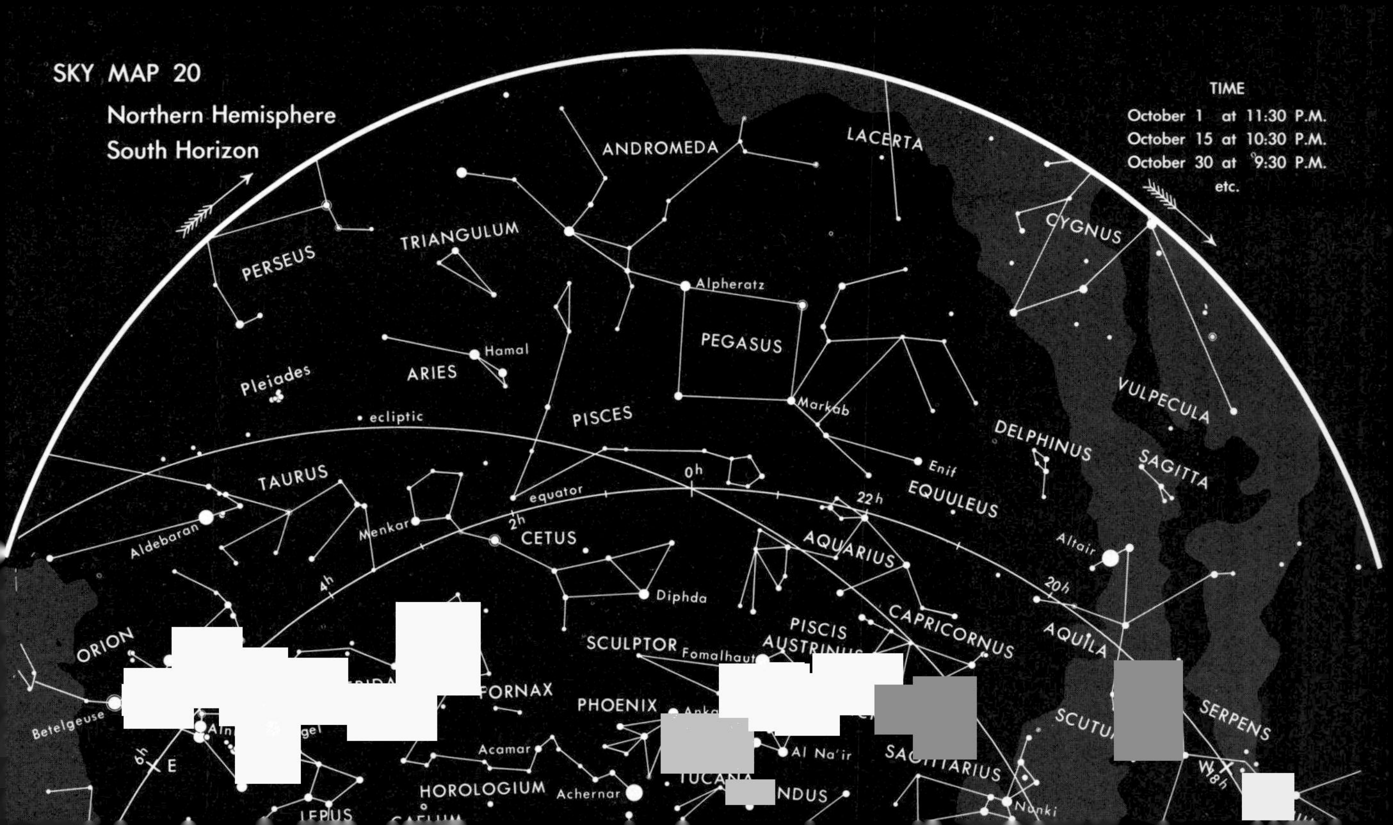
SKY MAP 20
Northern Hemisphere
South Horizon
TIME
October 1 at 11:30 P.M.
October 15 at 10:30 P.M.
October 30 at 9:30 P.M.
etc.
ANDROMEDA
LACERTA
CYGNUS
PERSEUS
TRIANGULUM
Alpheratz
PEGASUS
Hamal
ARIES
Pleiades
VULPECULA
Markab
PISCES
ecliptic
DELPHINUS
SAGITTA
Enif
EQUULEUS
0h
22h
TAURUS
equator
2h
CETUS
Menkar
Aldebaran
AQUARIUS
Altair
20h
4h
Diphda
CAPRICORNUS
AQUILA
PISCIS
ORION
SCULPTOR
Fomalhaut
FORNAX
PHOENIX
Betelgeuse
Acamar
Al Na'ir
SERPENS
E
6h
HOROLOGIUM
Achernar
Nunki
W
18h
LEPUS

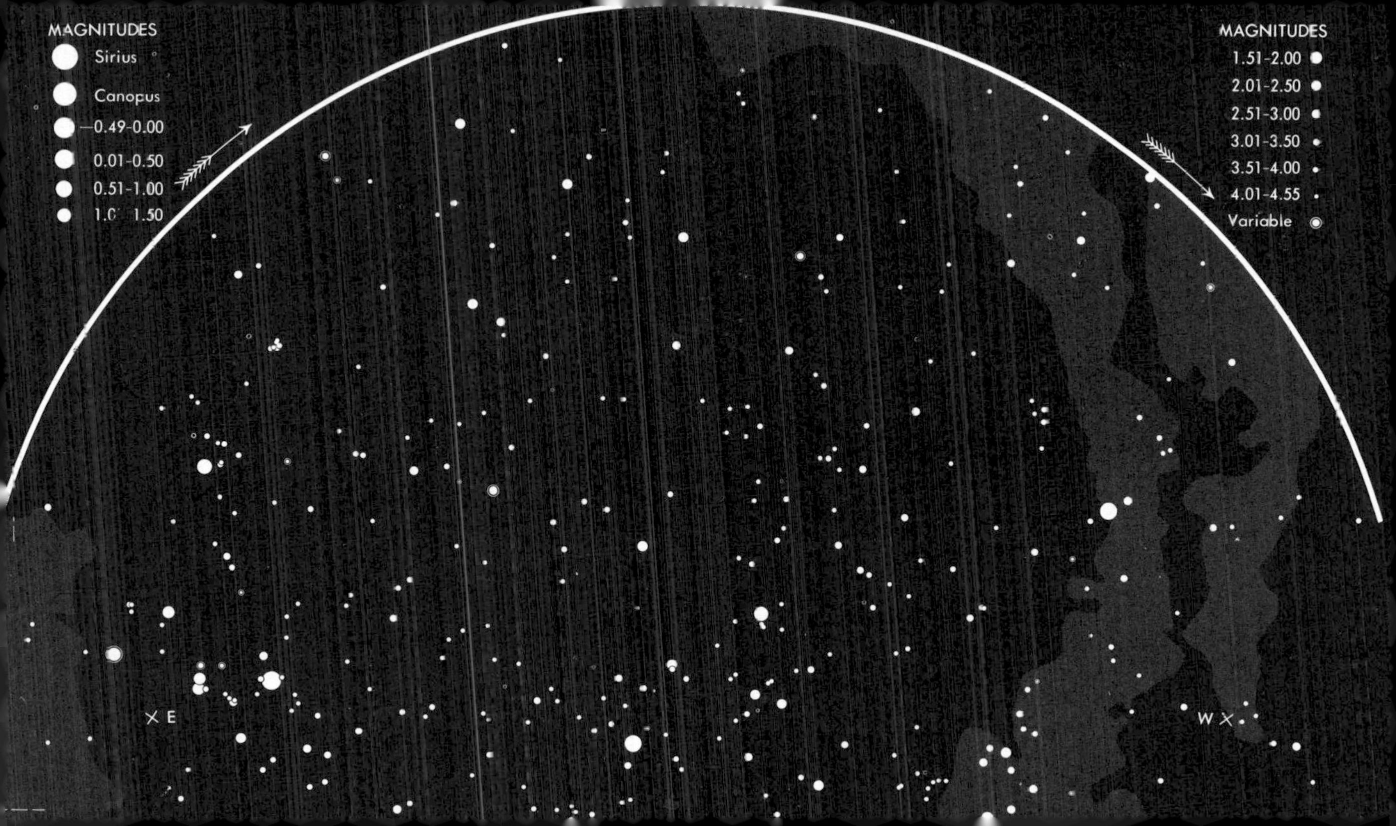
MAGNITUDES
Sirius
Canopus
—0.49-0.00
0.01-0.50
0.51-1.00
1.0' 1.50
MAGNITUDES
1.51-2.00
2.01-2.50
2.51-3.00
3.01-3.50
3.51-4.00
4.01-4.55
Variable
E
W

SKY MAP 21

Northern Hemisphere
North Horizon

TIME
November 1 at 11:30 P.M.
November 15 at 10:30 P.M.
November 30 at 9:30 P.M.
etc.

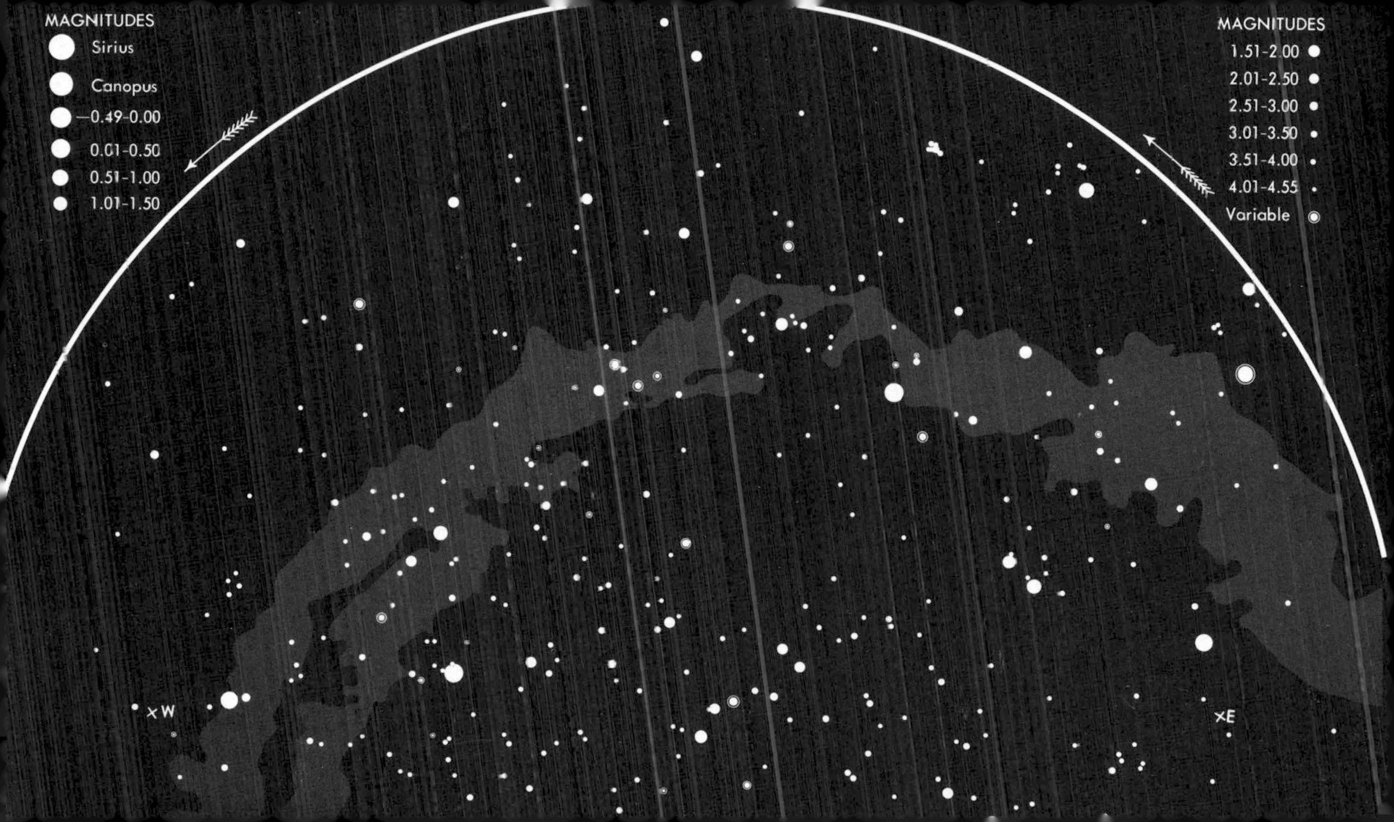
MAGNITUDES
Sirius
Canopus
—0.49-0.00
0.01-0.50
0.51-1.00
1.01-1.50
MAGNITUDES
1.51-2.00
2.01-2.50
2.51-3.00
3.01-3.50
3.51-4.00
4.01-4.55
Variable
×W
×E

SKY MAP 22

Northern Hemisphere
South Horizon

TIME
November 1 at 11:30 P.M.
November 15 at 10:30 P.M.
November 30 at 9:30 P.M.
etc.

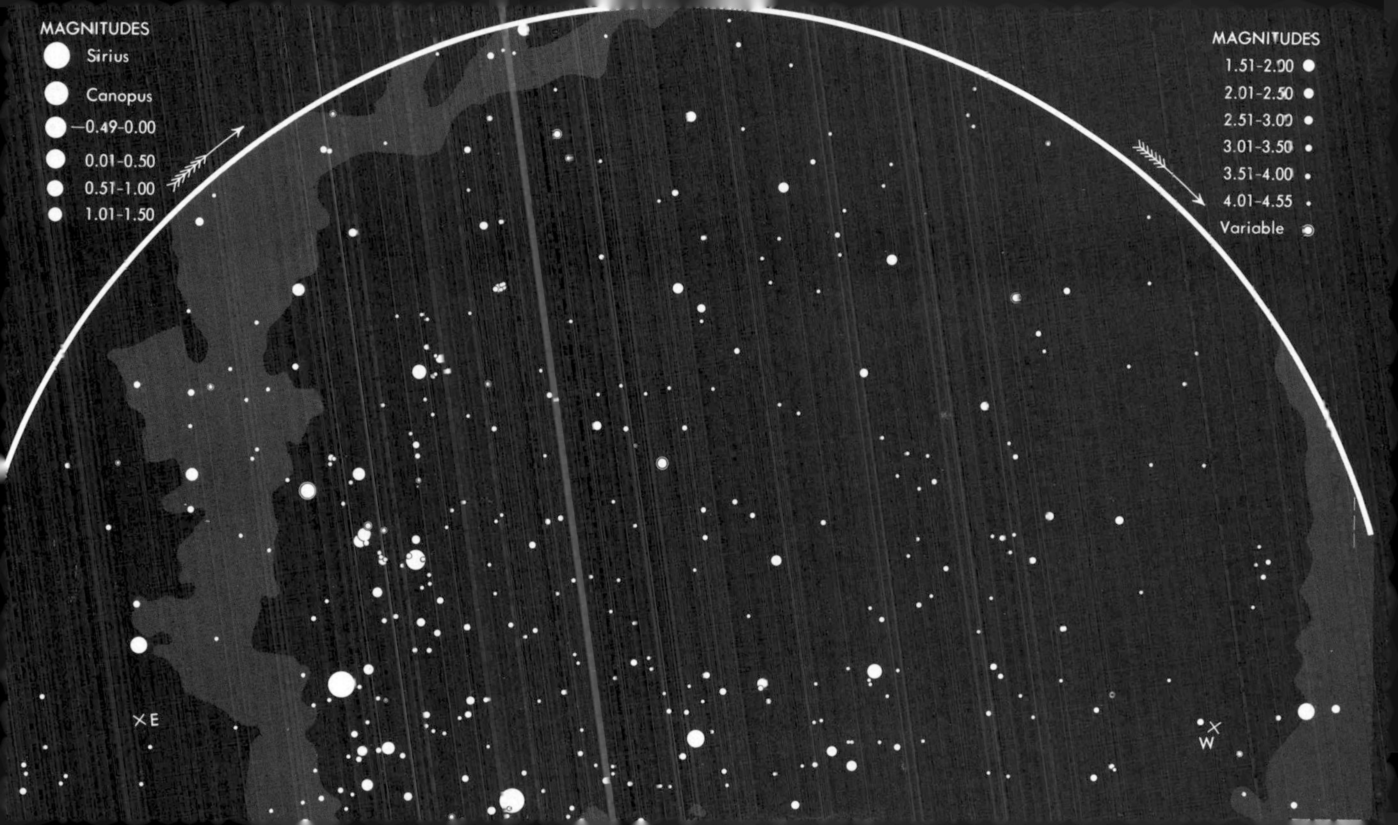
MAGNITUDES
Sirius
Canopus
−0.49–0.00
0.01–0.50
0.51–1.00
1.01–1.50
MAGNITUDES
1.51–2.00
2.01–2.50
2.51–3.00
3.01–3.50
3.51–4.00
4.01–4.55
Variable
E
W

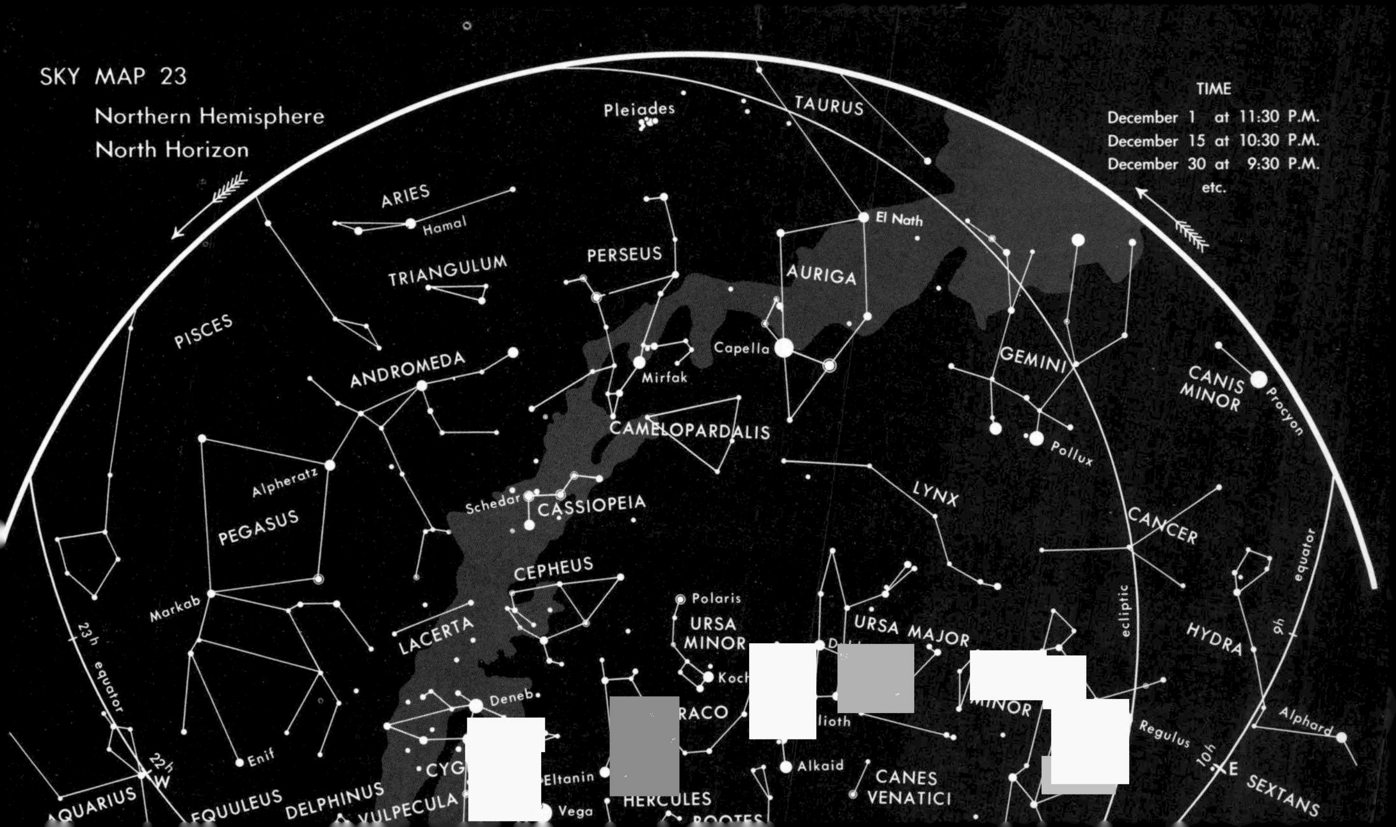

SKY MAP 23
Northern Hemisphere
North Horizon
TIME
December 1 at 11:30 P.M.
December 15 at 10:30 P.M.
December 30 at 9:30 P.M.
etc.
Pleiades
TAURUS
ARIES
Hamal
PERSEUS
El Nath
AURIGA
TRIANGULUM
PISCES
ANDROMEDA
Capella
Mirfak
GEMINI
CANIS MINOR
Procyon
Pollux
CAMELOPARDALIS
Alpheratz
Schedar
CASSIOPEIA
LYNX
PEGASUS
CANCER
CEPHEUS
Polaris
URSA MINOR
URSA MAJOR
ecliptic
equator
HYDRA
Markab
LACERTA
9h
23h
Deneb
Alphard
Enif
22h
W
Regulus
10h
E
Alkaid
CANES VENATICI
SEXTANS
Eltanin
Vega
HERCULES
AQUARIUS
EQUULEUS
DELPHINUS
VULPECULA

MAGNITUDES
Sirius
Canopus
—0.49-0.00
0.01-0.50
0.51-1.00
1.01-1.50
MAGNITUDES
1.51-2.00
2.01-2.50
2.51-3.00
3.01-3.50
3.51-4.00
4.01-4.55
Variable
W
E

SKY MAP 24

Northern Hemisphere

South Horizon

TIME

December 1 at 11:30 P.M.
December 15 at 10:30 P.M.
December 30 at 9:30 P.M.
etc.

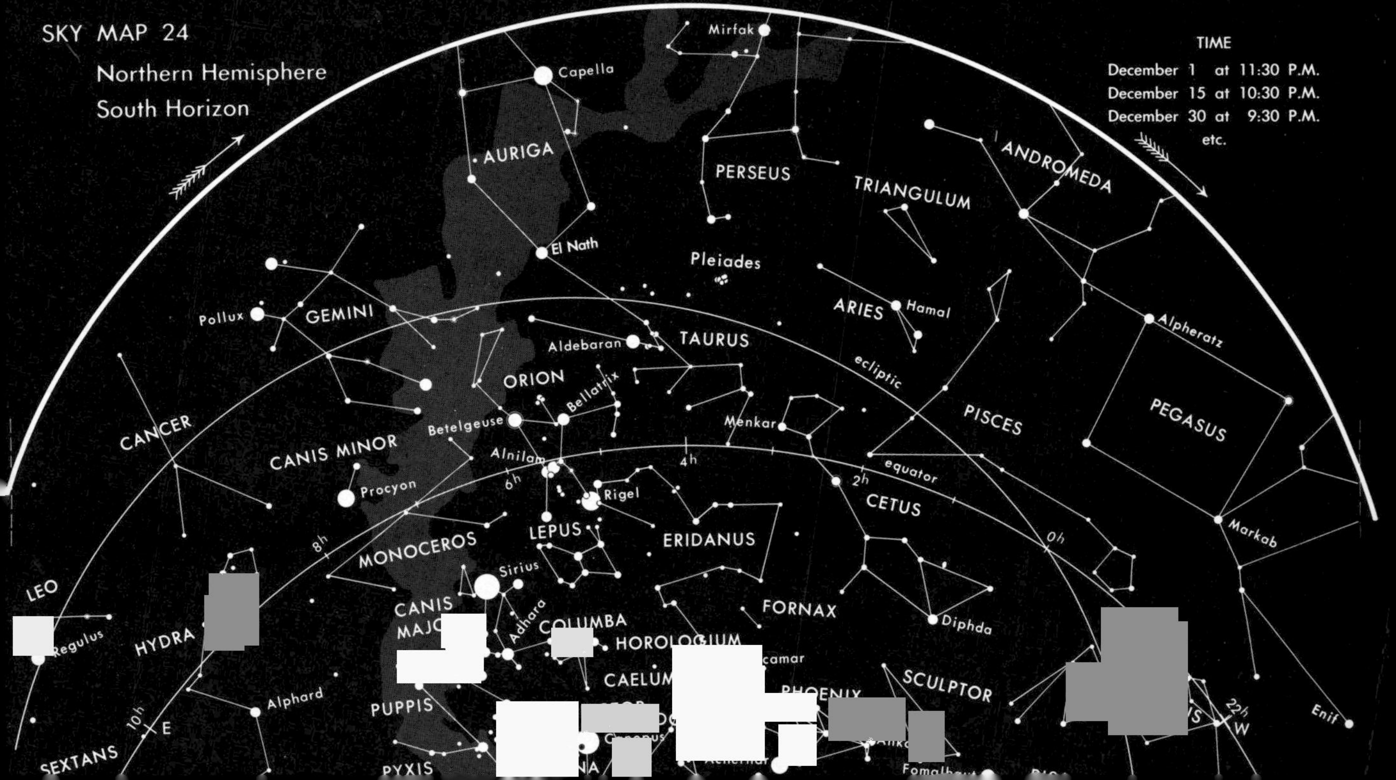

MAGNITUDES
Sirius
Canopus
−0.49–0.00
0.01–0.50
0.51–1.00
1.01–1.50
MAGNITUDES
1.51–2.00
2.01–2.50
2.51–3.00
3.01–3.50
3.51–4.00
4.01–4.55
Variable
E
W

SKY MAP 25

Southern Hemisphere

North Horizon

TIME

January 1 at 11:30 P.M.
January 15 at 10:30 P.M.
January 30 at 9:30 P.M.
etc.

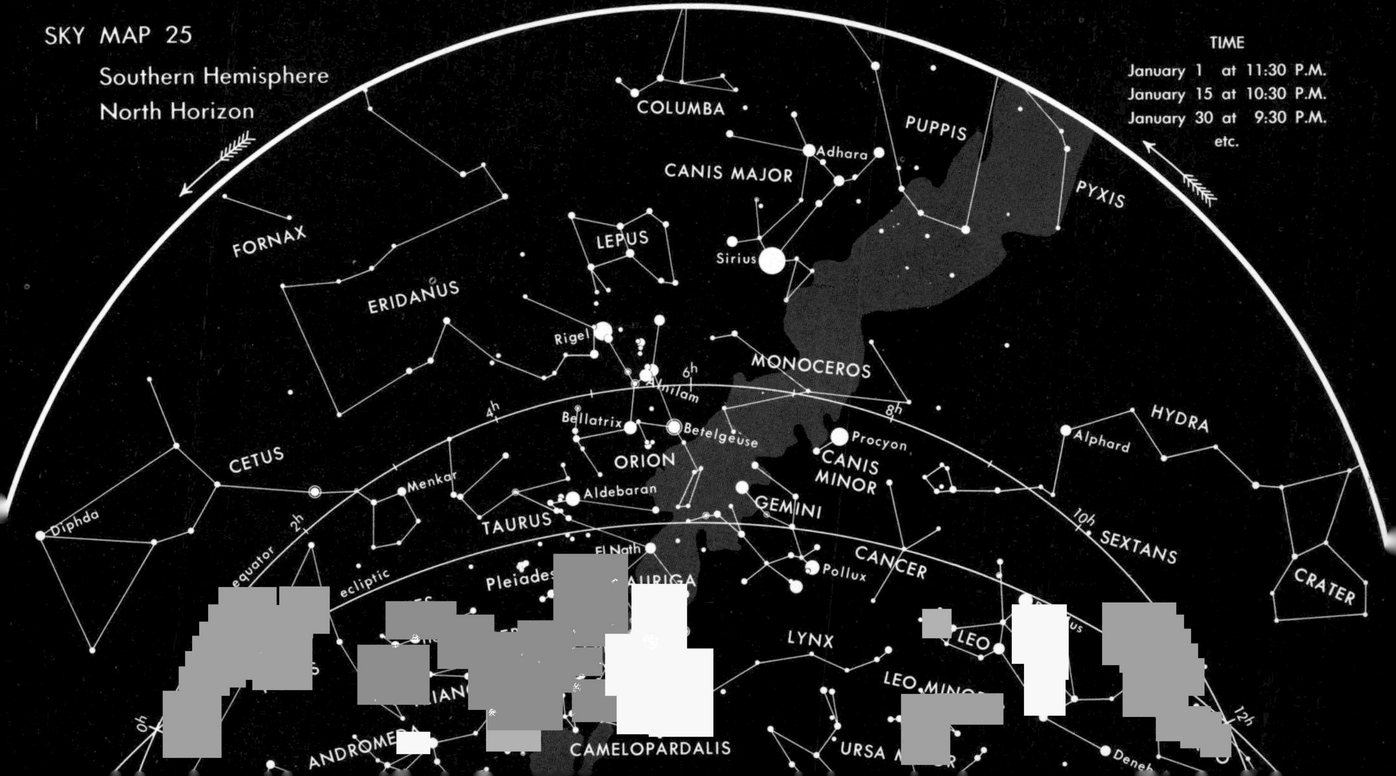

MAGNITUDES
Sirius
Canopus
−0.49-0.00
0.01-0.50
0.51-1.00
1.01-1.50
MAGNITUDES
1.51-2.00
2.01-2.50
2.51-3.00
3.01-3.50
3.51-4.00
4.01-4.55
Variable
W
E

SKY MAP 26

Southern Hemisphere

South Horizon

TIME

January 1 at 11:30 P.M.
January 15 at 10:30 P.M.
January 30 at 9:30 P.M.
etc.

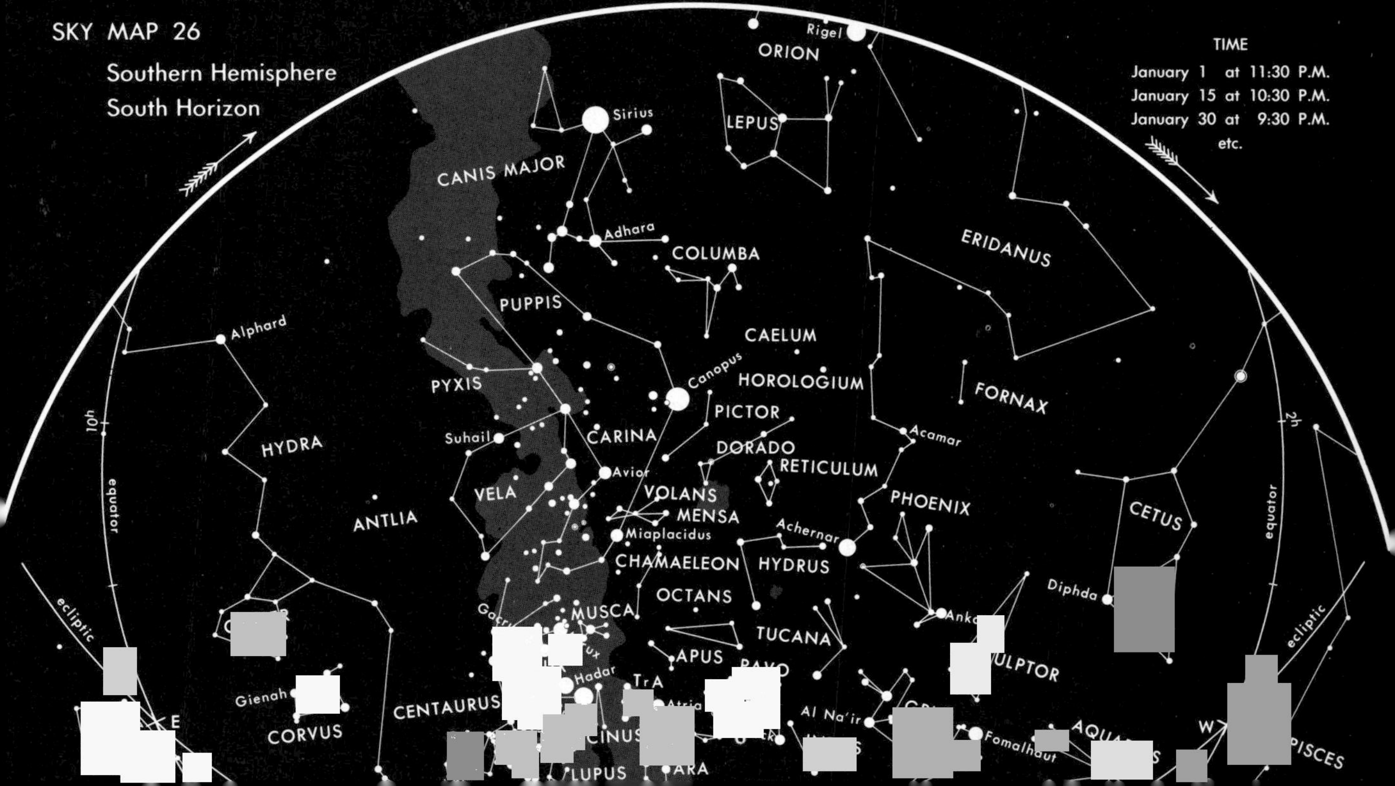

MAGNITUDES
Sirius
Canopus
−0.49-0.00
0.01-0.50
0.51-1.00
1.01-1.50
MAGNITUDES
1.51-2.00
2.01-2.50
2.51-3.00
3.01-3.50
3.51-4.00
4.01-4.55
Variable
E
W

SKY MAP 27

Southern Hemisphere
North Horizon

TIME
February 1 at 11:30 P.M.
February 15 at 10:30 P.M.
March 2 at 9:30 P.M.
etc.

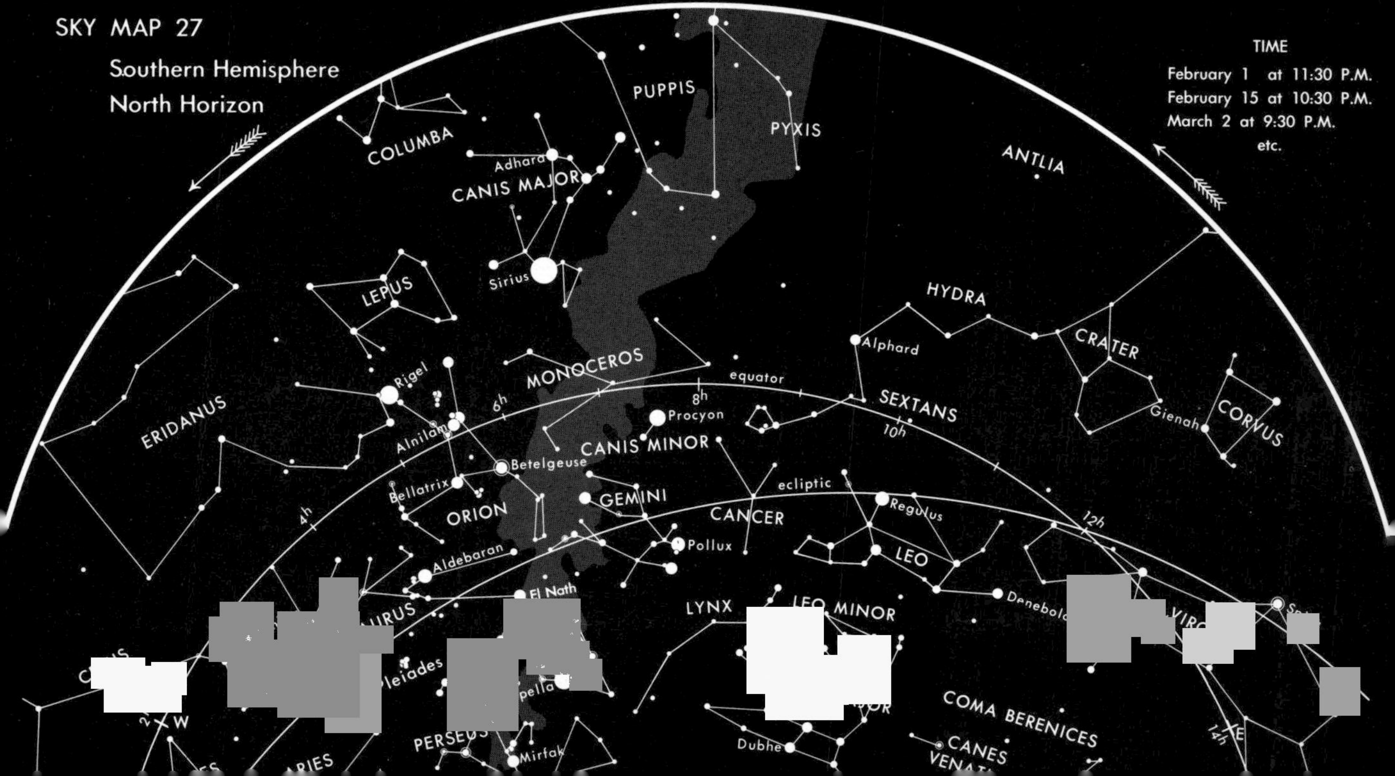

MAGNITUDES
Sirius
Canopus
−0.49-0.00
0.01-0.50
0.51-1.00
1.01-1.50
MAGNITUDES
1.51-2.00
2.01-2.50
2.51-3.00
3.01-3.50
3.51-4.00
4.01-4.55
Variable
W
E

SKY MAP 28

Southern Hemisphere

South Horizon

TIME

February 1 at 11:30 P.M.
February 15 at 10:30 P.M.
March 2 at 9:30 P.M.
etc.

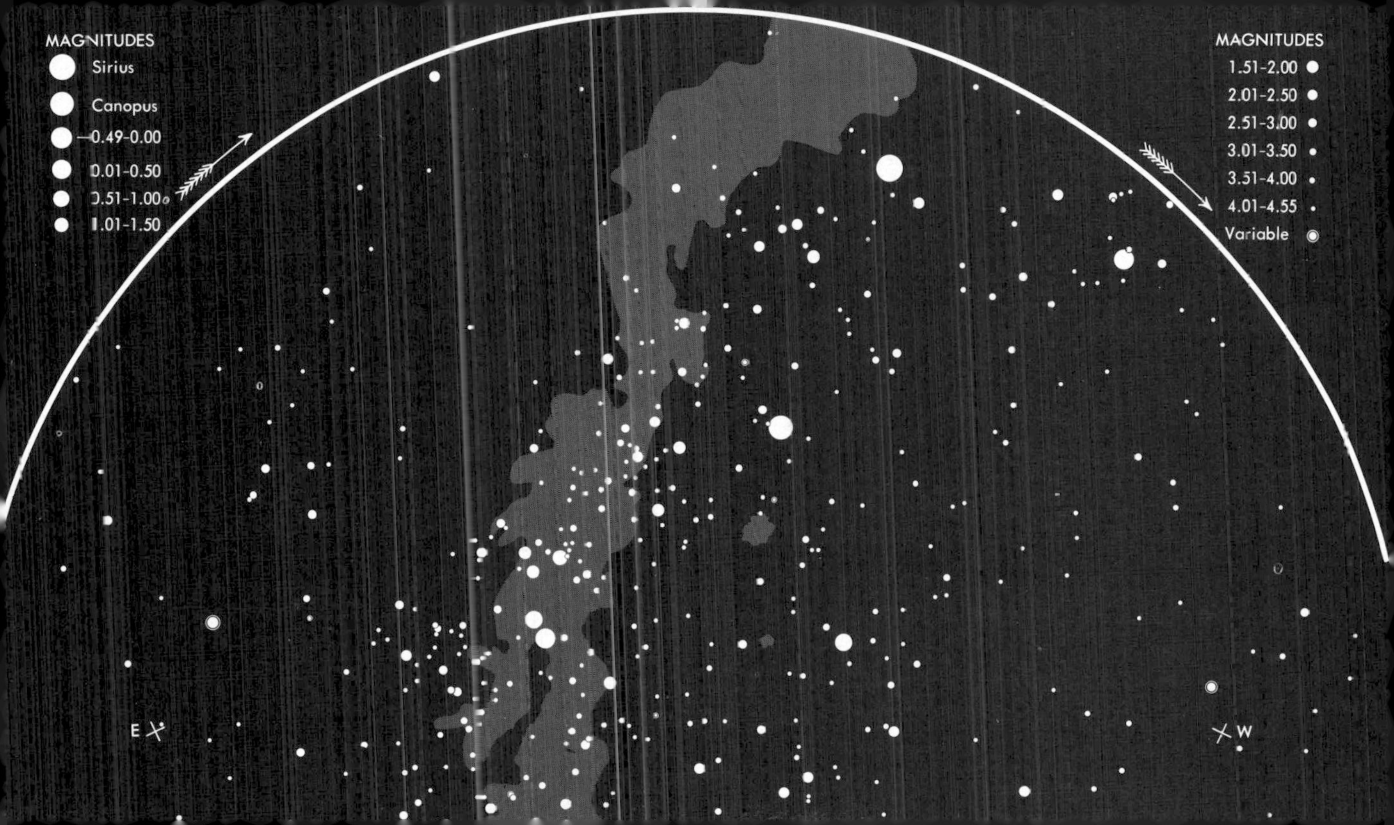
MAGNITUDES
Sirius
Canopus
−0.49-0.00
0.01-0.50
0.51-1.00
1.01-1.50
MAGNITUDES
1.51-2.00
2.01-2.50
2.51-3.00
3.01-3.50
3.51-4.00
4.01-4.55
Variable
E
W

SKY MAP 29

Southern Hemisphere
North Horizon

TIME
March 1 at 11:30 P.M.
March 15 at 10:30 P.M.
March 30 at 9:30 P.M.
etc.

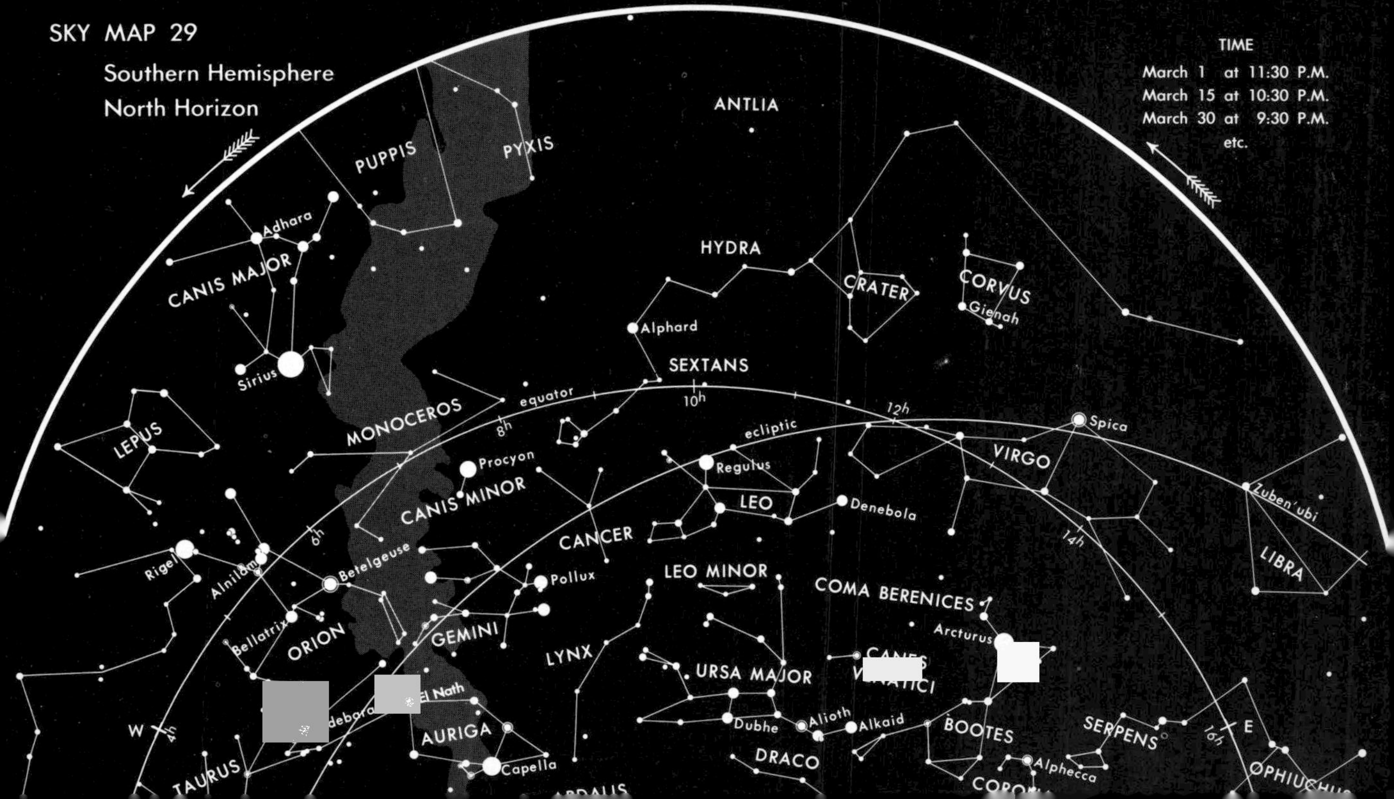

MAGNITUDES
Sirius
Canopus
—0.49-0.00
0.01-0.50
0.51-1.00
1.01-1.50
MAGNITUDES
1.51-2.00
2.01-2.50
2.51-3.00
3.01-3.50
3.51-4.00
4.01-4.55
Variable
W
E

SKY MAP 30
Southern Hemisphere
South Horizon
TIME
March 1 at 11:30 P.M.
March 15 at 10:30 P.M.
March 30 at 9:30 P.M.
etc.
Alphard
CRATER
HYDRA
Gienah
CORVUS
ANTLIA
PYXIS
MONOCEROS
CANIS MAJOR
Spica
VIRGO
PUPPIS
Suhail
VELA
Adhara
Sirius
14h
ecliptic
CARINA
ORION
6h
CENTAURUS
CRUX
Gacrux
Avior
COLUMBA
Menkent
Acrux
VOLANS
Canopus
Alnilam
MUSCA
equator
Zuben' ubi
Hadar
Miaplacidus
Bellatrix
Rigil Kent
PICTOR
Rigel
CHAMAELEON
CAELUM
LEPUS
LIBRA
CIRCINUS
DORADO
APUS
LUPUS
RETICULUM
ERIDANUS
Aust.
W

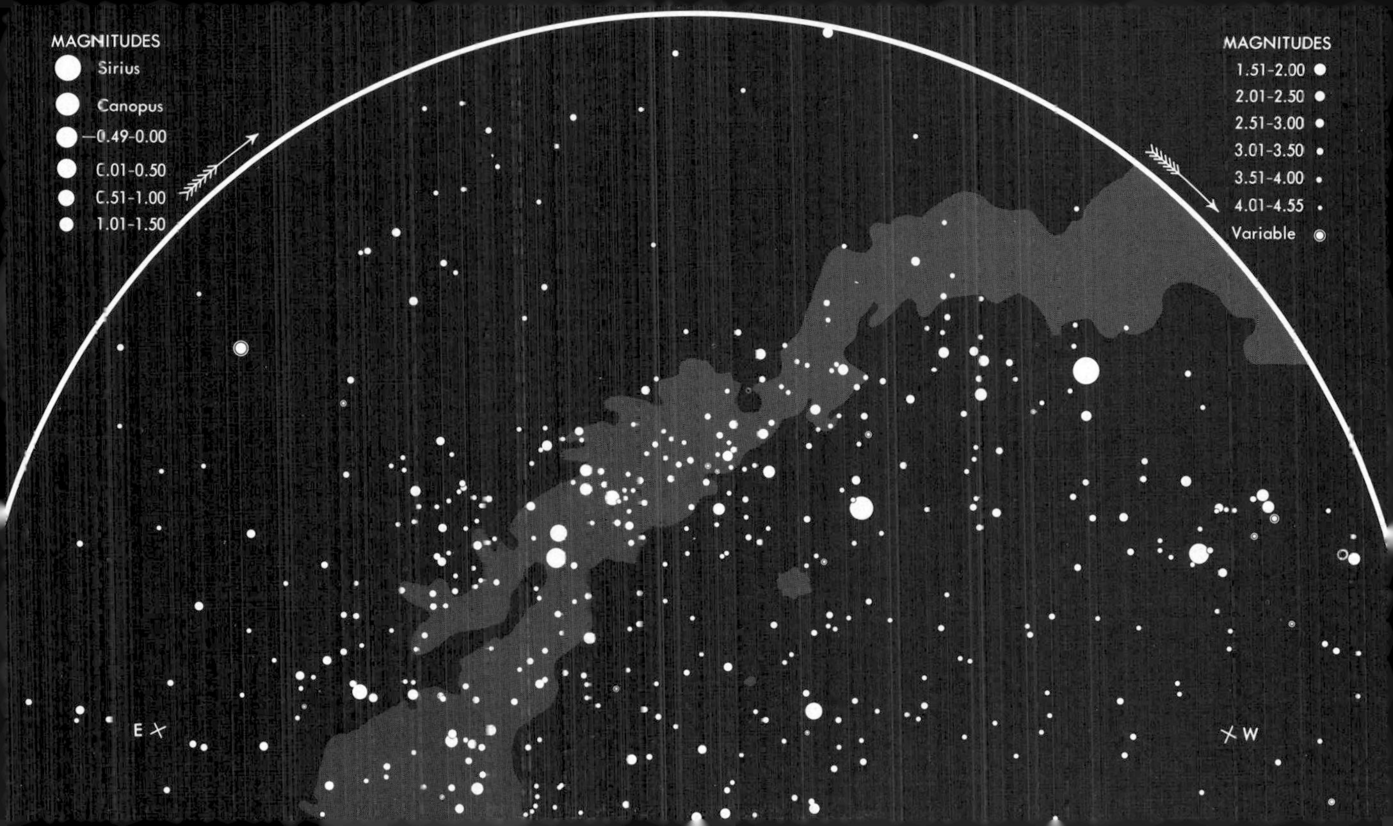
MAGNITUDES
Sirius
Canopus
—0.49-0.00
0.01-0.50
0.51-1.00
1.01-1.50
MAGNITUDES
1.51-2.00
2.01-2.50
2.51-3.00
3.01-3.50
3.51-4.00
4.01-4.55
Variable
E
W

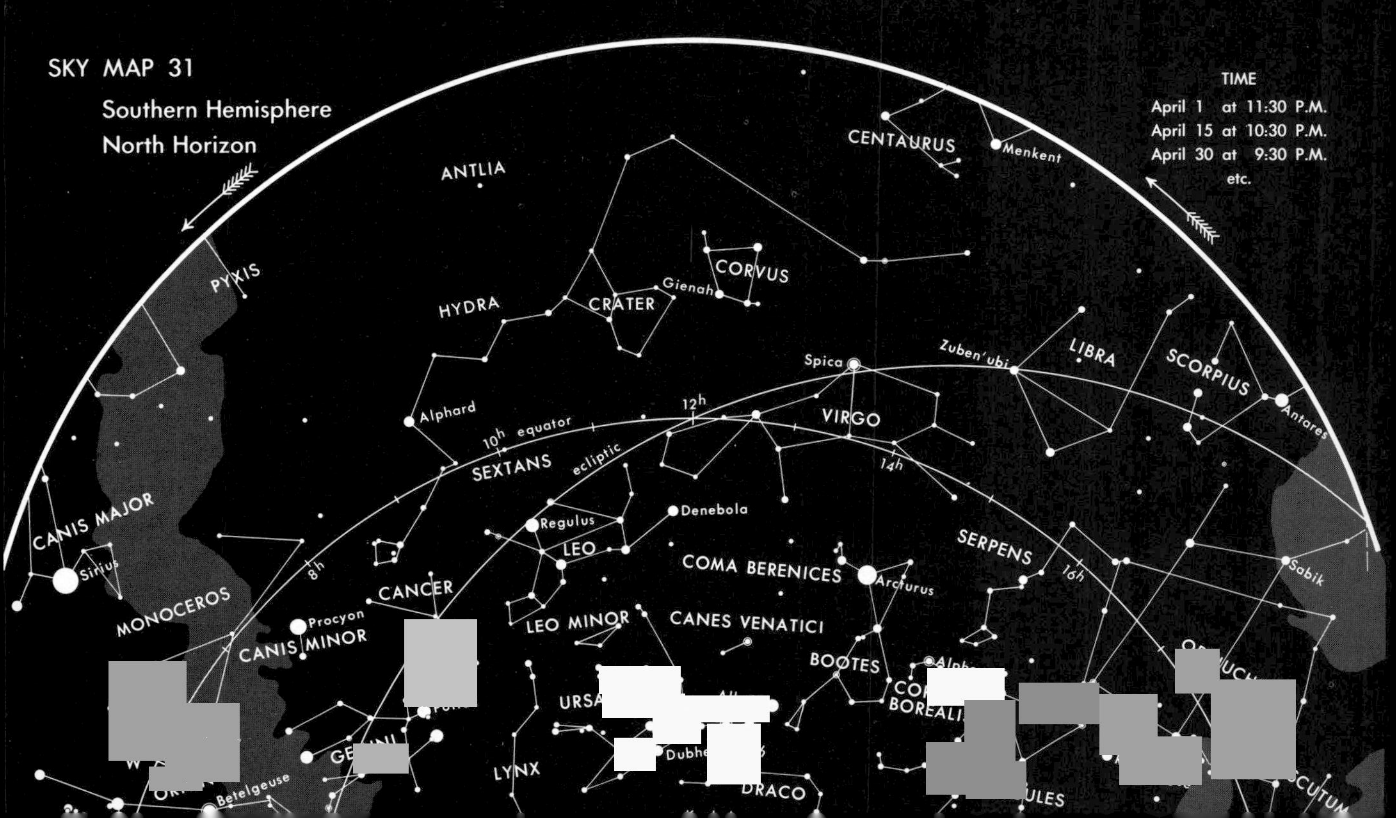
SKY MAP 31
Southern Hemisphere
North Horizon
TIME
April 1 at 11:30 P.M.
April 15 at 10:30 P.M.
April 30 at 9:30 P.M.
etc.
CENTAURUS
Menkent
ANTLIA
PYXIS
HYDRA
CRATER
CORVUS
Gienah
Spica
Zuben'ubi
LIBRA
SCORPIUS
Antares
Alphard
10h equator
12h
14h
16h
8h
SEXTANS
ecliptic
VIRGO
Regulus
LEO
Denebola
CANIS MAJOR
Sirius
MONOCEROS
Procyon
CANIS MINOR
CANCER
LEO MINOR
COMA BERENICES
Arcturus
CANES VENATICI
SERPENS
Sabik
BOOTES
BOREALI
URSA
LYNX
Dubhe
DRACO
Betelgeuse

MAGNITUDES
Sirius
Canopus
—0.49-0.00
0.01-0.50
0.51-1.00
1.01-1.50
MAGNITUDES
1.51-2.00
2.01-2.50
2.51-3.00
3.01-3.50
3.51-4.00
4.01-4.55
Variable
W
E

SKY MAP 32

Southern Hemisphere

South Horizon

TIME

April 1 at 11:30 P.M.
April 15 at 10:30 P.M.
April 30 at 9:30 P.M.
etc.

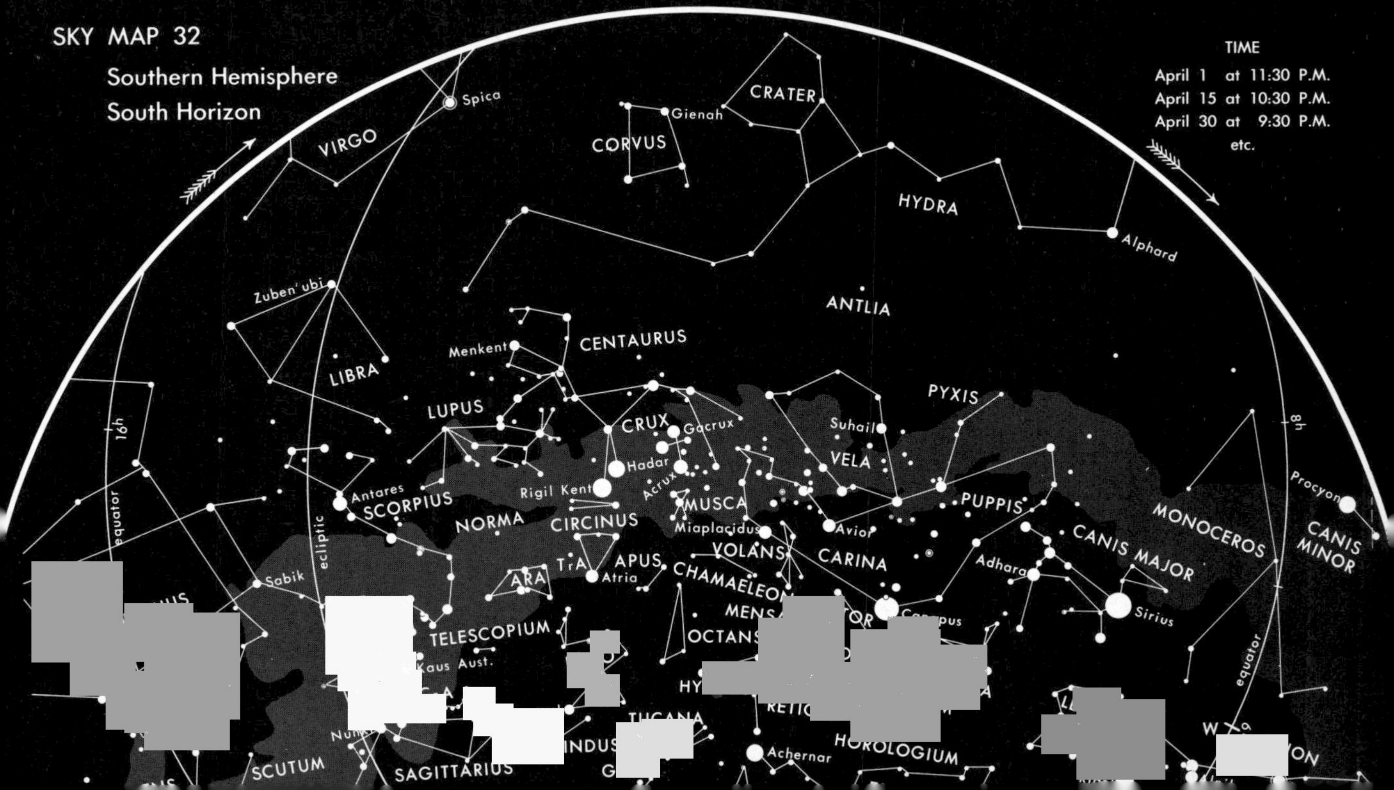

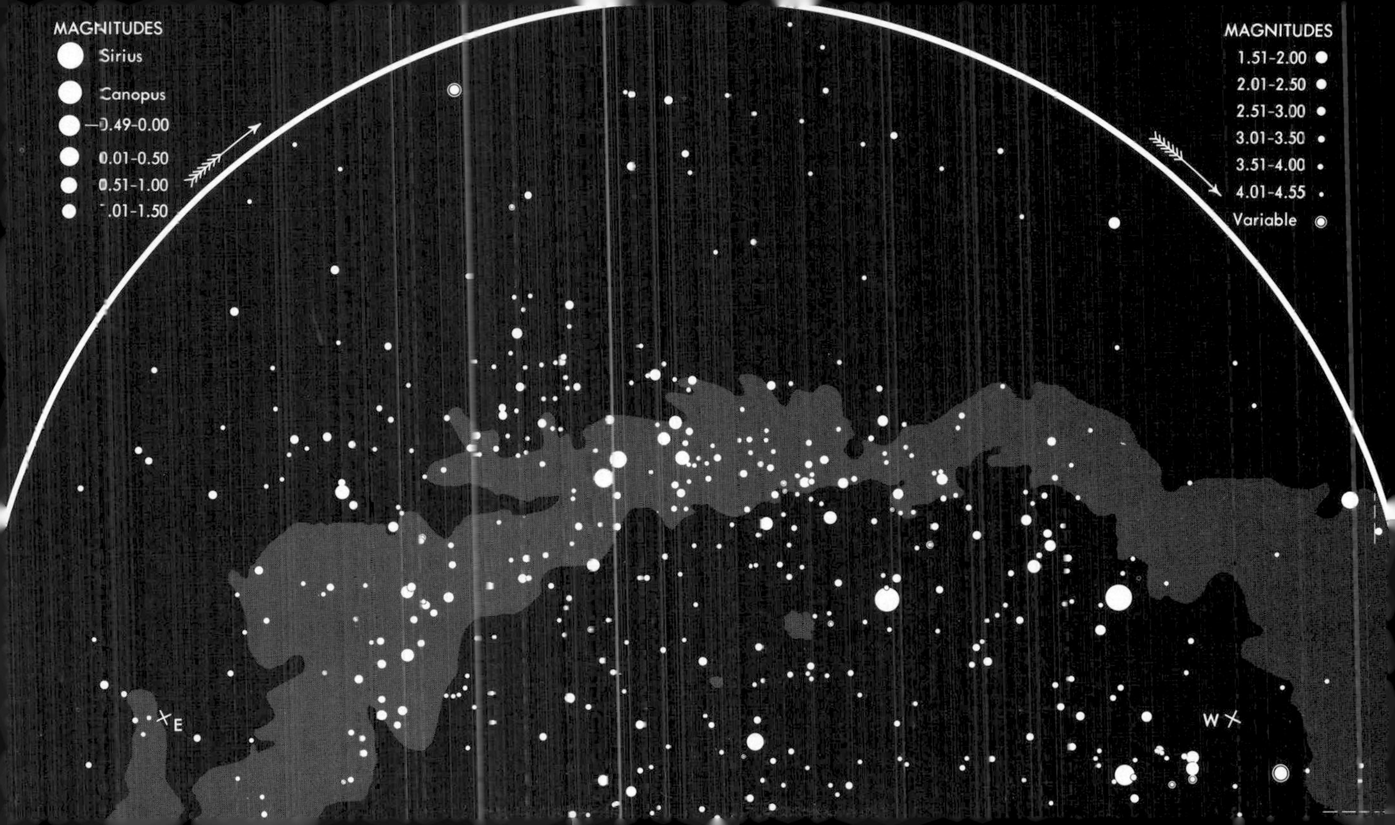
MAGNITUDES
Sirius
Canopus
—0.49-0.00
0.01-0.50
0.51-1.00
1.01-1.50
MAGNITUDES
1.51-2.00
2.01-2.50
2.51-3.00
3.01-3.50
3.51-4.00
4.01-4.55
Variable
E
W

SKY MAP 33

Southern Hemisphere

North Horizon

TIME

May 1 at 11:30 P.M.

May 15 at 10:30 P.M.

May 30 at 9:30 P.M.

etc.

MAGNITUDES
Sirius
Canopus
−0.49–0.00
0.01–0.50
0.51–1.00
1.01–1.50
MAGNITUDES
1.51–2.00
2.01–2.50
2.51–3.00
3.01–3.50
3.51–4.00
4.01–4.55
Variable
W
E

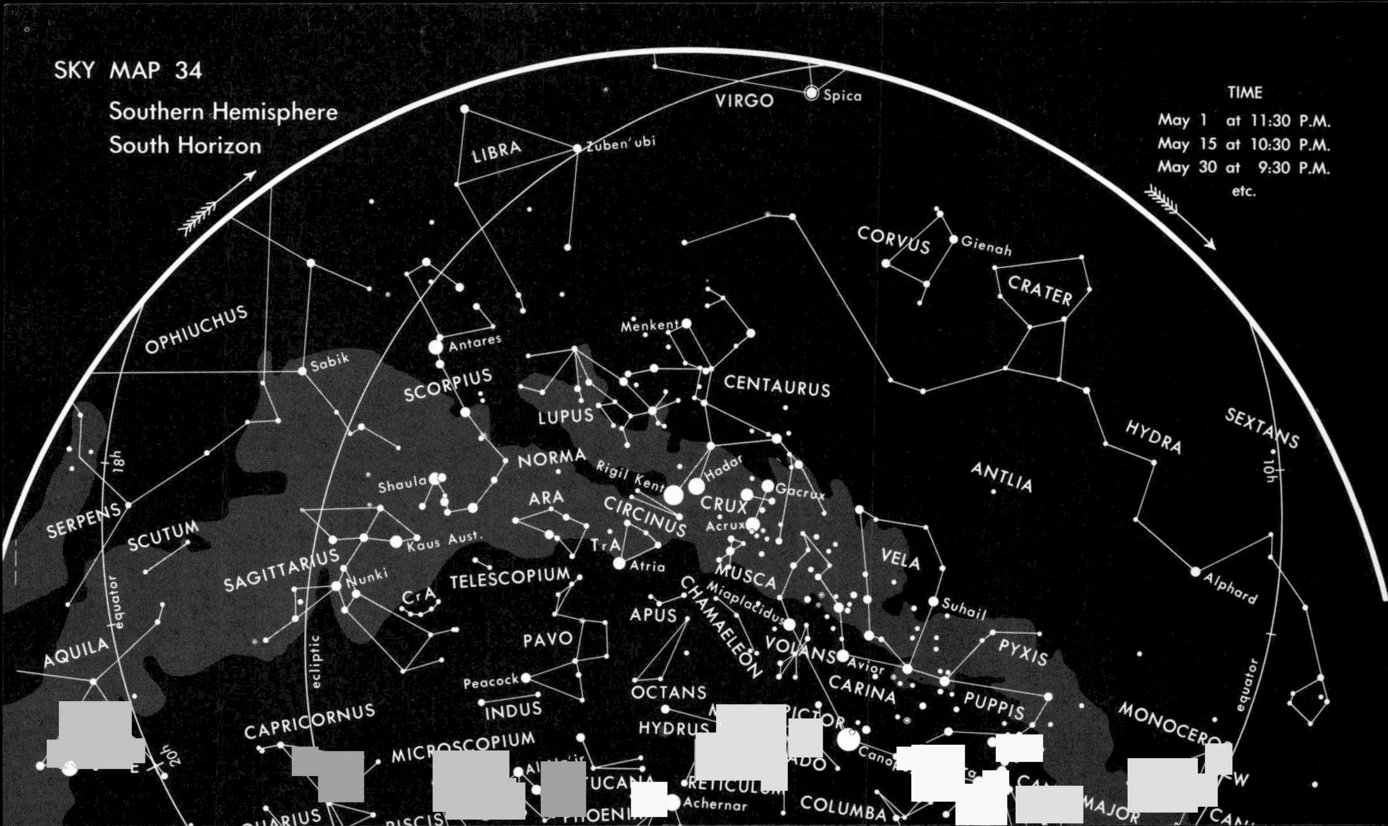
SKY MAP 34
Southern Hemisphere
South Horizon
TIME
May 1 at 11:30 P.M.
May 15 at 10:30 P.M.
May 30 at 9:30 P.M.
etc.
VIRGO
Spica
LIBRA
Zuben'ubi
CORVUS
Gienah
CRATER
OPHIUCHUS
Antares
Menkent
CENTAURUS
Sabik
SCORPIUS
LUPUS
HYDRA
SEXTANS
ANTLIA
18h
10h
NORMA
Rigil Kent
Hadar
Gacrux
CRUX
Acrux
SERPENS
Shaula
ARA
CIRCINUS
SCUTUM
Kaus Aust.
TrA
Atria
VELA
Alphard
SAGITTARIUS
TELESCOPIUM
MUSCA
Nunki
CrA
APUS
Miaplacidus
CHAMAELEON
Suhail
equator
AQUILA
PAVO
VOLANS
Avior
PYXIS
ecliptic
Peacock
OCTANS
CARINA
PUPPIS
INDUS
HYDRUS
MONOCEROS
CAPRICORNUS
Canopus
20h
MICROSCOPIUM
TUCANA
RETICULUM
COLUMBA
Achernar
MAJOR

MAGNITUDES
Sirius
Canopus
−0.49-0.00
0.01-0.50
0.51-1.00
1.01-1.50
MAGNITUDES
1.51-2.00
2.01-2.50
2.51-3.00
3.01-3.50
3.51-4.00
4.01-4.55
Variable
E
W

SKY MAP 35

Southern Hemisphere

North Horizon

TIME

June 1 at 11:30 P.M.
June 15 at 10:30 P.M.
June 30 at 9:30 P.M.
etc.

MAGNITUDES
Sirius
Canopus
−0.49–0.00
0.01–0.50
0.51–1.00
1.01–1.50
MAGNITUDES
1.51–2.00
2.01–2.50
2.51–3.00
3.01–3.50
3.51–4.00
4.01–4.55
Variable
W
E

SKY MAP 36

Southern Hemisphere

South Horizon

TIME

June 1 at 11:30 P.M.

June 15 at 10:30 P.M.

June 30 at 9:30 P.M.

etc.

MAGNITUDES
Sirius
Canopus
—0.49-0.00
0.01-0.50
0.51-1.00
1.01-1.50
MAGNITUDES
1.51-2.00
2.01-2.50
2.51-3.00
3.01-3.50
3.51-4.00
4.01-4.55
Variable
E
W

SKY MAP 37

Southern Hemisphere

North Horizon

TIME

July 1 at 11:30 P.M.

July 15 at 10:30 P.M.

July 30 at 9:30 P.M.

etc.

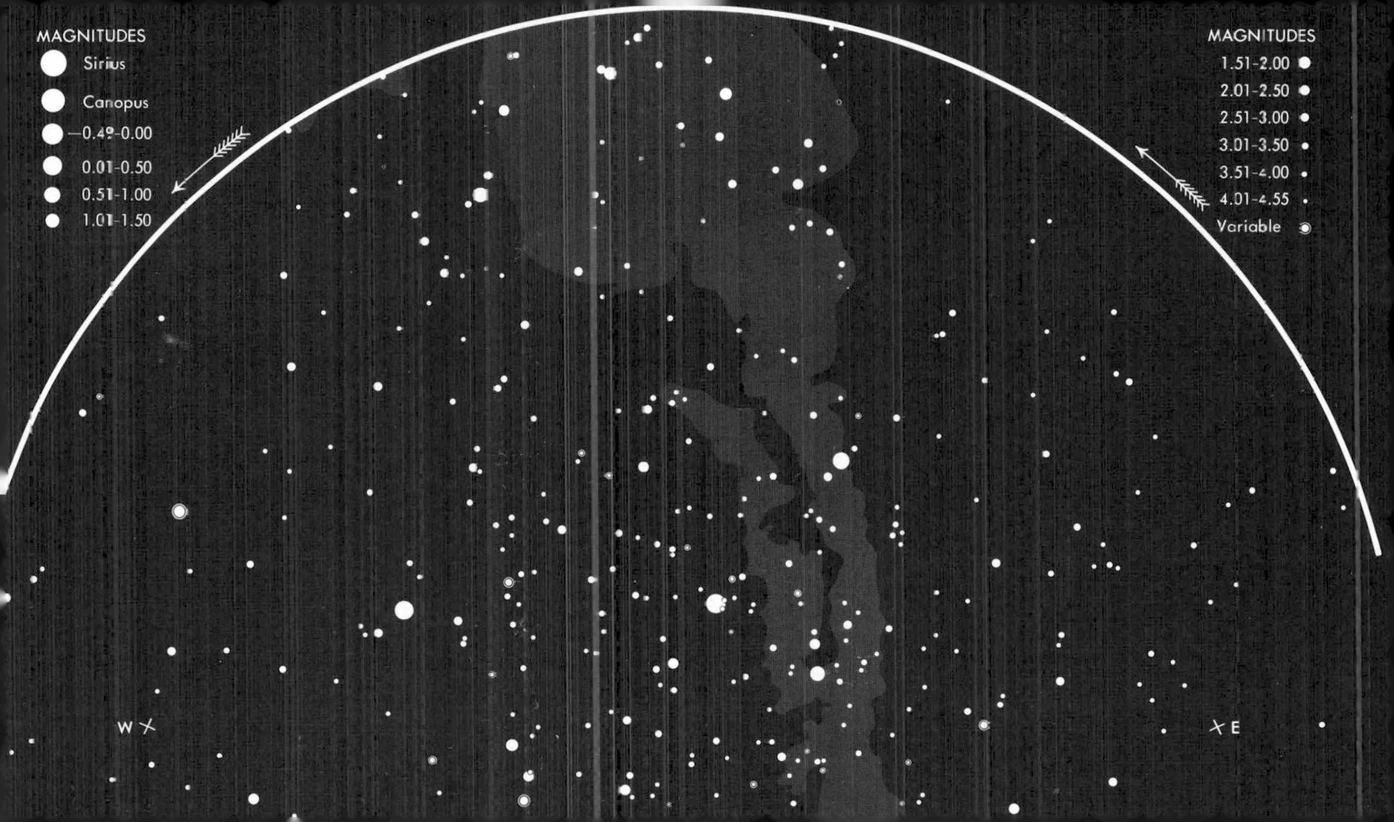
MAGNITUDES
Sirius
Canopus
−0.49–0.00
0.01–0.50
0.51–1.00
1.01–1.50
MAGNITUDES
1.51–2.00
2.01–2.50
2.51–3.00
3.01–3.50
3.51–4.00
4.01–4.55
Variable
W
E

SKY MAP 38

Southern Hemisphere

South Horizon

TIME

July 1 at 11:30 P.M.
July 15 at 10:30 P.M.
July 30 at 9:30 P.M.
etc.

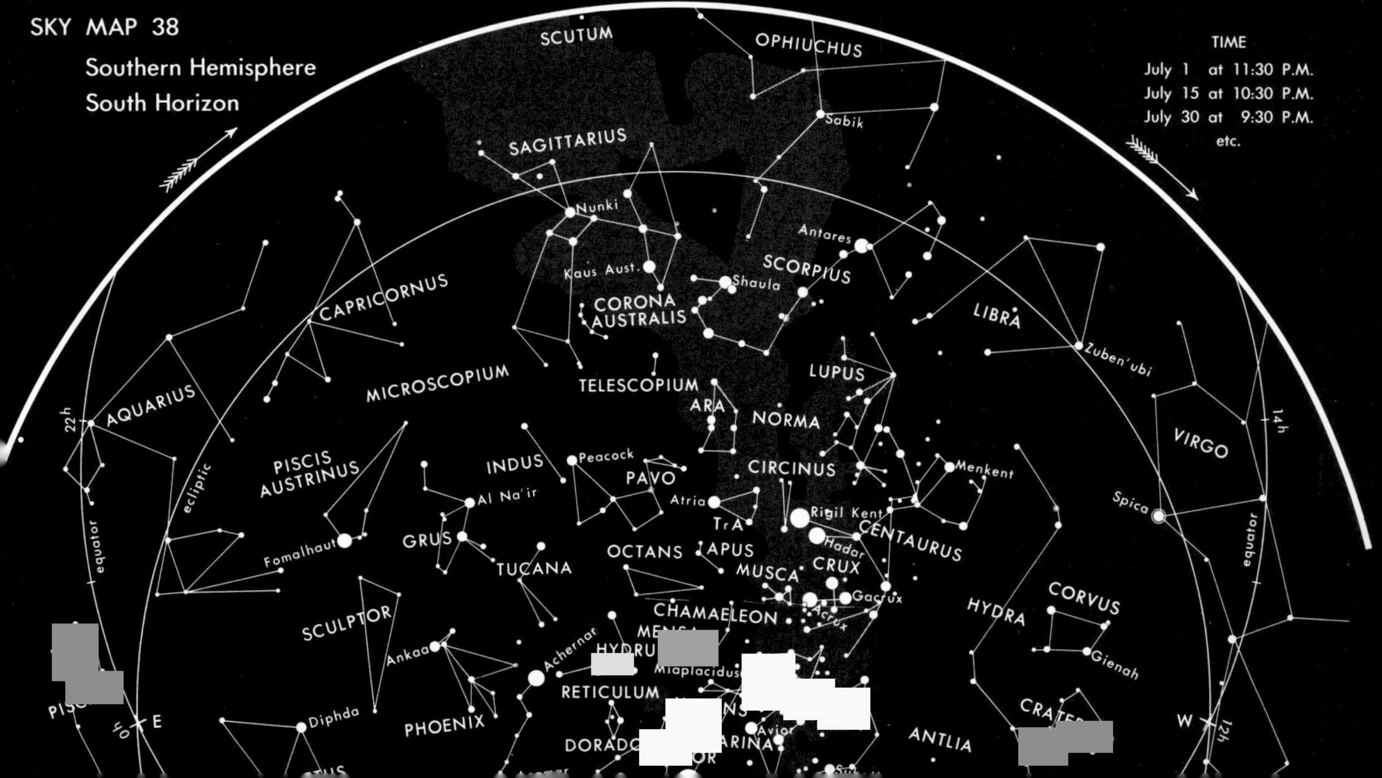

MAGNITUDES
Sirius
Canopus
−0.49–0.00
0.01–0.50
0.51–1.00
1.01–1.50
MAGNITUDES
1.51–2.00
2.01–2.50
2.51–3.00
3.01–3.50
3.51–4.00
4.01–4.55
Variable
E
W

SKY MAP 39

Southern Hemisphere

North Horizon

TIME

August 1 at 11:30 P.M.
August 15 at 10:30 P.M.
August 30 at 9:30 P.M.
etc.

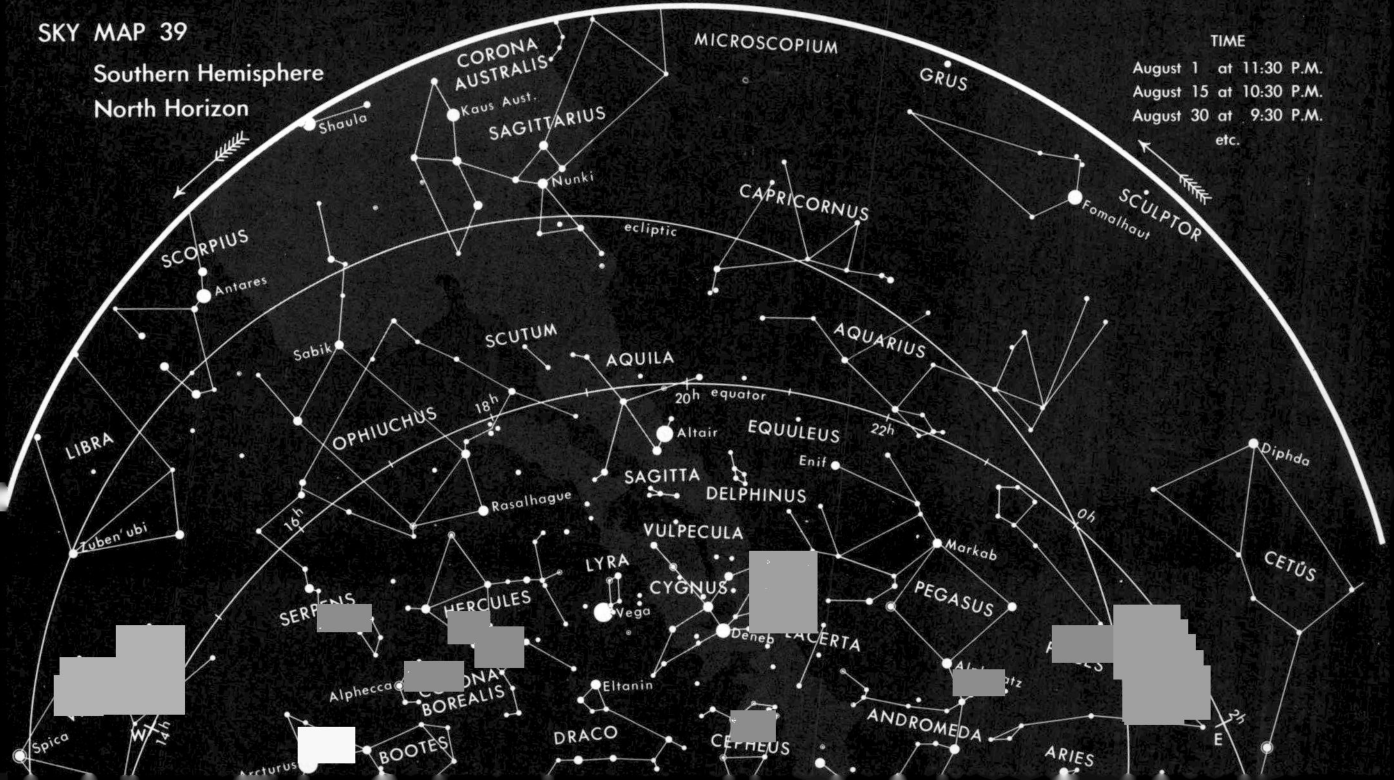

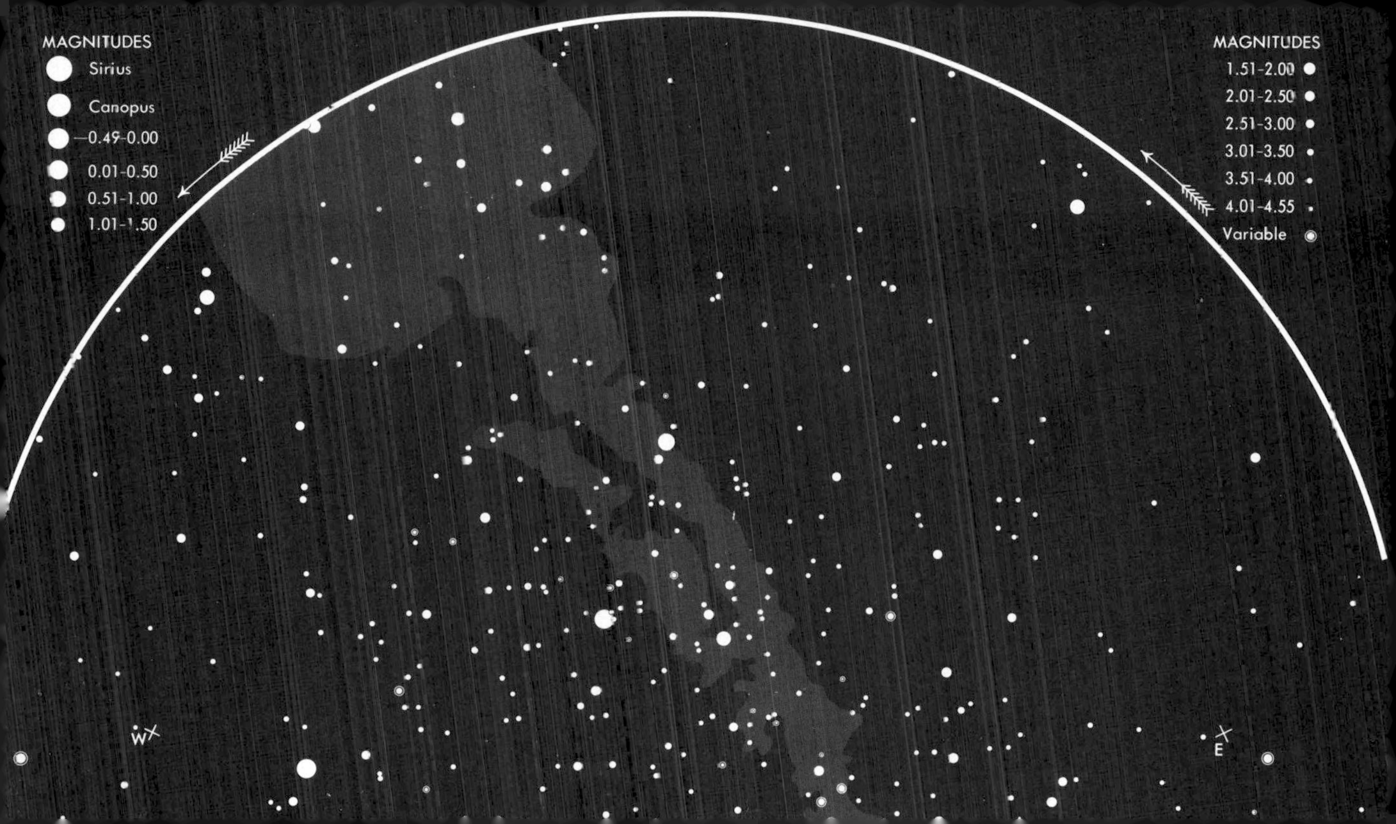

MAGNITUDES
Sirius
Canopus
—0.49-0.00
0.01-0.50
0.51-1.00
1.01-1.50
MAGNITUDES
1.51-2.00
2.01-2.50
2.51-3.00
3.01-3.50
3.51-4.00
4.01-4.55
Variable
W
E

SKY MAP 40

Southern Hemisphere

South Horizon

TIME	
August 1	at 11:30 P.M.
August 15	at 10:30 P.M.
August 30	at 9:30 P.M.
etc.	

MAGNITUDES
Sirius
Canopus
−0.49–0.00
0.01–0.50
0.51–1.00
1.01–1.50
MAGNITUDES
1.51–2.00
2.01–2.50
2.51–3.00
3.01–3.50
3.51–4.00
4.01–4.55
Variable
E
W

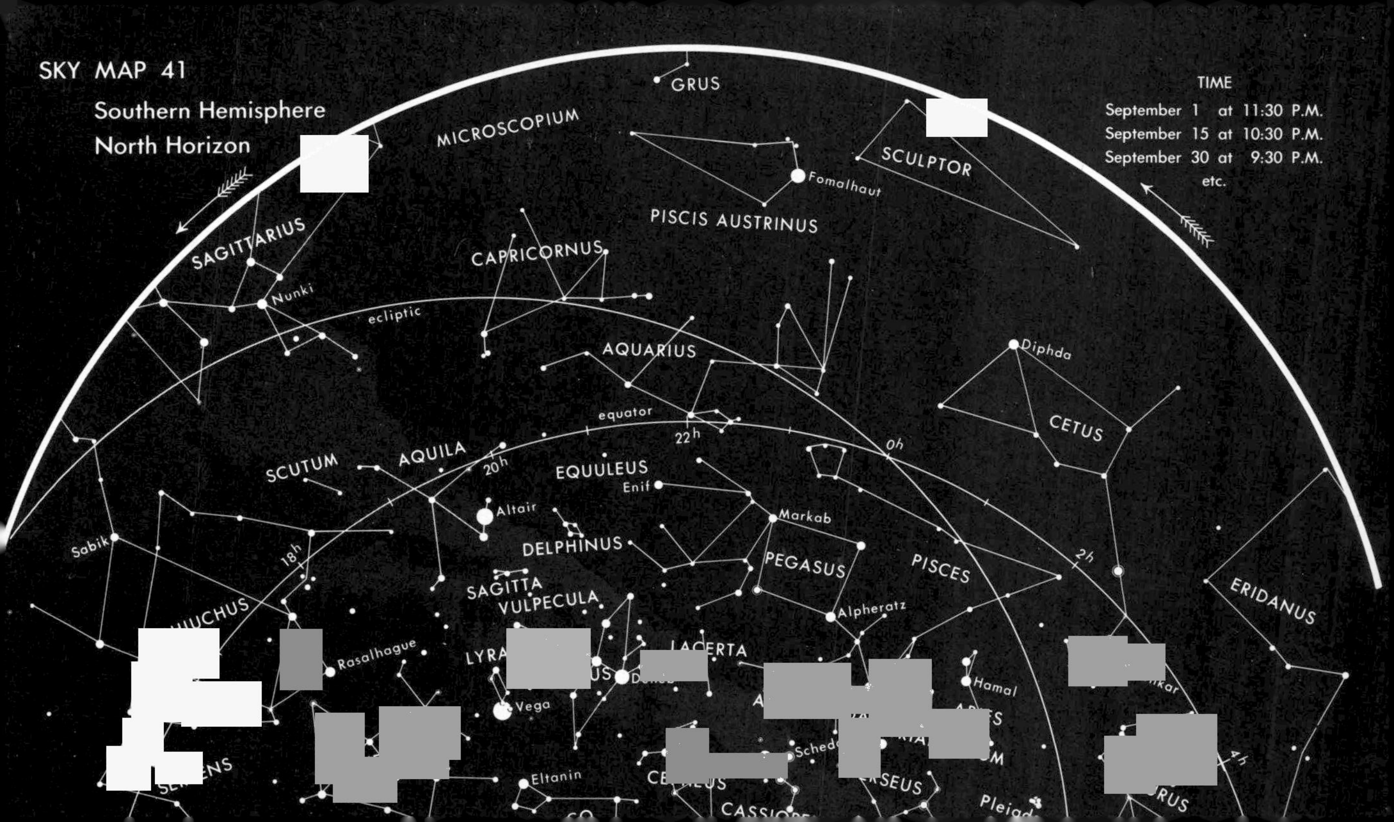

SKY MAP 41
Southern Hemisphere
North Horizon
TIME
September 1 at 11:30 P.M.
September 15 at 10:30 P.M.
September 30 at 9:30 P.M.
etc.
GRUS
MICROSCOPIUM
SCULPTOR
Fomalhaut
PISCIS AUSTRINUS
SAGITTARIUS
Nunki
CAPRICORNUS
ecliptic
AQUARIUS
Diphda
CETUS
equator
22h
0h
20h
18h
2h
4h
SCUTUM
AQUILA
EQUULEUS
Enif
Markab
Altair
DELPHINUS
PEGASUS
PISCES
Sabik
SAGITTA
VULPECULA
Alpheratz
ERIDANUS
LACERTA
Rasalhague
Hamal
Vega
Eltanin

MAGNITUDES
Sirius
Canopus
−0.49–0.00
0.01–0.50
0.51–1.00
1.01–1.50
MAGNITUDES
1.51–2.00
2.01–2.50
2.51–3.00
3.01–3.50
3.51–4.00
4.01–4.55
Variable
W
E

SKY MAP 42

Southern Hemisphere
South Horizon

TIME

September 1 at 11:30 P.M.
September 15 at 10:30 P.M.
September 30 at 9:30 P.M.
etc.

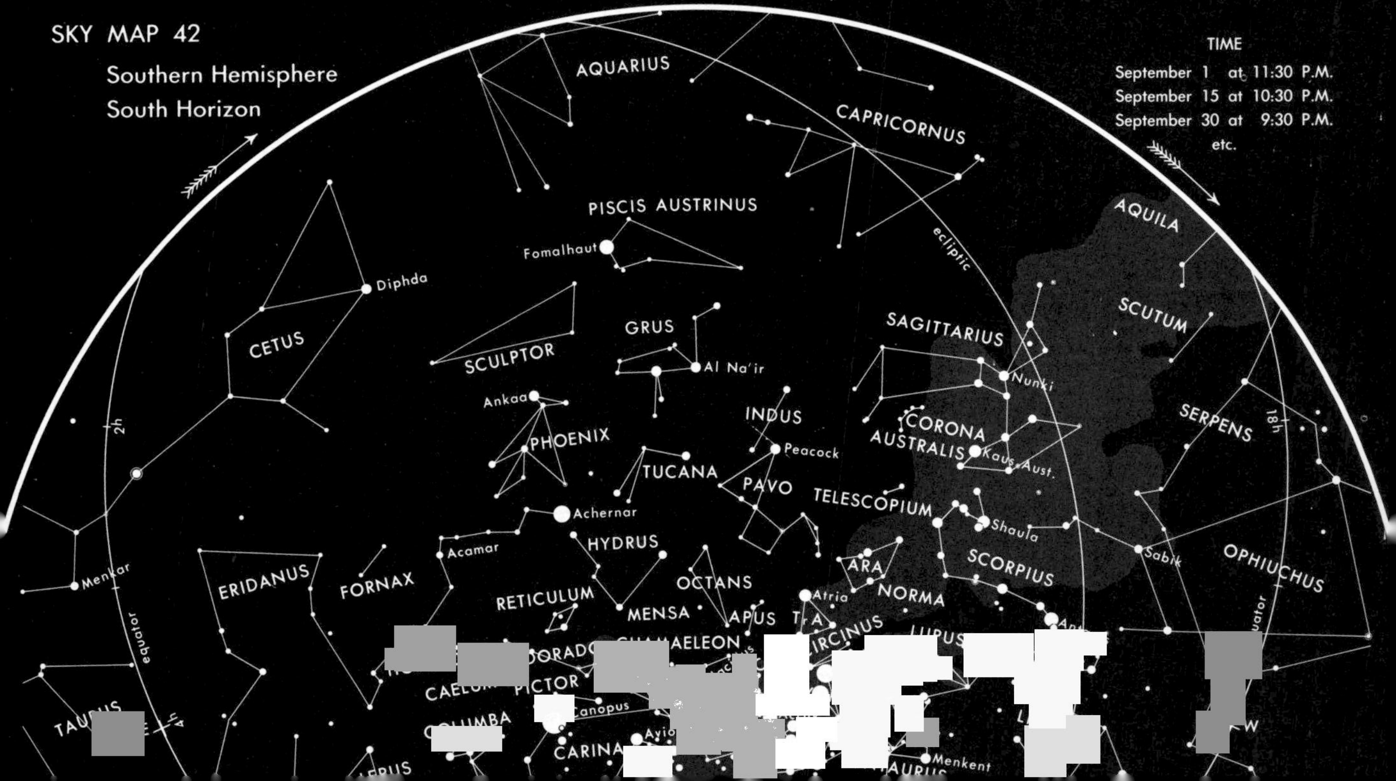

MAGNITUDES
Sirius
Canopus
−0.49–0.00
0.01–0.50
0.51–1.00
1.01–1.50
MAGNITUDES
1.51–2.00
2.01–2.50
2.51–3.00
3.01–3.50
3.51–4.00
4.01–4.55
Variable
E
W

SKY MAP 43

Southern Hemisphere
North Horizon

TIME

October 1 at 11:30 P.M.
October 15 at 10:30 P.M.
October 30 at 9:30 P.M.
etc.

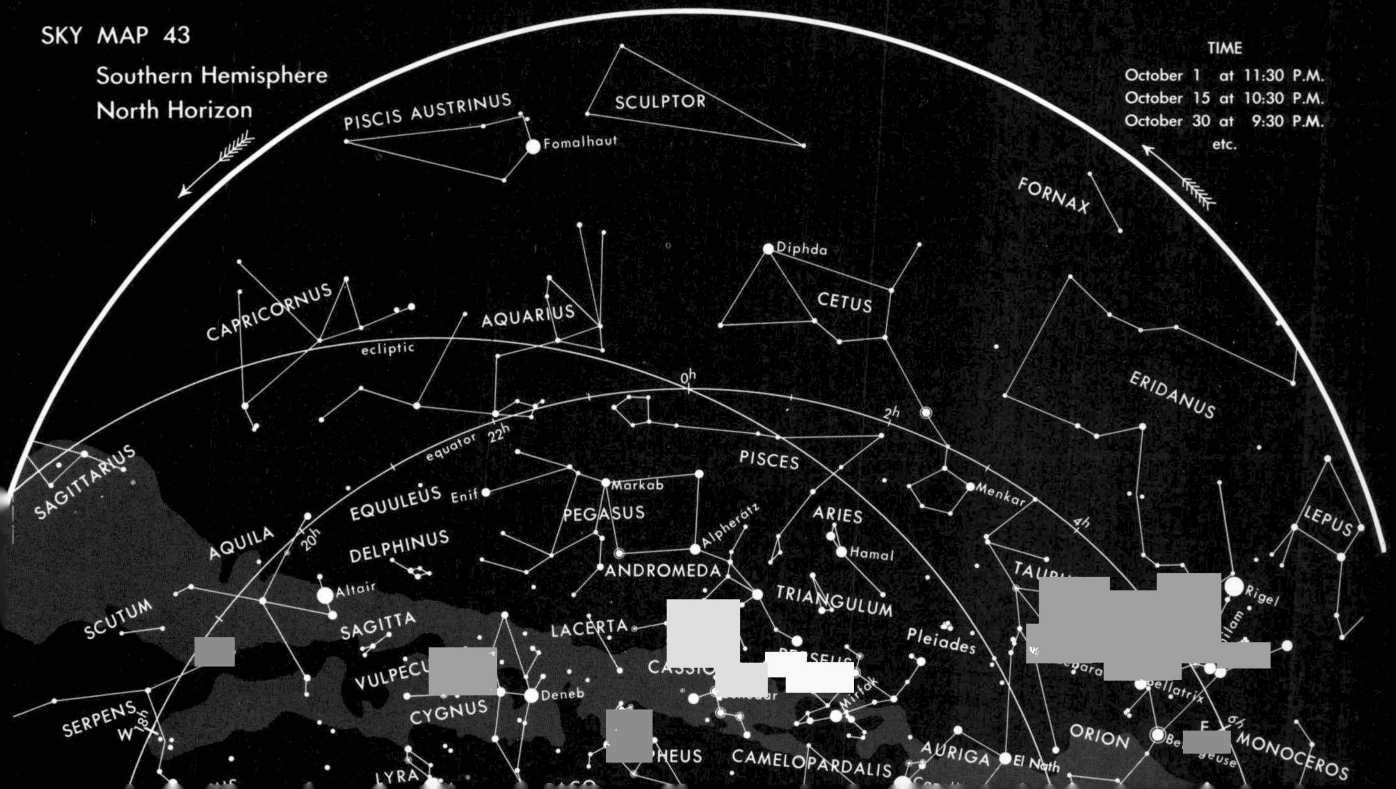

MAGNITUDES
Sirius
Canopus
—0.49-0.00
0.01-0.50
0.51-1.00
1.01-1.50
MAGNITUDES
1.51-2.00
2.01-2.50
2.51-3.00
3.01-3.50
3.51-4.00
4.01-4.55
Variable
W
E

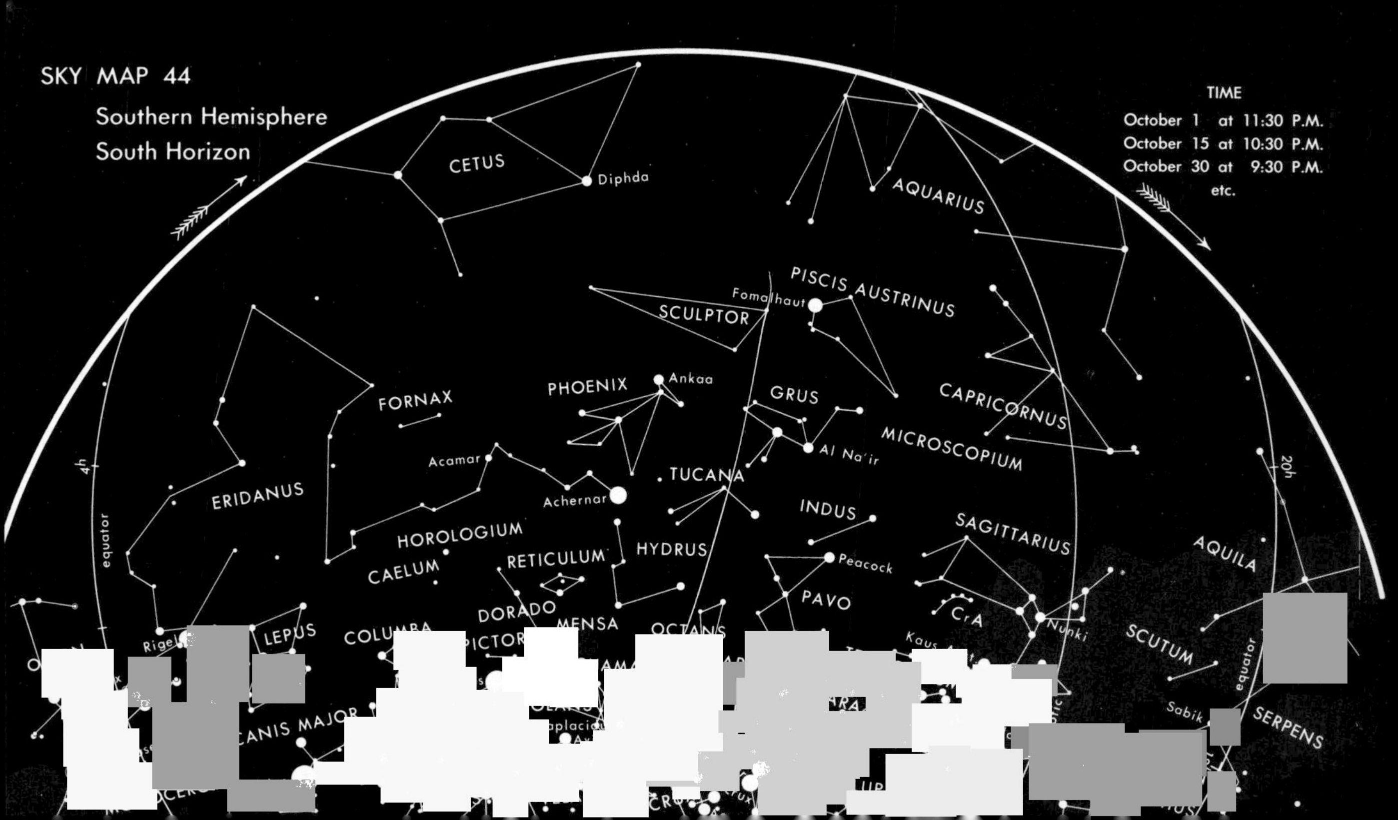

SKY MAP 44
Southern Hemisphere
South Horizon
TIME
October 1 at 11:30 P.M.
October 15 at 10:30 P.M.
October 30 at 9:30 P.M.
etc.
CETUS
Diphda
AQUARIUS
PISCIS AUSTRINUS
Fomalhaut
SCULPTOR
CAPRICORNUS
MICROSCOPIUM
PHOENIX
Ankaa
GRUS
Al Na'ir
FORNAX
Acamar
Achernar
TUCANA
INDUS
SAGITTARIUS
AQUILA
ERIDANUS
HOROLOGIUM
CAELUM
RETICULUM
HYDRUS
Peacock
PAVO
CrA
Nunki
SCUTUM
Sabik
SERPENS
DORADO
MENSA
OCTANS
COLUMBA
LEPUS
Rigel
CANIS MAJOR
equator
4h
20h

MAGNITUDES
Sirius
Canopus
—0.49-0.00
0.01-0.50
0.51-1.00
1.01-1.50
MAGNITUDES
1.51-2.00
2.01-2.50
2.51-3.00
3.01-3.50
3.51-4.00
4.01-4.55
Variable
E
W

SKY MAP 45

Southern Hemisphere
North Horizon

TIME

November 1 at 11:30 P.M.
November 15 at 10:30 P.M.
November 30 at 9:30 P.M.
etc.

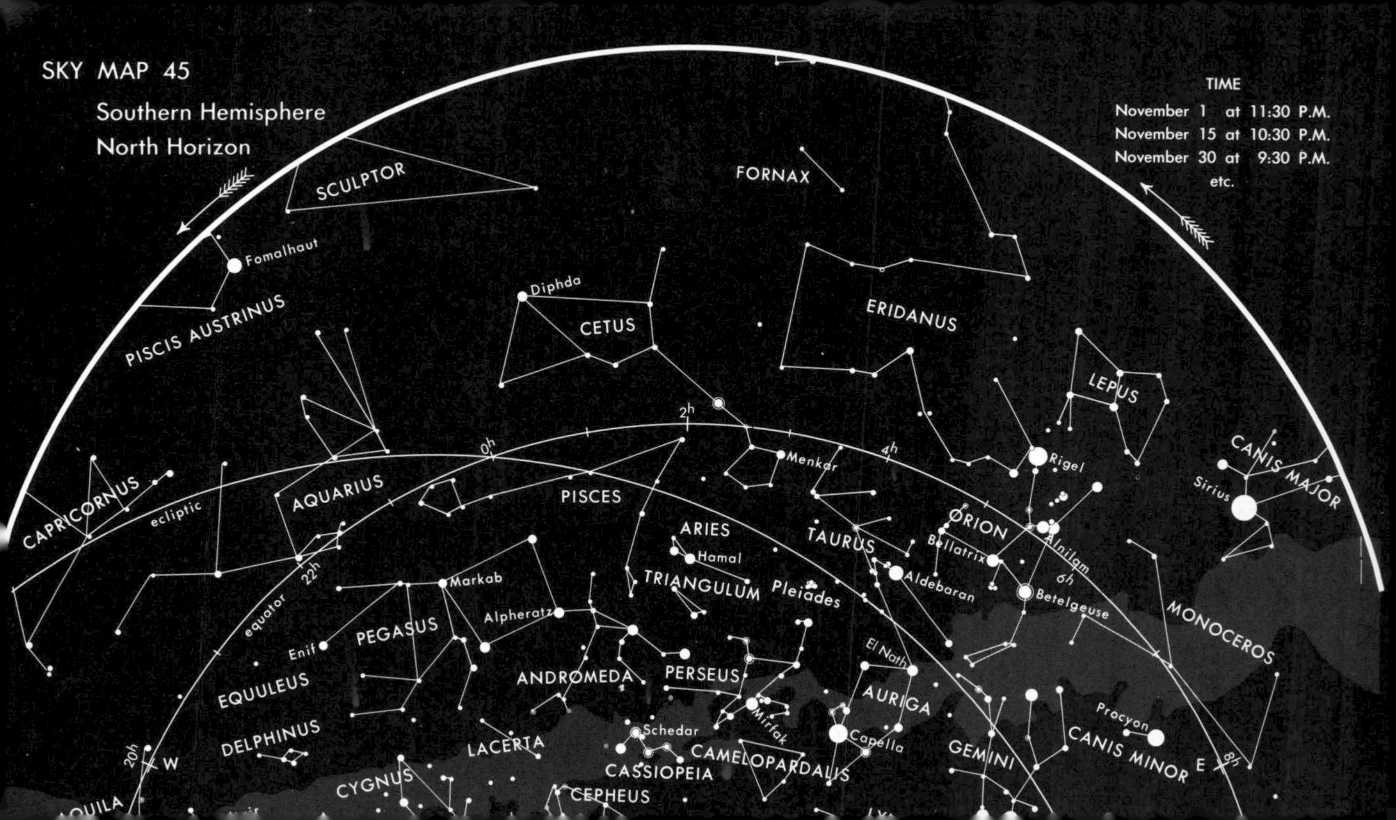

MAGNITUDES
Sirius
Canopus
−0.49-0.00
0.01-0.50
0.51-1.00
1.01-1.50
MAGNITUDES
1.51-2.00
2.01-2.50
2.51-3.00
3.01-3.50
3.51-4.00
4.01-4.55
Variable
W
E

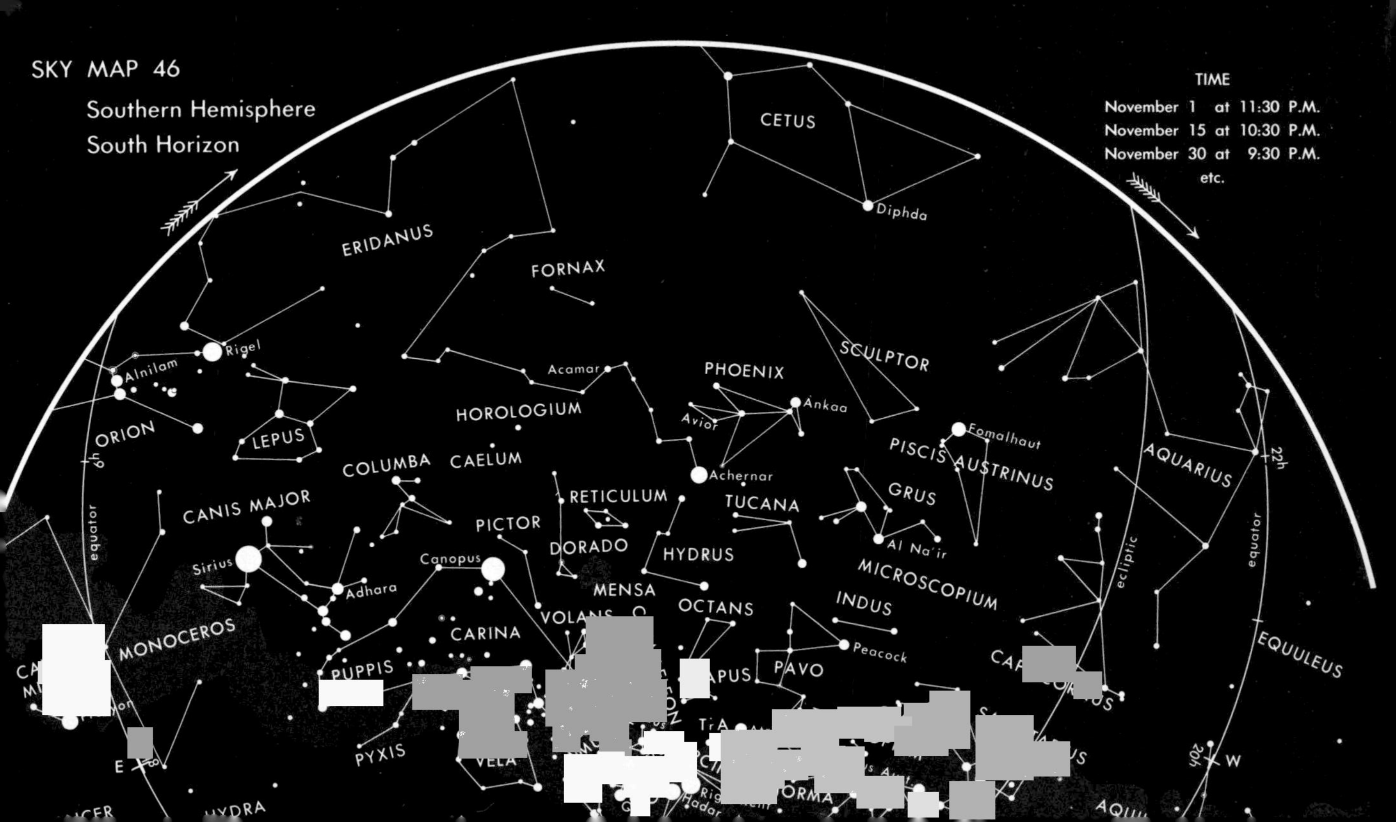
SKY MAP 46
Southern Hemisphere
South Horizon
TIME
November 1 at 11:30 P.M.
November 15 at 10:30 P.M.
November 30 at 9:30 P.M.
etc.
CETUS
Diphda
ERIDANUS
FORNAX
SCULPTOR
Rigel
Alnilam
ORION
LEPUS
Acamar
HOROLOGIUM
PHOENIX
Ankaa
Avior
Fomalhaut
PISCIS AUSTRINUS
AQUARIUS
22h
COLUMBA
CAELUM
Achernar
RETICULUM
TUCANA
GRUS
CANIS MAJOR
PICTOR
DORADO
HYDRUS
Al Na'ir
Sirius
Canopus
Adhara
MENSA
OCTANS
MICROSCOPIUM
INDUS
VOLANS
CARINA
MONOCEROS
PUPPIS
Peacock
PAVO
APUS
PYXIS
VELA
EQUULEUS
equator
ecliptic
6h
20h
W
E

MAGNITUDES
Sirius
Canopus
−0.49–0.00
0.01–0.50
0.51–1.00
1.01–1.50
MAGNITUDES
1.51–2.00
2.01–2.50
2.51–3.00
3.01–3.50
3.51–4.00
4.01–4.55
Variable
E
W

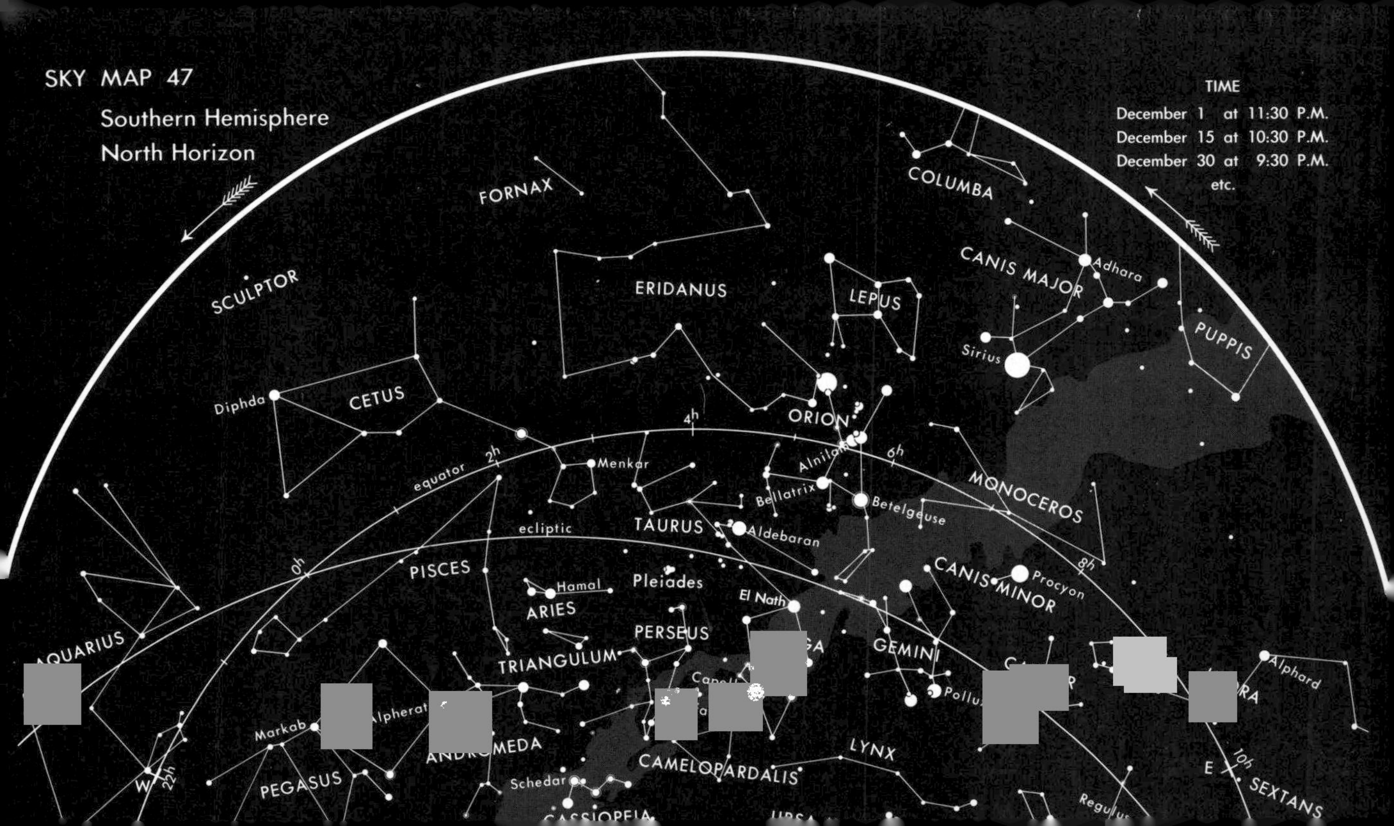

SKY MAP 47
Southern Hemisphere
North Horizon
TIME
December 1 at 11:30 P.M.
December 15 at 10:30 P.M.
December 30 at 9:30 P.M.
etc.
FORNAX
SCULPTOR
ERIDANUS
COLUMBA
CANIS MAJOR
Adhara
PUPPIS
LEPUS
Sirius
CETUS
Diphda
ORION
Alnilam
Bellatrix
Betelgeuse
Menkar
equator
ecliptic
2h
4h
6h
8h
0h
10h
22h
MONOCEROS
TAURUS
Aldebaran
CANIS MINOR
Procyon
PISCES
Hamal
ARIES
Pleiades
El Nath
PERSEUS
GEMINI
Pollux
Alphard
TRIANGULUM
Markab
PEGASUS
Schedar
CAMELOPARDALIS
LYNX
E
W
SEXTANS

MAGNITUDES
Sirius
Canopus
−0.49-0.00
0.01-0.50
0.51-1.00
1.01-1.50
MAGNITUDES
1.51-2.00
2.01-2.50
2.51-3.00
3.01-3.50
3.51-4.00
4.01-4.55
Variable
W
E

SKY MAP 48

Southern Hemisphere
South Horizon

TIME
December 1 at 11:30 P.M.
December 15 at 10:30 P.M.
December 30 at 9:30 P.M.
etc.

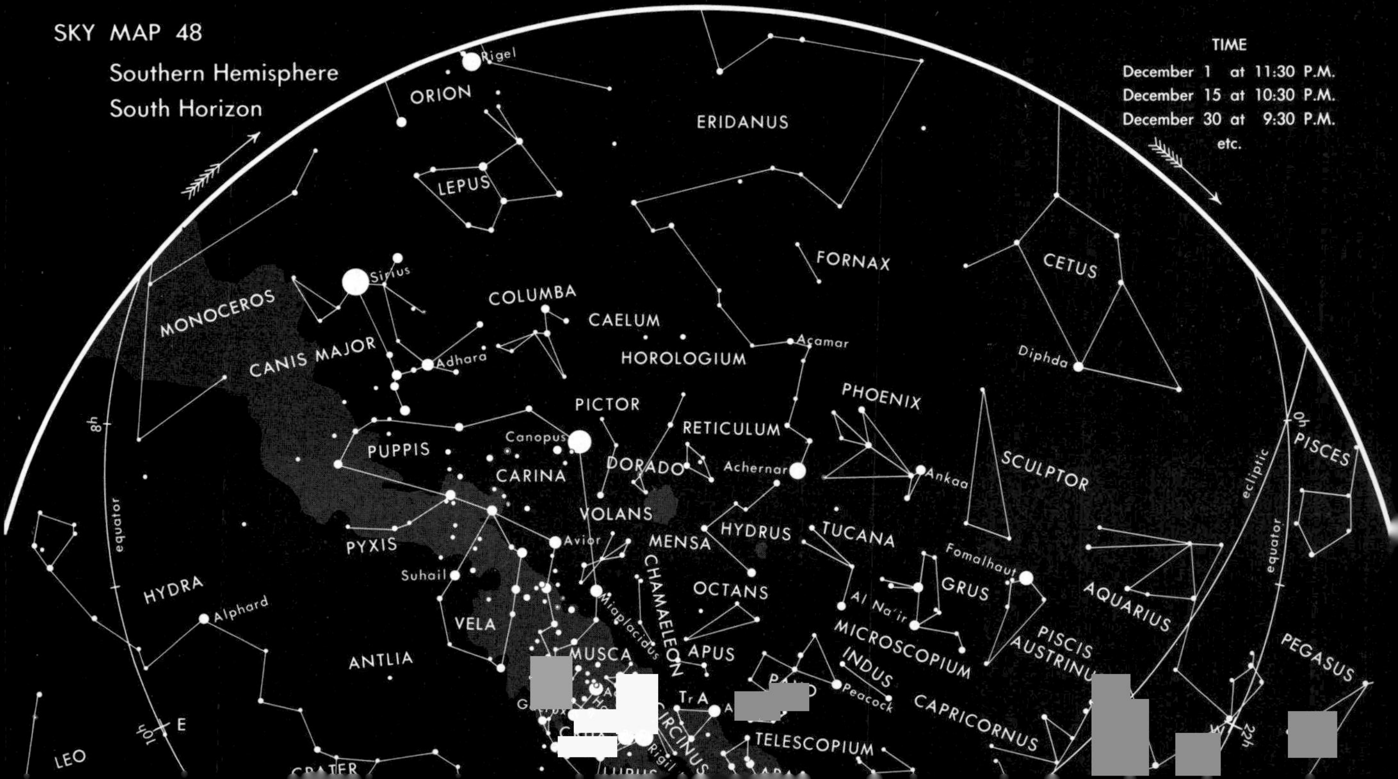

MAGNITUDES
Sirius
Canopus
−0.49–0.00
0.01–0.50
0.51–1.00
1.01–1.50
MAGNITUDES
1.51–2.00
2.01–2.50
2.51–3.00
3.01–3.50
3.51–4.00
4.01–4.55
Variable
E
W

III

Pathways in the Sky

To FIND your way among the stars, choose some familiar pattern as a starting point and gradually work your way from one star group to another. The endpaper on the inside front cover illustrates useful ways of tracking from one constellation to another. The green arrows direct the eye to significant features of various constellations. Miscellaneous triangles of bright stars, indicated by dash lines, help you to check your findings and extend your knowledge of the sky.

The Big Dipper (Ursa Major) is perhaps the best place to begin, because it is easily recognized and also because it lies above the horizon for so large a portion of the globe. The Pointers, the two stars in the front of the Dipper, are the most dependable guides to start with. In the northerly direction they indicate the all-important Pole Star. In the opposite direction they point toward the constellation of Leo, with its conspicuous asterism, the Sickle. From the Pointers proceed directly to the Pole Star and then turn a sharp right angle toward the right and you will encounter the prominent 1st-magnitude star Capella, in the constellation of Auriga. At Capella, angle across the constellation of Gemini toward Procyon, in Canis Minor.

Now return to the Big Dipper, to the star that lies between the bowl and the bend in the Dipper handle. Proceed again to Polaris and continue straight on for an equal distance on the opposite side. There you will encounter the W-shaped figure, Cassiopeia. Along a curved line, as indicated, lie the "four C's" (Camelopardalis, Cassiopeia, Cepheus, and Cygnus), in alphabetical order. Deneb and Vega form the base of an isosceles triangle, with Altair in the vertex.

Return again to Polaris and draw a line from that star through Caph (β Cassiopeiae) and extend it south to Alpheratz (α Andromedae) and the eastern edge of the Great Square of Pegasus. The rear feet of the Winged Horse rest on Aquarius. Just south of the Square lies an asterism, the Circlet, a delicate ring of stars marking the head of the western fish in Pisces. To the east of Andromeda lies the associated constellation Perseus, the hero rushing to save the lady chained to the rock. A line drawn southeastward from the legs of Andromeda encounters, in turn, Triangulum, Aries, and the Head of Cetus. The Knot of Pisces lies just west of the Neck of Cetus.

Again return to Polaris and note the Guards, the bright pair of stars at the end of the bowl of the Little Dipper. Draw a line from Polaris near the eastern Guard star and extend it southward. It will indicate another delicate circle of stars, the well-known constellation Corona Borealis, the Northern Crown. Note how the body of Draco seems to hold the Little Dipper, Ursa Minor in its folds, its head well marked by the asterism, the Lozenge. Just to the south of the head of Draco lies the constellation Hercules, with its distinctive asterism, the Keystone. Note how the head of Hercules and the head of the other giant, Ophiuchus, are essentially touching. Hercules, holding a bow in his outstretched hand, has just launched an arrow, Sagitta, toward the two birds, Cygnus and Aquila, both of whom seem to have escaped. Lyra, on old sky maps, also appears as a bird — Vultur, the Vulture.

Sagittarius, with its conspicuous asterism, the Milk Dipper, lies south of Aquila. Directly south of Ophiuchus lies Scorpius, with its bright red star Antares and the sharp sting beneath this giant's right foot. To the west of Scorpius lies Libra, designated by the ancients as the Claws of the Scorpion.

Return again to the north and from the Big Dipper extend the handle along an *arc* to *Arc*turus and continue this line in a *spike* to Spica. Note that Spica and Arcturus form the base of an isosceles triangle with its vertex at Denebola (β Leonis). Regulus, Procyon, and Alphard (α Hydrae) form a right-angle triangle. Procyon, Sirius, and Betelgeuse form another. And, as indicated in Table 6, "Asterisms," the stars Aldebaran, Capella, Castor, Pollux, Procyon, Sirius, Rigel, Bellatrix, and Betelgeuse make up a large and conspicuous asterism sometimes called the Heavenly G.

Orion continually fights Taurus, the Bull. Lepus, the Hare, lies just beneath Orion's foot.

To continue the pathways into the southern half of the sky, turn to the endpaper on the inside back cover. The Belt stars of Orion point toward the brightest star in the sky, Sirius. The River, Eridanus, rises near Rigel in the foot of Orion and wends its tortuous way south via the 1st-magnitude star Achernar. Canopus, in Carina, lies directly south of Sirius. Rigel, Sirius, and Canopus form a distinct right-angled triangle.

Canopus, Achernar, and Rigil Kent form another triangle that encloses a smaller triangle, formed by the Large Magellanic Cloud, the Small Magellanic Cloud, and the south celestial pole. As previously noted, there is no South Star analogous to Polaris.

On the other side of the pole, a line from the foot of Sagittarius through Pavo also points directly to the south celestial pole. Note the closeness of the four birds, Pavo, Tucana, Phoenix, and Grus. From the north, water flows from Aquarius into the mouth of Piscis Austrinus, the Southern Fish.

A somewhat irregular line drawn southward from the tail of Scorpius crosses Ara, Triangulum Australe, and eventually leads to the South Pole. Ara lies east of Lupus. Note, also, that Centaurus lies just south of Corvus and the Southern Triangle slightly east of Centaurus. The conspicuous pair of 1st-magnitude stars, α and β Centauri, point toward the nearby Southern Cross. This will help distinguish the true cross from the somewhat larger "False Cross" formed by ι and ϵ Carinae and δ and κ Velorum. The False Cross also lies in the Milky Way and its arms roughly parallel those of Crux.

When you have become familiar with the pathways just described, you will easily find a number of constellations, sufficient to provide sky marks for recognizing the positions of others. There is no royal road to learning the sky. To master each new area, you must continually check and recheck your path, especially from season to season, as the stars slowly change their places relative to the sun.

A résumé of the sky paths and additional helpful hints appear in Chapters IV and VI.

IV

Order and System of the Constellations

THE SKY was man's first picture book. Primitive races the world over have whiled away the hours of night looking at the slowly moving heavenly pageant. The stars suggested to them vague outlines of familiar animals and men. Gradually, as man's intellect grew, he expanded his relationship with the universe through an intricate legendry of the constellations. The myths we have inherited from the ancients enrich our appreciation for the starry patterns associated with them.

Rarely does the constellation resemble its supposed terrestrial prototype so clearly that recognition is immediate and beyond question. Orion, a major exception, does suggest a giant figure. But the Big Dipper, though a remarkable replica, is a modern asterism, part of Ursa Major, which more nearly resembles a long-tailed mouse than a lumbering bear.

How, then, can one expect to learn the sky as a whole, with its 88 distinct constellations, if each individual star group is independent of all the others? Fortunately, the sky is not quite as disorderly as a casual inspection would indicate. Indeed, all 88 constellations fall into only 8 separate and distinctive families, with a small amount of overlapping. We shall consider each family in turn, giving a brief commentary on each constellation. Chapter III provided a geometrical association between the groups; this chapter indicates a logical association based on mythology or other factors.

A. The Ursa Major Family

1. Ursa Major (the Great Bear) and its familiar asterism, the Big Dipper (known to the British as the Plow, the Wain, or Wagon), is the best skymark for starting constellation study. In the latitude of the United States, the Dipper never sets. Ursa Major rides highest in the early evenings of spring and lowest in the fall; legends of North American Indians imply that the animal is looking for a place to lie down, preparatory to winter hibernation.

The 7 stars are lettered α, β, γ, δ, ϵ, ζ, η from bowl to handle. Their respective names are: Dubhe, Merak, Phecda, Megrez, Alioth, Mizar, and Alkaid. Dubhe and Merak, at the fore of the

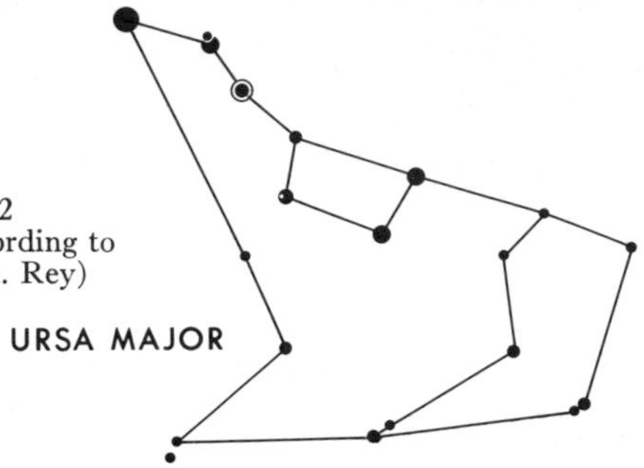

Fig. 2
(according to
H. A. Rey)

bowl, are often called the Pointers, because they point to the Pole Star (Polaris), the star at the end of the tail of

2. Ursa Minor (the Little Bear). This group is sometimes termed the Little Dipper because of a resemblance to an old-fashioned cream ladle or gravy spoon. The two brightest stars in the bowl of the dipper are often called the Guardians of the Pole or, more simply, the Guards.

3. Draco (the Dragon) is one of the few constellations that resemble the objects for which they are named. The tail starts north of the Big Dipper handle. The body turns northward enveloping Ursa Minor in a fold and then deviates abruptly south to a conspicuous lozenge-shaped head, near Hercules. Mizar, the star at the bend of the Dipper handle, Polaris, and the head of Draco form an approximate equilateral triangle.

4. Canes Venatici (the Hunting Dogs) consists of two bright stars south of, but still within, the bend of the Dipper handle. These dogs, yapping at the heels of Ursa Major, are a modern formulation by the astronomer Hevelius.

5. Boötes (the Bear Driver, sometimes called the Herdsman) lies south and east of Ursa Major. (Pronounce both *o*'s of Boötes, as in "coordinate.")

6. Coma Berenices (Berenice's Hair) imaginatively depicts the shorn locks of beautiful Berenice, an Egyptian queen. A line from Polaris extended through Alioth points toward this group, which lies just south of Canes Venatici. Actually, Coma is an open star cluster, more beautiful when seen with field glasses than with the naked eye.

7. Corona Borealis (the Northern Crown), a delicate circlet of stars, lies between Boötes and Hercules.

8. Camelopardalis (the Giraffe) is situated in a region where stars are few, between Polaris and Auriga.

9. Lynx (the Lynx), likewise in a barren region, forms a sort of fence in front of Ursa Major, whose Pointer stars, extended in

the direction opposite to Polaris, indicate Leo, a zodiacal constellation.

10. Leo Minor (the Smaller Lion) lies just north of Leo. Leo Minor, Lynx, and Camelopardalis are modern astronomical images providing identification for a few faint stars unattached to the older groups.

B. The Zodiacal Family

Twelve constellations, one for each month of the year, make up the Zodiac. Memorize them in order if you can. The following rhyme may help:

The Ram, the Bull, the Heavenly Twins,
And next the Crab, the Lion shines,
The Virgin and the Scales,
The Scorpion, Archer, and the Goat,
The man who holds the watering pot
And Fish with glittering tails.

11. Leo (the Lion), whose asterism, the Sickle, forms an arched mane, really looks like the king of beasts.

12. Virgo (the Virgin) lies south of Coma Berenices.

13. Libra (the Scales) was once regarded as the claws of

14. Scorpius (the Scorpion), whose curving back and poised sting form a realistic group on the edge of the Milky Way. Antares, the red luminary of the group, signifies "rival of Mars."

15. Sagittarius (the Archer), represents a Centaur. Its characteristic asterism, the Milk Dipper, appears in the beautiful star clouds of the southern Milky Way.

16. Capricornus (the Sea Goat),

17. Aquarius (the Water Carrier), and

18. Pisces (the Fish) lie in a region of the sky devoted largely to water and denizens of the sea. Most maps picture the Fish tied together with a ribbon and bowknot; I suggest that the ribbons indicate streams of water pouring from the mouths of live fish and that the bow denotes the splash against the body of Cetus. (For further information, refer to F below, "The Heavenly Waters.")

19. Aries (the Ram) represents the animal famous for its golden fleece, the goal of the Argonautic expedition. The asterism, the Northern Fly, hovers appropriately over the rump of Aries.

20. Taurus (the Bull), with long curving horns and fiery Aldebaran for an eye, is a magnificent constellation. The V-shaped cluster, the Hyades, marks the head of Taurus and the Pleiades cluster lies in the shoulder. Both clusters are beautiful objects seen either with field glasses or a small telescope. The Bull continuously backs away from the advancing Orion.

21. Gemini (the Twins) is the constellation of the mythological Castor and Pollux. It occurs in a densely populated region of the northern Milky Way.

22. Cancer (the Crab) is associated with the Hercules family. Astronomically, Cancer is significant for its delicate cluster Praesepe, or the Beehive.

C. The Perseus Family

We turn northward to consider the numerous constellations associated with Perseus.

23. Cassiopeia (the Lady of the Chair) is a W-shaped group lying on a line drawn from Alioth through Polaris and extended an equal distance on the opposite side of the pole.

24. Cepheus (royal consort of Cassiopeia) lies to the north and west of the Queen, whose boast that she was fairer than Juno evoked the wrath of the sea nymphs. They sent a sea monster (Cetus) to ravage the coast and banished Cassiopeia to the sky, where she hangs head downward half of the time, learning humility. Neptune demanded that her daughter, Andromeda, be chained to a rock as a sacrifice to Cetus. Perseus, flying in on Pegasus the winged horse, rescued the maiden,

25. Andromeda.

26. Perseus is a clearly delineated figure. His body extends approximately parallel to the Milky Way. The right leg, indicated by a number of bright stars, stretches almost to the Pleiades. The left leg is not quite so noticeable.

27. Pegasus (see also Eridanus).

28. Cetus (the Whale or Sea Monster) also belongs to the family I call "Heavenly Waters."

29. Auriga (the Charioteer), in my opinion, represents Neptune, the sea god often shown driving his chariot drawn by sea horses.

30. Lacerta (the Lizard) is an inconspicuous modern group

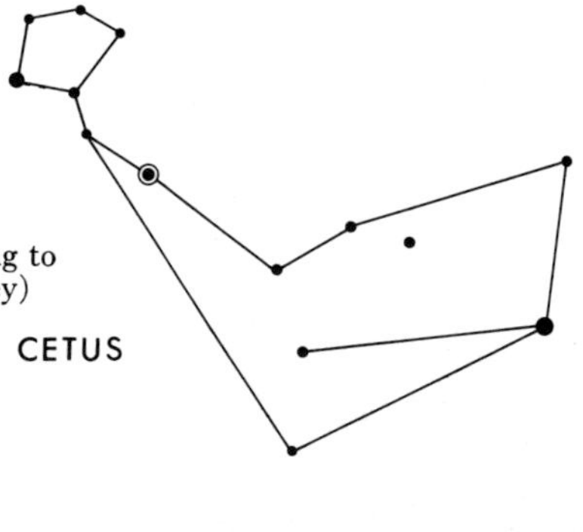

Fig. 3 (according to H. A. Rey)

lying between Cepheus on the north and Pegasus on the south.

31. Triangulum (the Triangle) contains some interesting star fields.

D. The Hercules Family

Again we turn north, where we find a number of star groups related to the legends of Hercules.

32. Hercules. Historians have frequently wondered why Hercules appears to be kneeling. I think the answer is simple: he holds a bow in his outstretched arm, one foot rests on the head of Draco. Many ancient carvings and paintings show him thus.

33. Sagitta (the Arrow) flies from the bow, directed toward Aquila, Lyra, and Cygnus, which seems to me to represent the Stymphalian Birds of one of Hercules' labors.

34. Aquila (the Eagle).

35. Lyra (the Lyre) often appears in early descriptions as a vulture and sometimes as a tortoise. Mercury invented the lyre by placing strings across the back of a tortoise shell. Lyra has a distinctive geometrical pattern consisting of a parallelogram and an equilateral triangle with the bright star Vega at one vertex.

36. Cygnus (the Swan) is also known as the Northern Cross. The arrow has also narrowly missed another creature,

37. Vulpecula (the Fox), which lies just south of Cygnus.

38. Hydra (the Sea Serpent) probably represents the Lernean Hydra that Hercules encountered on one of his labors. The head is an asterism. The body straggles almost interminably south and eastward across the sky, below Virgo. With Regulus and Procyon, Alphard (Solitary One in the Sea Serpent), forms a right triangle. At least two zodiacal constellations—Leo (the Nemean Lion) and Cancer—also figure in the labors of Hercules. And Eridanus may be the river that the giant used to flush out the Augean stables.

39. Sextans (the Sextant), a modern group,

40. Crater (the Cup [of Bacchus]), and

41. Corvus (the Crow) appear to stand on the body of Hydra lying between the Sea Serpent and the zodiacal groups Leo and Virgo.

42. Ophiuchus (the Serpent Holder), wrestling with

43. Serpens (the Serpent), is a fitting associate for Hercules. The two giants lie with their heads almost in contact, their bodies extending in opposite directions. Ophiuchus appears to tread on the Scorpion, whose sting is striking toward the giant's right leg. The Milky Way, running southward from Ophiuchus through Scorpius,

44. Scutum (the Shield), Sagittarius, and Centaurus, is spectacular.

45. Centaurus (the Centaur) also figured in the legends of Hercules. Sagittarius is also a Centaur, remember.

46. Lupus (the Wolf) has undergone many transformations in history. In my opinion, the original figure represented the Erymanthian boar, which Hercules hunted in company with the Centaurs.

47. Corona Australis (the Southern Crown),

48. Ara (the Altar), and

49. Triangulum Australe (the Southern Triangle) lie on the southern fringe of the Milky Way.

50. Crux (the Southern Cross) consists of four stars, two of the 1st magnitude. It lies in the Milky Way, on the southern border of Centaurus. A dark starless patch of cosmic dust, known as the Coalsack, is silhouetted against the Milky Way.

E. The Orion Family

51. The giant Orion, battling Taurus (the Bull) and followed by his two dogs, is one of the most striking figures of the sky. Betelgeuse (pronounced *Bet-el-gerz*) is the red star in his right shoulder. Rigel is the bright blue star in the left leg. The central star of the belt is Alnilam (String of Pearls). From Orion's belt hangs a well-defined dagger, sometimes regarded as the handle of Venus' Mirror, an asterism formed in conjunction with the belt and the star in the left hip. The mirror is diamond-shaped. Scorpius, on the opposite side of the sky, is sometimes credited with stinging Orion to death. The Belt stars point toward Sirius, the brightest star in the sky, in

52. Canis Major (the Larger Dog).

53. Canis Minor (the Smaller Dog), having one 1st-magnitude star, Procyon, forms an equilateral triangle with Sirius and Betelgeuse. The dog appears to stand on the back of

54. Monoceros (the Unicorn), which gallops behind Orion in a brilliant region of the winter Milky Way. Orion appears to be standing on

55. Lepus (the Hare).

F. The Heavenly Waters

A large portion of the heavens consists of watery wastes, a sort of celestial aquarium inhabited by creatures natural to that medium. The zodiacal constellations Pisces, Aquarius, and Capricornus lie near the northern border of the celestial sea.

56. Delphinus (the Dolphin) and

57. Equuleus (the Little Horse, which I interpret to be a small sea horse) lie between Pegasus and Aquila. The delicate diamond

of Delphinus forms the asterism known as Job's Coffin. Cetus (the Whale or Sea Monster) lies just south of Pisces and Aries. The latter group, portraying the golden fleece, took part in a well-known aquatic adventure, the Argonautic expedition.

58. Eridanus (the River) rises near Rigel and meanders southward to its terminus at the bright star Achernar. The rear hoofs of Pegasus strike over the water jar in Aquarius, thus giving an astronomical basis for the association of Pegasus with the mythical spring Hippocrene, supposed to have burst forth at a blow of the horse's hoofs. The stream from Aquarius flows southward to be engulfed by

59. Piscis Austrinus (the Southern Fish). On the edge of the sea, as if wrecked upon the shore, rests the great ship Argo, subdivided into four constellations:

60. Carina (the Keel),
61. Puppis (the Stern),
62. Vela (the Sails), and
63. Pyxis (the Mariner's Compass).
64. Columba (the Dove) flies near the stern of the Ship.

G. The Bayer Group

The foregoing constellations were known to the ancients. When the astronomer Johann Bayer early in the 17th century delineated the star groups near the South Pole of the sky, he continued the concept of sea and land, suitably introducing sea creatures in the southern extension of the waters:

65. Hydrus (the Water Snake),

66. Dorado (the Goldfish), with the Large Magellanic Cloud within the constellation borders, and

67. Volans (the Flying Fish), on the southern edge of Carina. On the shore we encounter five Bayer birds and an Indian:

68. Apus (the Bird of Paradise),

69. Pavo (the Peacock),

70. Grus (the Crane; appropriately just south of Piscis Austrinus),

71. Phoenix (the Phoenix),

72. Tucana (the Toucan), with the Small Magellanic Cloud, and

73. Indus (the Indian). Although the name is masculine, Flamsteed drew a female figure. Perhaps the group refers to the Amazonian queen Hippolyta, whose golden girdle was the objective of one of the labors of Hercules. Might not Corona Australis be the girdle itself?

The two remaining Bayer animals lie south of Carina (the Keel [of Argo]). They are:

74. Chamaeleon (the Chameleon), a lizard with tongue extended toward the neighboring minor constellation,

75. Musca (the Fly, which Bayer originally designed as a bee). Someone, perhaps Halley of comet fame, later changed the designation.

H. The La Caille Family

I consider the 75 groups listed above as the major constellations. There are, however, 13 others, conceived by the astronomer La Caille to fill in the star-poor regions between the Bayer and other groups. With one exception (Mensa) they represent scientific equipment or instruments. They break up the pattern of related families. Nevertheless, since they are currently recognized groups, I must include them in the list of 88 constellations.

76. Norma (et Regula), the Level (and Ruler).
77. Circinus (the Compasses).
78. Telescopium (the Telescope).
79. Microscopium (the Microscope).
80. Sculptor (the Sculptor's Apparatus).
81. Fornax (the Furnace).
82. Caelum (the Graving Tool).
83. Horologium (the Clock).
84. Octans (the Octant), site of the south celestial pole.
85. Mensa (Table Mountain at Capetown, site of La Caille's observatory).
86. Reticulum (the Net).
87. Pictor (the Easel).
88. Antlia (the Air Pump).

I. The Milky Way

The Milky Way, neither a constellation nor an asterism, deserves special mention, if only because the luminous band with which it rings the entire heavens has always struck awe into the wondering mind of man. The diffused, hazy strip divides the sky into two hemispheres. Telescopes show that the glow comes from millions of stars, individually too faint to be seen with the naked eye but whose collective brightness is plainly visible.

The Milky Way system, more properly termed the Galaxy, contains about 100,000,000,000 stars, of which our sun is one. The stars are arranged in a flattened spiral, resembling the galaxy NGC 628 (see Fig. 17g, p. 134). The diameter of our galaxy is 100,000 light-years and its thickness about 10,000 light-years. The center, which lies in the direction of Sagittarius, is some 30,000 light-years away. The thin, watch-shaped distribution of the stars accounts for the concentration of stars toward the long diameter. The Milky Way slowly revolves, one turn requiring about 200,000,000 years for a star at the distance of the sun from the galactic center. We do not see the separate spiral arms clearly because of our relatively poor location inside the Galaxy.

Even a small telescope resolves the Milky Way haze into swarms of stars. Many of the more interesting telescopic objects lie close to the Milky Way: the star clusters (both open and globular), the gaseous nebulae (both diffuse and planetary), and the great dust clouds that stand out in dark silhouette against the brighter Milky Way background. One of the most conspicuous of these dark patches is the Northern Coalsack, in Cygnus.

In Cepheus and Cassiopeia, the Milky Way comes closest to the north celestial pole. Perseus, Auriga, and Gemini display many interesting fields and beautiful clusters. Some dark obscurations occur between Gemini and the bright fields of Monoceros.

The Galaxy grows brighter through Puppis, Vela, and Carina. Crux, the Southern Cross, has the most pronounced sharp-edged black spot, the Coalsack, which the Bushmen of Africa call the Old Bag.

Centaurus, Norma, Lupus, Scorpius, and Sagittarius are magnificent to the unaided eye as well as when seen through a telescope. This region, in the direction of the galactic center, abounds in nebulae and clusters. The small constellation Scutum is also spectacular. Through Ophiuchus and Aquila we see the Milky Way cut in two by a black rift of obscuring dust. The rift continues from the northern side of Sagitta on into Vulpecula and southern Cygnus.

The distribution of galaxies, sometimes called "external galaxies" (labeled "cg" on the Photographic Atlas Charts, Chap. VI) clearly shows the effect of the obscuring dust. Most of these objects lie quite far from the plane of the Milky Way, where the dust tends to thin out. The Large and Small Magellanic Clouds, which look like patches of Milky Way torn loose and set elsewhere in the sky, are also independent galaxies, the nearest to us of all such objects.

V

The Nature of the Stars and Nebulae

THE ANCIENT ASTRONOMERS distinguished two types of stars, the fixed stars that form the constellations and the wandering stars that today we recognize as planets. The planets are members of our own solar system and derive their luminosity from reflected sunlight. The fixed stars are in reality suns — hot balls of glowing gas. Most of the stars visible to the naked eye are even brighter than our sun. They appear faint to us only because of their enormous distances. Light, traveling at a speed of 186,000 miles a second, takes 8 minutes to reach us from the sun. From the very nearest stars, the double star called Rigil Kent (α Cen) and its faint companion Proxima, light rays take about 4 years to reach the earth. Table 7 lists the positions, magnitudes, spectral types, and distances (in light-years) of the brightest stars. These are taken from the *Yale Catalogue of Bright Stars* by Schlesinger and Jenkins. A *light-year* is the distance that light travels in a year, approximately six million million (6,000,000,000,000) miles. Since light can traverse a distance equal to the circumference of the earth in only 1/7 of a second, celestial distances clearly are enormous.

Stars differ from one another in both size and surface temperature. The hotter a star is, the brighter will be its surface. Hence a small, hot star can in fact radiate as much light as a large, cool one. The diameters of stars range from several hundred times that of our sun to approximately one one-hundredth of the solar value. Betelgeuse is so large that it would fill the orbit of the earth, whereas Antares would extend beyond the earth's orbit as far as Mars. At the other extreme we have found a few very faint stars, invisible to the naked eye, which are scarcely larger than the earth.

Table 7 lists 85 stars above magnitude 2.90 (expressed by numerically *lower* numbers; see p. 117) in order of brightness, plus one fainter star, Acamar, a navigational star. The first column gives the astronomical designation, a Greek letter followed by the abbreviation of the constellation (see Table 5).

The second column gives the common name of the star. We have inherited most of these names from the Arabic. During the Middle Ages, science all but disappeared from the central Mediterranean region. In the Western world only the Arabs continued to observe and chart the motions of the planets. Our only records of the important star catalog by Ptolemy of Alexandria come from the hybrid Arabic translation *Almagest* (or *The Syntaxis*, which might be more freely rendered as "The Ordering of Nature"). The Arabs preserved the *Almagest*, with its locations of the stars in the various figures, and they gave to these stars the abbreviated Arabic names; many were translations of the cumbersome notations developed by the Greek scientist Hipparchus, sometimes called the father of modern astronomy. For example, *Denebola* denotes the "tail of the lion," *Denebkaitos*, "tail of the sea monster," *Rasalhague*, the "head of the serpent charmer," and so on.

Although columns 1 and 2 serve to identify the star, the positions given in columns 3 and 4 serve as a more precise identification. This system of celestial coordinates is analogous to latitude and longitude on the earth. The heavens appear to revolve about two fixed points, the celestial poles, and we can draw a great circle midway between them. From the pole we draw a line through the star, perpendicular to the equator. This is the star's *hour circle*. We divide the equator into 24 hourly segments, with the *vernal equinox* (the point where the sun — which follows the *ecliptic*, see p. 329 — crosses the equator in the spring) as zero. The numbers increase eastward and we may further subdivide each hour into minutes and seconds. *Right ascension* (RA), then, is merely the arc measured along the equator, from the vernal equinox to the foot of the star's hour circle and *declination* (Dec), the angle from the equator, north (+) or south (−), along the hour circle to the star (see also p. 329). Thus, right ascension is akin to terrestrial longitude and declination akin to terrestrial latitude. Since the vernal equinox slowly moves along the equator, the positions of stars also may change. As explained in Chapter I, unless otherwise indicated the positions here given are for the epoch 1950.

Column 5 gives the brightness of the star. The larger the number denoting the magnitude, the fainter is the star. Zero or 1st magnitude indicates some of the brightest stars. Still brighter are those of *negative* magnitude, like Sirius, whose magnitude is −1.58.

Each unit of magnitude signifies a difference in brightness of 2.512 times. Thus a star of magnitude 1 is 2.512 times brighter than one of magnitude 2; similarly, a star of magnitude 2 is 2.512 times brighter than one of magnitude 3. To follow the ratio further: a star of magnitude 1 is $2.512 \times 2.512 \times 2.512 \times 2.512 \times 2.512 = 100$ times brighter than a 6th-magnitude star just visible to the naked eye. A difference of 5 magnitudes implies a 100-fold brightness ratio. This scale continues to still fainter objects. A star of the 21st magnitude is $21 - 6 = 15$ magnitudes fainter than a star at the naked-eye limit. It is, therefore, $100 \times 100 \times 100$, or 1,000,000 times too faint to be visible to the naked eye.

At the other range of brightness, the sun is of magnitude −26.7, the full moon of magnitude −12.5, Venus at brightest −4.3, and of Jupiter at opposition -2.3. Astronomers also recognize a scale of photographic magnitudes. Since the ordinary photographic plate, unless corrected with proper filters, is more sensitive than the human eye to blue light, blue stars are photographically brighter and red stars fainter than they are on the visual scale.

Column 6 gives the *spectral type* of the star, and column 7 its *distance in light-years*, which is the number of years it takes light traveling at a speed of six million million miles per year to reach us.

The color of a star provides a good index of its surface temperature. The spectrum, as we call the rainbow band of color that results when light passes through a glass prism, is an even better gauge. The spectrum indicates the chemical composition of the atmosphere of the star and the conditions of temperature and pressure prevailing in the hot stellar gas. The correspondence between the type of spectrum, color, and surface temperature on the absolute Centigrade scale appears in Table 8. The spectral type, designated by the letter and numbers in the first column, conveys more exact information than does the color determination we make by eye. To describe a star, the seven most common varieties of spectrum (O, B, A, F, G, K, and M) have been subdivided on a decimal system. Thus, A5 is halfway between A0 and F0. Astronomers use the terms "early" or "late" to indicate the relative position of a star in the tabulation. A few cool stars have spectra falling outside this classification. These are the peculiar red variables of classes R, N, and S. At the other extreme of temperature are the very hot stars, usually surrounded by extensive globes of gas, denoted by the letter W.

The variegated colors to be found among the brighter stars enhance the interest of looking at the sky either with the eye or a telescope. Although the color differences are subtle, not glaring like advertisement signs, they should be immediately evident to anyone who is not color-blind. Moreover, the colors form an additional check on constellation identification. Hence, use the spectral classification of the brighter stars given in Table 7 as a rough guide to the color you will expect to find. Stars of Class A usually appear almost white, though they seem to have a tinge of green when coupled with a companion of later spectral class.

Many stars vary in brightness — some almost imperceptibly, others by a large amount. There are two main causes of such variation: one star may eclipse another temporarily by circling about it in a regular orbit; or actual changes may take place in the condition of the star's outer layers. Table 9 lists the known *variables* with maxima brighter than magnitude 6.0. The data are from the standard *Catalog of Variable Stars* by the Russian astronomers B. V. Kukarkin and P. P. Parenago, and others.

Column 1 gives the name of the variable. When the star does not already have some accepted name or Greek letter symbol, we distinguish it by one of the capital letters between R and Z, followed by the name of the constellation (often abbreviated) in which the star appears. Some constellations contain more than nine variables, however. We must then use a system of paired letters from R through Z (RR, RS, etc., SS, ST, etc.) and, when these combinations have been used up, pair A through Q in the same manner. (Following ZZ, we have AA, AB to AZ, BB to BZ, and finally QQ, QR to QZ.) This system of lettering will take care of 334 stars in each constellation. When the alphabet is used up, additional variables receive number designations beginning with V (V335, V336, and so on).

Columns 2 and 3 give the positions, column 4 specifies the *magnitude range*, column 5 the *period of variation* when known, column 6 the spectrum, and column 8 the *type of variability*. The recognized types of variable stars and the symbols we use to indicate each class are described below.

E — Eclipsing stars in general.

EA — Eclipsing stars of the Algol type, whose light is nearly constant between eclipses.

EB — Eclipsing stars of the Beta Lyrae type. In this class, two stars are so close together that they are almost in contact, and the force of their mutual gravitation makes them egg-shaped. The changing aspects of the stars themselves, as well as their eclipses, result in a continuous variation of the light.

Ell — Similar to the stars of the EB type, but we attribute the variation in their brightness mainly to the fact that the revolving stars are elliptical, rather than to the fact that they eclipse one another.

N — Novae or "new" stars. These stars apparently explode catastrophically, expelling their outer shells with enormous force. They may increase their brilliance by a million times or more, usually within a day or two. Thereafter the star fades slowly, returning to about the same magnitude it had prior to the outburst. If you should chance to find a bright star in a part of the sky where it doesn't belong, notify the nearest observatory immediately, after first checking to see whether one of the bright planets could be responsible. A few novae, such as P Cygni and η Carinae, are still visible, several hundred years after their outburst.

RN — Repeating novae. Among these are T Pyx, RS Oph, and T CrB, which have had several outbursts. Watch these stars for further nova behavior and report any unusual activity.

Nl — Nova-like variables, irregular in outburst, but clearly explosive in behavior.

UG — Prototype of this class, U Gem or SS Cyg, stars also showing a nova-like behavior, except that the range in brightness is smaller (2 or 3 magnitudes) and the intervals between successive explosions are about two months, with considerable fluctuations.

Flare Stars — Usually red dwarfs, which show occasional, short-lived, irregular brightenings. Our sun is probably a flare star, but the flares do not produce an appreciable increase in the total light.

R CrB — An unusual star that may suddenly and without warning drop from its normal magnitude of about 6 to as faint as 14, a decrease of some 1600 times in luminosity. The reason for the variation is not understood.

LP — Long-period variables. Many red-giant stars show a periodic large variation in magnitude, with periods ranging from several months to almost two years, the majority being somewhat under a year. The best-known example is Mira (o Cet), which ranges from magnitude 2 to 10, apparently by both shrinking and cooling, so that it becomes less luminous.

Cep — The Cepheids. These variables take their name from β Cephei and δ Cephei, both of which (particularly the latter) typify the class. They are probably related to the long-period variables, but are hotter. Their pulsations are not quite so pronounced and the periods are shorter.

β C — Hot, pulsating stars of extremely short period.

Cl — Cluster variables. A type of pulsating star intermediate between ordinary Cepheids and β Cephei, found chiefly in globular clusters.

α CV — A variable of early type and very small range in magnitude, perhaps related to the Cepheids. α CVn, the prototype, shows many peculiar spectral characteristics.

δ Sct — A short-period variable of small-magnitude range. δ Sct is the prototype star.

SR — Semiregular variables, usually red giants.

I — Irregular variables. These stars exhibit erratic behavior, both in the degree of light variation and in their periods.

Still other types of variables exist, but they are less important than those listed above.

The amateur will find that observation of such objects is an interesting and rewarding activity. The American Association of Variable Star Observers (AAVSO, 187 Concord Avenue, Cambridge, Massachusetts 02138), headquarters for such observers, invites interested persons to join the society. Watching night after night for the privilege of seeing the irregular variable SS Cygni ascend sharply to maximum brilliance is a favorite pastime of observers of variable stars.

Figures 4 through 9 are representative variables of selected types. The charts are adapted from those published by the AAVSO, with positions advanced to 1950. The numbers beside various stars are the magnitudes of stars that may be used for comparison with the variable. The decimal point is omitted because it might be mistaken for a star. Hence, 67 signifies magnitude 6.7, 104 indicates magnitude 10.4, and so on. These maps are adapted for use with field glasses or low-power telescope, magnification up to 25 times.

Many stars are double or multiple — that is, they form an actual *physical system* in orbital revolution around their common

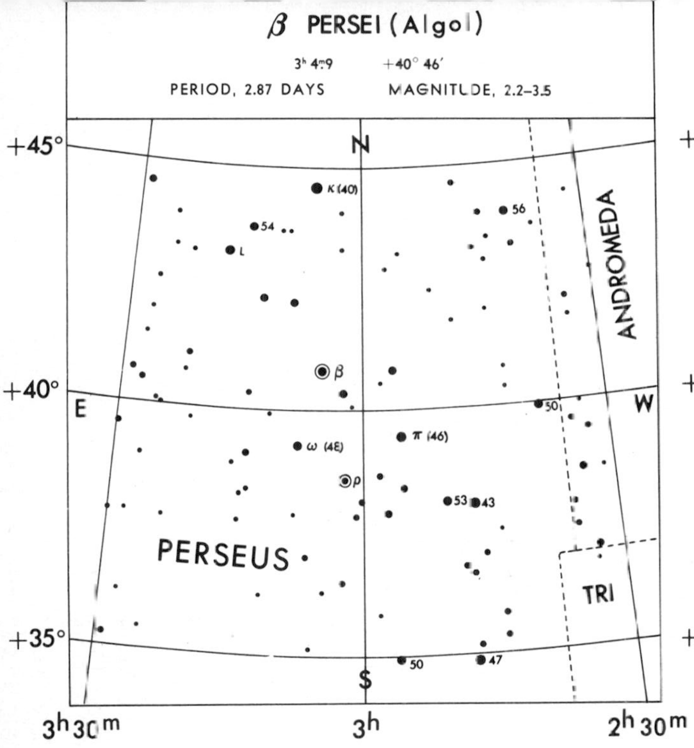
β PERSEI (Algol)
3h 4m9 +40° 46′
PERIOD, 2.87 DAYS MAGNITUDE, 2.2–3.5
+45°
+40°
+35°
N
E
W
S
κ (40)
54
56
β
ω (4E)
π (46)
ρ
53
43
50
47
PERSEUS
ANDROMEDA
TRI
3h 30m
3h
2h 30m

Fig. 4

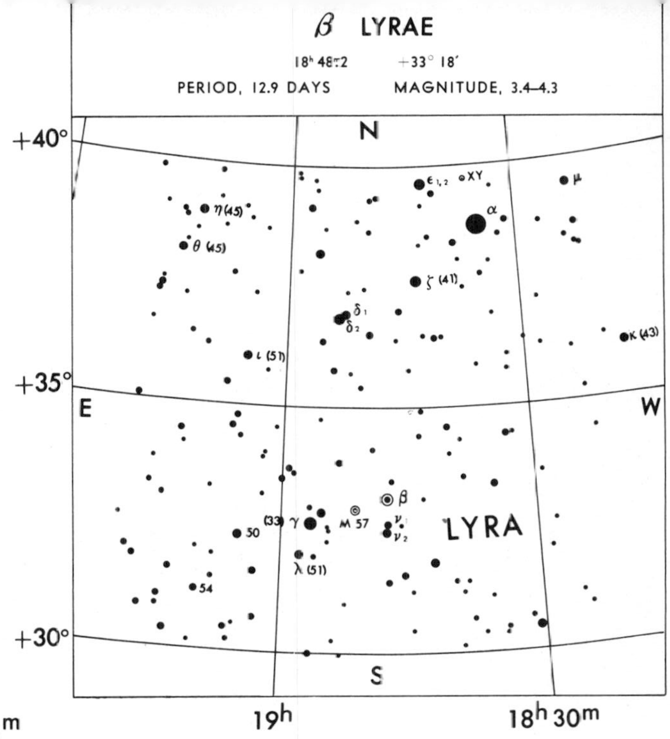
β LYRAE
18h 48m2 +33° 18′
PERIOD, 12.9 DAYS MAGNITUDE, 3.4–4.3
+40°
+35°
+30°
N
E
W
S
ε1,2
XY
α
μ
η (45)
θ (45)
ζ (41)
δ1
δ2
κ (43)
ι (51)
β
(33) γ
M 57
ν1
ν2
50
λ (51)
54
LYRA
19h
18h 30m

Fig. 5

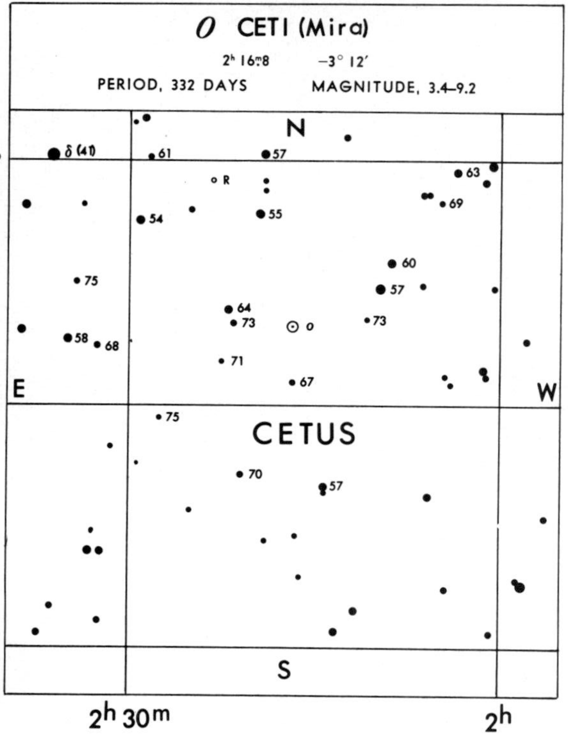
ο CETI (Mira)
2h 16m.8 −3° 12′
PERIOD, 332 DAYS MAGNITUDE, 3.4–9.2
N
S
E
W
CETUS
0°
−5°
−10°
2h 30m
2h
δ (41)
61
57
63
R
69
54
55
60
75
57
64
73
o
73
58
68
71
67
75
70
57

Fig. 6

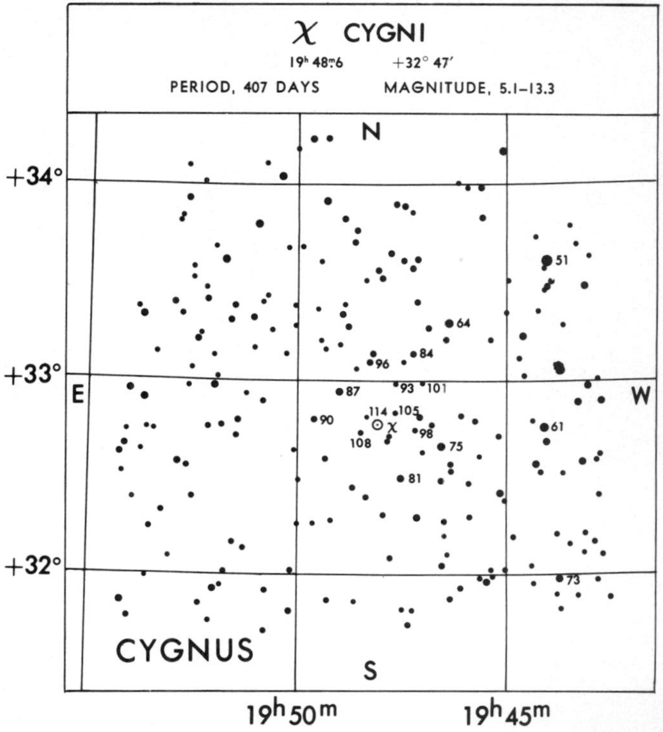
χ CYGNI
19h 48m.6 +32° 47′
PERIOD, 407 DAYS MAGNITUDE, 5.1–13.3
N
S
E
W
CYGNUS
+34°
+33°
+32°
19h 50m
19h 45m
51
64
96
84
87
93
101
114
105
90
χ
98
108
75
61
81
73

Fig. 7

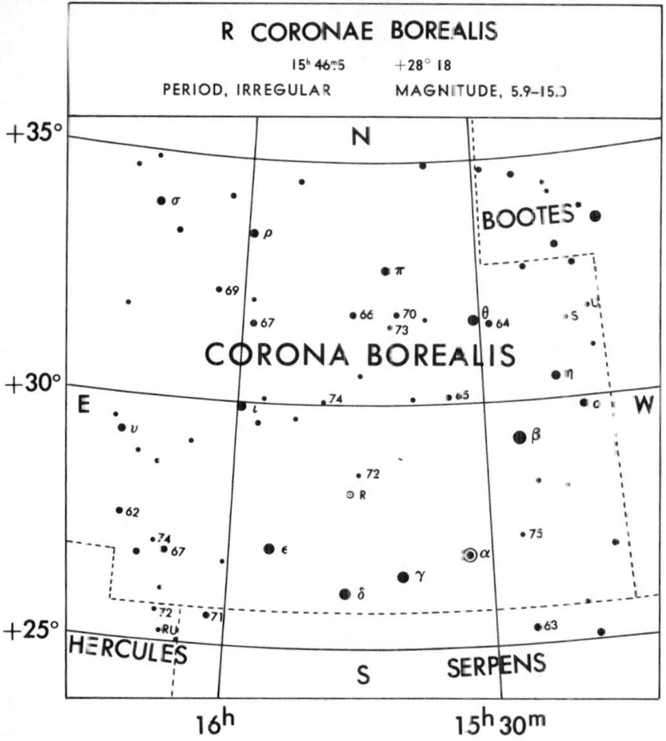
R CORONAE BOREALIS
15h 46m5
+28° 18
PERIOD, IRREGULAR
MAGNITUDE, 5.9–15.0
N
S
E
W
+35°
+30°
+25°
16h
15h 30m
BOOTES
CORONA BOREALIS
HERCULES
SERPENS
69
67
66
70
73
64
74
72
R
62
74
67
75
72
RU
71
63

Fig. 8

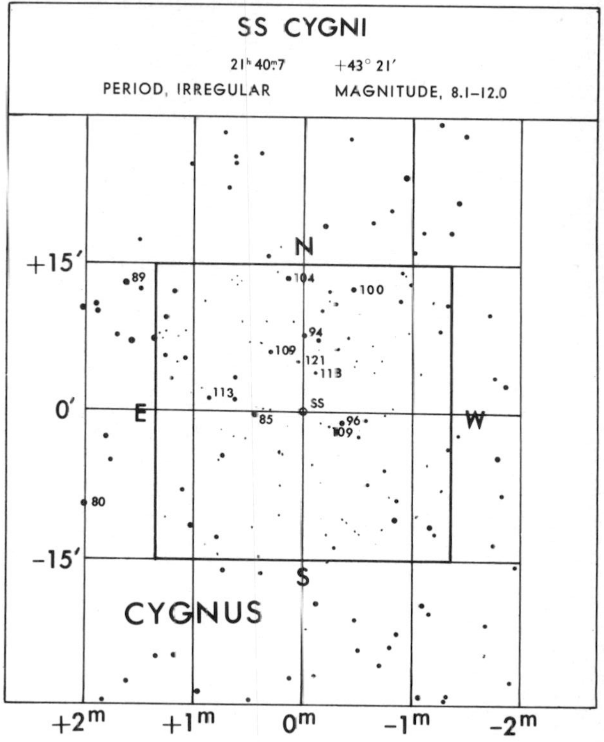
SS CYGNI
21h 40m7
+43° 21′
PERIOD, IRREGULAR
MAGNITUDE, 8.1–12.0
N
S
E
W
+15′
0′
−15′
+2m
+1m
0m
−1m
−2m
89
104
100
94
109
121
113
113
SS
85
96
109
80
CYGNUS

Fig. 9

Fig. 10. Double cluster h and χ in Persei

center of gravity. Tables 10 and 11 list representative doubles over the entire sky. The numbers in the first column of Table 10, used to designate the doubles from declinations (see p. 117) 90° to −30°, are taken from Robert G. Aitken's *New General Catalogue of Double Stars*. The magnitudes of both members of the pairs appear in column 4. The fainter component is sometimes called a *comes*. The *position angles* (PA) specify the angle for 1950, as measured eastward from the brighter star as a center. The *separation* (in seconds of arc) and the distance (in light-years) are given in columns 6 and 7. Special comments relative, for example, to color contrasts or *periods of revolution* when known, appear in the captions for the Photographic Atlas Charts of the sky. Table 11 extends this information to declinations from −30° to −90°; the data come from the *Southern Double Star Catalog* by R. T. A. Innes and others. The numbers with the prefix *b* have been assigned arbitrarily.

The next category of stars includes more elaborate stellar families, the so-called *open clusters* (oc), large, loose aggregations of stars that appear to the eye, and even to the smaller telescopes, as mere patches of light. The double cluster h and χ Persei is an outstanding case in point (see Fig. 10). Examples are listed in Table 12. The data for this table are from *Star Clusters* by Harlow Shapley. The first column lists the identifying number of the cluster, as given in J. L. E. Dreyer's *New General Catalogue.* Alphabetical prefixes to the number indicate the source to be:

1. Dreyer's original *NGC* or his supplementary catalogs (I or II),
2. Melotte (Mel), an observer who catalogued mainly southern objects,
3. Harvard (H), or
4. an early catalog compiled by Messier (prefix M followed by a number, as in M35).

The Messier assignments are often used for the brighter objects. Columns 4, 5, and 6 give, respectively, the diameter in minutes of arc, the approximate number of stars, and the magnitude of the 5th brightest star of the group (Mag. 5★). Other remarks will be found in the captions for the Photographic Atlas Charts of the sky.

Still more organized and compact than the open clusters are the *globular star clusters* (gc) such as those listed in Table 13, also designated by their NGC numbers. These objects contain many thousands of stars in a centrally condensed group. Columns 4, 5, and 6 respectively give the diameter (in minutes of arc), the total magnitude of the cluster, and its distance in light-years. The data are from W. Lohmann, "Die Entfernungen der Kugelförmigen Sternhaufen," and Shapley, *Star Clusters*. Special comments appear in the captions to the Photographic Atlas Charts. Small telescopes show most globular clusters as unresolved haze. Larger telescopes reveal them as magnificent groups of brilliant stars (see Fig. 11).

Fig. 11. Globular star cluster, ω Centauri

The Milky Way marks a part of the sky where the stars lie thickly concentrated in a volume shaped roughly like a pocket watch, among immense quantities of gas and dust (see p. 114). In the vicinity of hot, bright stars (usually of spectral types O and B), the gas glows or the dust clouds dimly reflect light from imbedded stars. Irregular patches of bright or dark nebulosity result, the *diffuse galactic nebulae* (gn), so called because they generally occur in or close to the galactic belt. Column 1 of Table 14 gives the NGC number (see above for explanation of alphabetical prefixes to numbers). Columns 4, 5, and 6 list respectively the magnitude of the brightest star involved, the approximate dimensions of the luminous patch, and the distance from the earth in light-years. Many of these nebulae are spectacular objects for the small telescope. The captions to the Photographic Atlas Charts contain additional references. The tabular data are from Cederblad, "Studies of Bright Diffuse Galactic Nebulae." Figures 13–16 portray characteristic examples of diffuse nebulae.

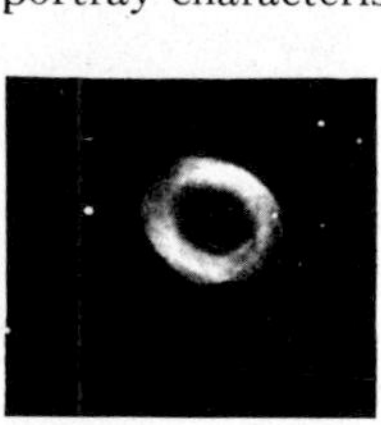

Fig. 12. Ring Nebula in Lyra

The heavens also contain a number of ellipsoidal or spherical gas clouds excited to luminescence by a hot, central star. These objects have received the name *planetary nebulae* (pn), because their small disks display some resemblance to planets (Fig. 12). These nebulae may be the remnants of stars that exploded long ago — ancient novae. Indeed, one of them, the Crab Nebula, listed in Table 14 as gn 1952, is almost certainly the debris of a star that exploded

Fig. 13. The Pleiades: low power, above; high power showing nebulosity, below

Fig. 14. The great Orion Nebula

Fig. 15. North American Nebula

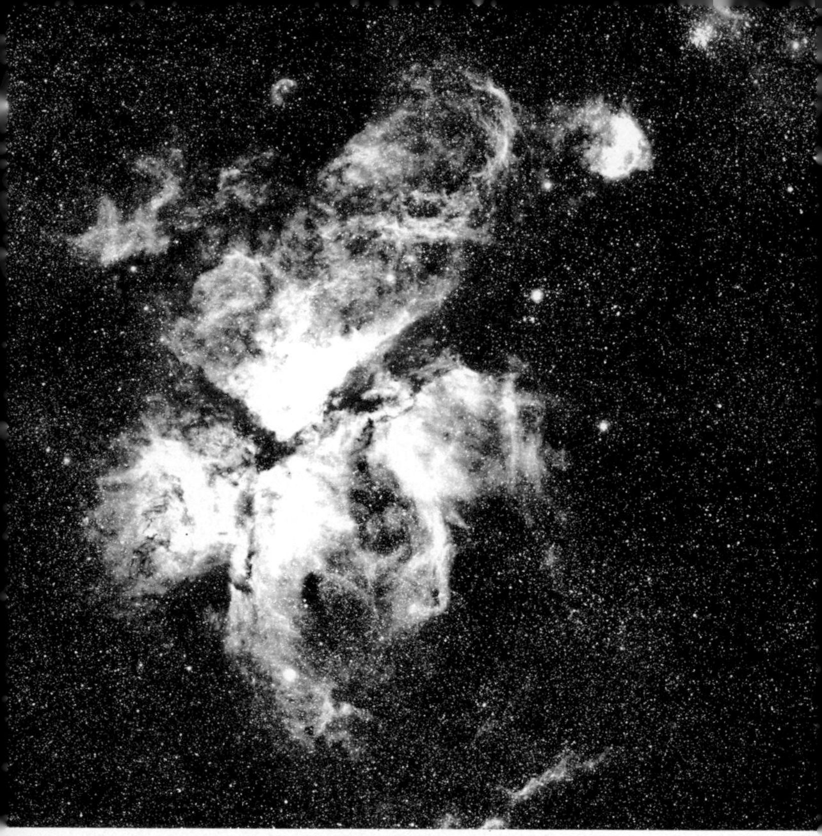

Fig. 16. Region of η Carinae

A.D. 1054, as the Chinese recorded in their annals of the time. Table 15 gives the basic data for the planetaries (pn); see page 126 for explanation of alphabetical prefix to numbers. Columns 5 and 6 present respectively the magnitudes of the nebula and of the central star. For additional comments see the captions to the Photographic Atlas Charts. The data are from *Gaseous Nebulae and Novae* by Vorontsov-Velyaminov.

The Milky Way, or Galaxy (see pp. 114–15), consists of some 100 billion stars, probably arranged in a spiral pattern, with an over-all diameter of about 100,000 light-years. Large telescopes reveal thousands of galaxies (designated by eg, for *external galaxies*), some apparently similar to our own. The wide variety of forms suggest an evolutionary or, at least, a family relationship. Figure 17, *a* through *i*, presents examples of the various classes of galaxy, according to the scheme devised by E. P. Hubble and given in Table 16. Additional comments on special objects appear in

Chapter VI on the Photographic Atlas Charts. The data are from Harlow Shapley and Adelaide Ames, "A Survey of the External Galaxies Brighter than the Thirteenth Magnitude." The galaxies have distances measured in millions of light-years, but the exact scale is still not fully settled. Figures 18–20 show additional well-known examples of galaxies.

Figs. 17a–i. GALAXIES. In these captions: E signifies elliptical; S, spiral; SB, barred spiral; and I, irregular; numbers 0–5 indicate increasing eclipticity. Objects with large unresolved centers are designated by a, intermediate by b, and very small by c.

Fig. 17a (above). NGC 4486, E0 (200-inch)
Fig. 17b. NGC 205, dwarf E5 (resolved 200-inch)

Fig. 17c. NGC 5866, S0 edgewise (60-inch)
Fig. 17d. NGC 3031, Sb (200-inch)

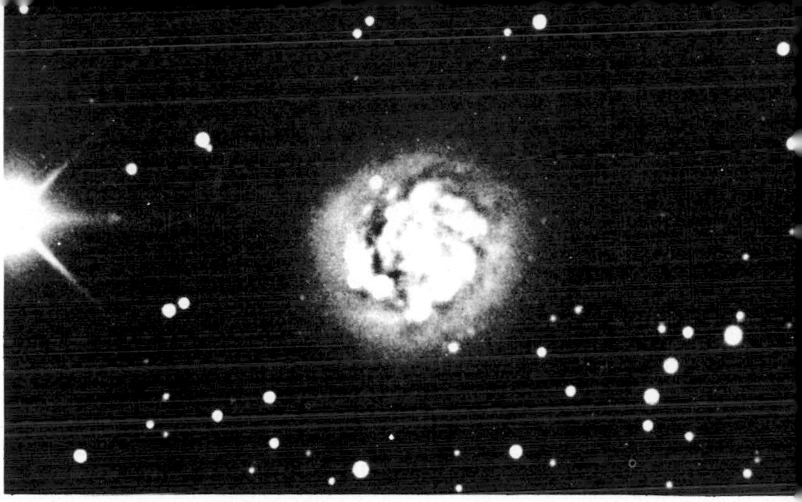

Fig. 17e. NGC 278, Sb (60-inch)
Fig. 17f. NGC 891, Sb edgewise (60-inch)

Fig. 17i. NGC 6822, I (100-inch)

Fig. 17g (opposite, above). NGC 628, Sc (200-inch)
Fig. 17h (opposite, below). NGC 1300, SBb (200-inch)

Fig. 20. Andromeda Nebula, Sb

'ig. 18 (opposite, above). Small Magellanic Cloud, I
'ig. 19 (opposite, below). M 51 in Canes Venatici, Sc

VI

The Photographic Atlas Charts

THE HAND-DRAWN Sky Maps are ideal for identifying the constellations and checking the positions of stars brighter than magnitude 4.55. One often wishes to check areas of the sky for fainter stars, perhaps even below magnitude 6.0, the normal limit of naked-eye visibility. Below magnitude 4.55 the number of stars multiplies so rapidly that photographic charts are the most satisfactory method of portrayal. The atlas that follows consists of prints from Harvard College Observatory's vast collection of plates.

Each photograph appears twice. The left-hand chart shows a negative print on which superposed lines connect the stars of the relevant constellations. The brighter stars and interesting objects selected as worthy of special attention carry identifying labels, listed in Tables 9 through 16. Most of these objects are visible only with a telescope.

Fifty-four charts cover the entire sky. Table 17 lists the Harvard plate number, the approximate right ascension and declination (see p. 117) for the center of each picture, the date of the photograph, and the exposure time. The left-hand chart also carries the constellation name and the official boundaries of the constellation (broken lines). Coordinate lines for each hour of RA and for every 10 degrees of Dec complete the picture. These coordinates are specifically for the year 1950, though they slowly move because of precession (see p. 329).

The right-hand chart is a positive print showing the sky approximately as it would look to the eye or a small telescope. The plates show stars down to nearly the limit of magnitude (12.5). One special warning is necessary. As previously noted, ordinary photographic plates are more sensitive to blue than to red light. In consequence, these Atlas Charts make red stars somewhat less brilliant and blue stars somewhat more brilliant than they appear to the eye (see p. 118). The captions below each pair of charts call attention to regions or objects of special interest in the area.

The following Atlas Charts and descriptive text furnish the reader with a pictorial road map to sky watching with binocular or telescope. The scale is small enough to permit naked-eye recognition of the brighter stars and constellations by comparison with the monthly Sky Maps. It is large enough for identification of the magnified star fields, especially with instruments of very low power.

Since power of a telescope is a relative term, let us see what this means. My 7 × 50 Bausch and Lomb binocular, magnifying 7 times, has a field of nearly 10 degrees. It covers, therefore, a circular area whose diameter is roughly one third that of an Atlas Chart. This is very low power indeed, but still useful for sweeping, since it is sufficient to resolve the Milky Way and larger star clusters. For the average amateur telescopes of 3- or 4-inch aperture, I use the term "very low power" to indicate magnification in the range 5–10 times, "low power" 10–30 times, "medium power" 30–80, "high power" 80–250, and "very high power" 250–1000. For further discussion see Chapter XI.

The astronomer often uses the term "difficult object" in the sense of its being difficult to see or detect. For a nebula, the term may signify low surface brightness; for a double star, a close pair near the limit of resolution of the telescope (see pp. 312–13).

The labeling of the charts has posed some difficulties, especially when the object is faint. The brighter stars, especially those forming part of the constellation outline are clearly indicated, since they usually lie in the breaks of the heavy lines. Otherwise, the labels have been set, in so far as possible, directly to the right of the object designated. Where crowding makes the use of this convention impossible, the label has been set to the left of the object with a short horizontal line pointing to the object.

Objects near the extreme right of the chart also have the short-line indicator, unless they are so bright that no ambiguity results to the object. In the most crowded areas, the labels sometimes run in a vertical direction. Again, as you read the label, the object is to the left.

All of the objects listed in Tables 9–16 appear on at least one of the charts, with the exception of VV Orionis, where the region was too crowded for the identification. The brighter stars are designated by Greek letters when such assignments have been made. Fainter stars carry the Bayer letters or Flamsteed numbers.

The designation of double stars posed something of a problem. In the strictest sense doubles are indicated by the "a" (Aitken) letters or the arbitrary "b" letters of the southern stars in Tables 10 and 11. But many of the brighter doubles are better known by their Greek-letter or other designation. On the Photographic Atlas Charts, therefore, only the fainter doubles carry the *a* or *b* numerical designation. If you wish to identify a particular given double in Table 9 or 10 and do not find the *a* or *b* number, you may assume that the star appears under some other designation. You can find out what the star is by noting the position of the star from the tables and locating it on the chart. Thus a 1477 is α Ursa Minoris (or Polaris), a 6175 is α Geminorum, and so on. Special references often appear in the captions to the various Atlas Charts.

The other objects listed in Tables 9–16 should cause no great problem. Remember, however, that the galaxies ("eg") are usually very faint and hence are often difficult to locate.

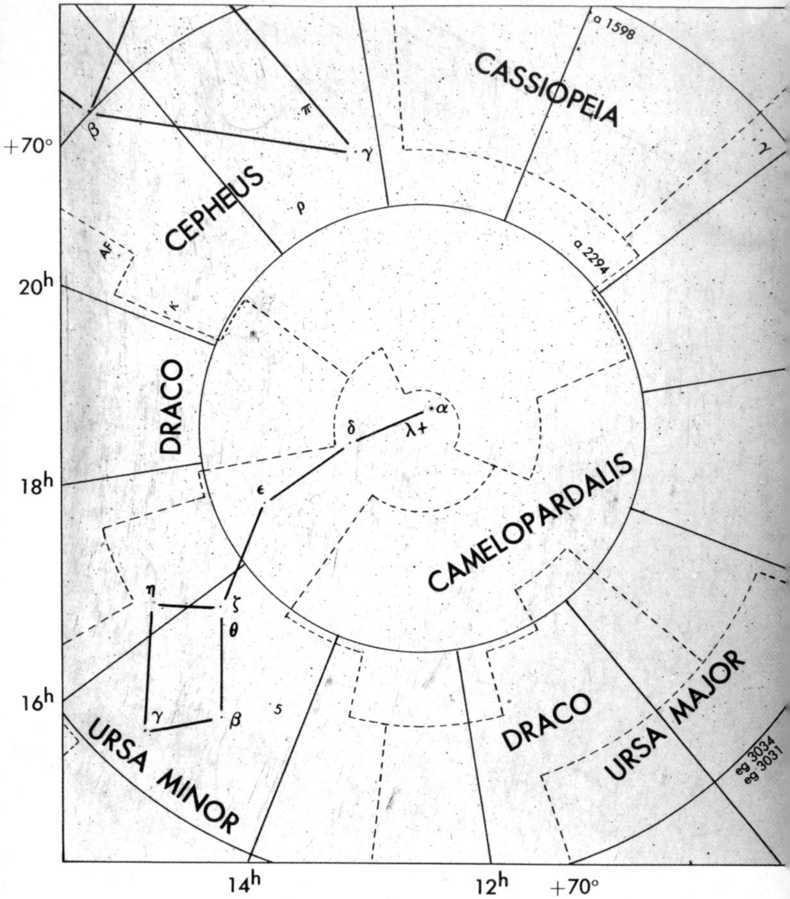

Atlas Chart 1. The area surrounding the north celestial pole contains relatively few bright stars. α U Mi, otherwise known as Polaris or the Pole Star, actually lies about 1° from the true pole (indicated by a +), and circles the pole once every 24 hours. Polaris is double, having a 9th-magnitude companion at a distance of 18″. A good 3-inch telescope will resolve the pair. The double a 2294 provides an interesting contrast of red and white.

eg 3031 and eg 3034 are interesting, though a 4-inch telescope is generally necessary to show them clearly, as hazy patches of light. The region of Camelopardalis is particularly devoid of stars, especially in the areas bordering Draco and Ursa Minor. The background becomes richer in Cepheus and Cassiopeia, however. Sweeping this region with a binocular or a small telescope is a rewarding experience.

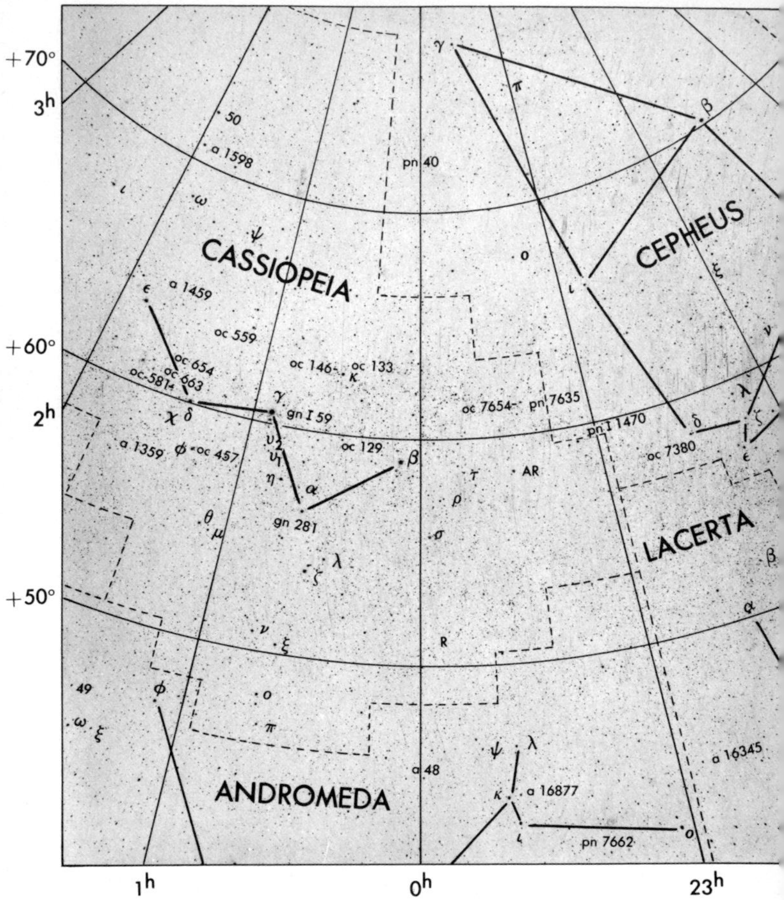

Atlas Chart 2. The Milky Way in the vicinity of Cassiopeia and Cepheus provides an interesting field for observation with field glasses, binocular, or telescope. Even opera glasses will disclose individual stars in the numerous hazy patches of stars that lie just below the limit of the eye's power to resolve. Most of these objects are open clusters. A few, however, prove to be gaseous nebulae, of the galactic or planetary types. The region around δ Cas is particularly rich. oc 581, also known as M 103, is fan-

shaped. It lies at a distance of approximately 6200 light-years. oc 654 and oc 7654 (also called M 52) are somewhat triangular in shape. oc 457 is richer than the average and one of the brightest in the sky.

gn I 59, near γ Cas, consists of two fans pointing northwest. gn 281 lies 1° east of α Cas. pn 7635 has a high total brightness, though its large size makes it seem faint, except with low telescopic power. Faintness makes pn 40 a difficult object.

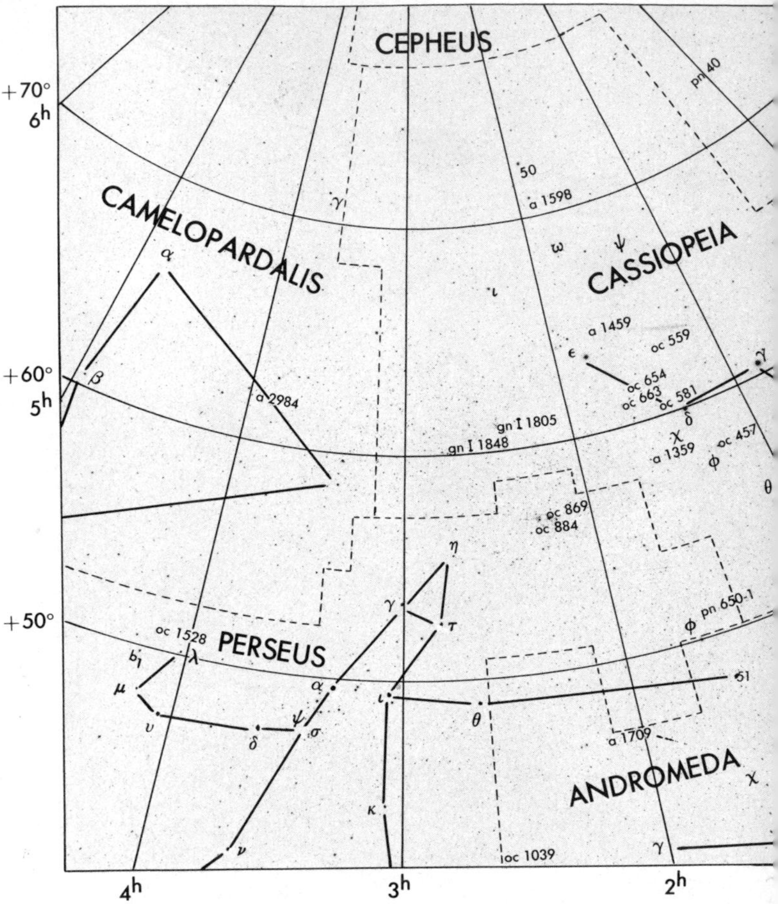

Atlas Chart 3. From Cassiopeia, the Milky Way sweeps southeastward through Perseus, revealing some of the most interesting areas of the northern sky. For a description of Cassiopeia, refer to Atlas Chart 2. The diffuse nebulae gn I 1805 and gn I 1848 probably have star clusters associated with them. The pair oc 869 and oc 884, in Perseus, also carry the designations h and χ, respectively (see also Fig. 10, p. 124). The naked eye can glimpse the condensations of the Milky Way as hazy patches. These

clusters, magnificent in a small telescope, lie at the estimated distances of 3900 and 4900 light-years.

To observe oc 1528 most effectively, use low telescopic power. The Milky Way background, in the vicinity of α Per, is especially beautiful. Use low magnification, or sweep with binocular.

The double pn 650-1 is large but faint, with irregular extensions to 157″. This object, also designated M 76, lies at an estimated distance of 8200 light-years.

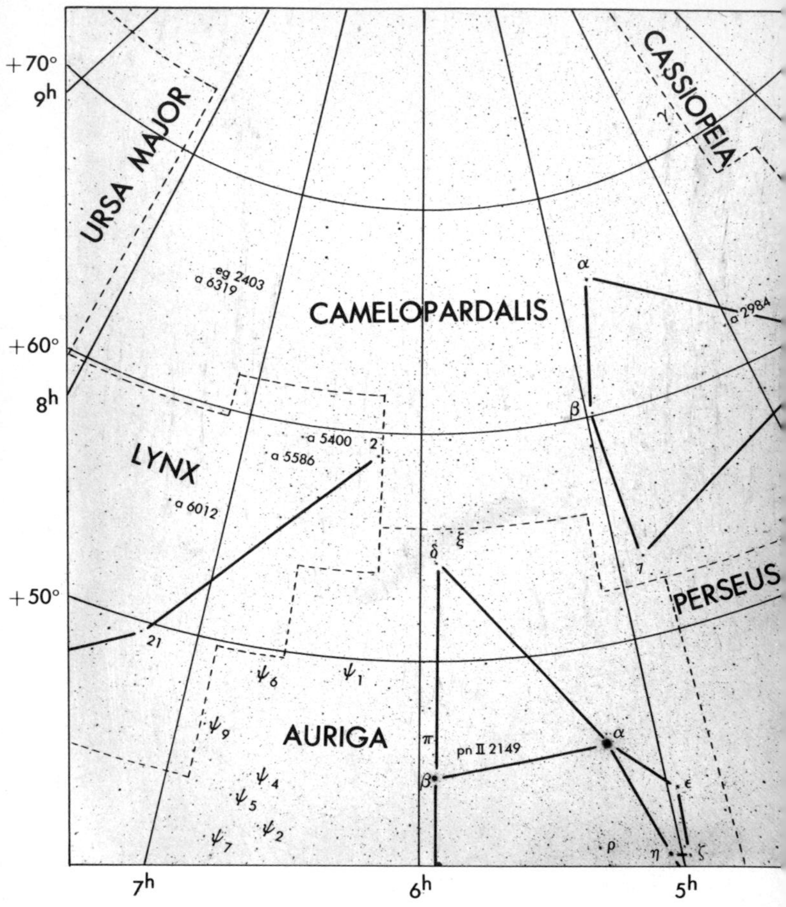

Atlas Chart 4. The region of Camelopardalis is relatively uninteresting, either with the naked eye or with the telescope. The richness of the sky background decreases rapidly as we leave the Milky Way. The double star a 2984 is actually multiple, part of a cluster. The star a 5400 is an interesting triple. The double a 6012 possesses several faint *comes*. The 1st-magnitude star Capella (α Aur) lies at a distance of 42 light-years. The spectroscope

shows that Capella is really double, each component being roughly 75 times brighter than our sun; they swing around one another every 104 days. Menkalinen (β Aur) is also a spectroscopic binary. The stars complete a revolution once every 4 days. Since they eclipse one another in the interval, the star appears to vary in brightness. pn II 2149 is a small oval ring of 10th magnitude.

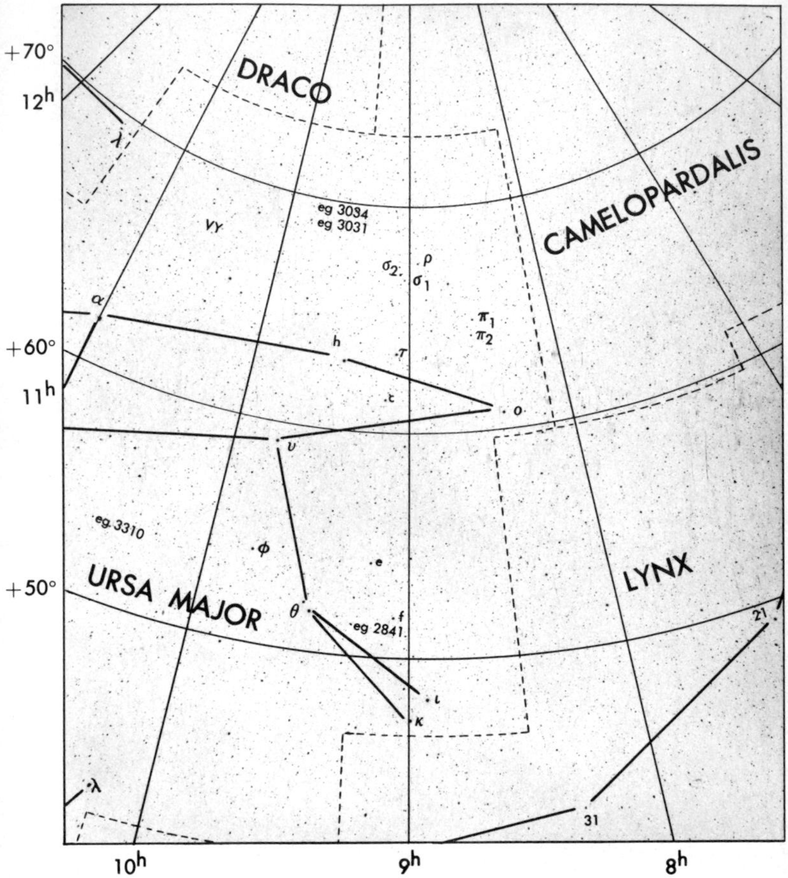

Atlas Chart 5. In the region of Ursa Major the stars and dust of the Milky Way thin out so that we can see the distant galaxies. Several bright examples of these objects occur in this region. eg 3031 is the beautiful spiral M 81. eg 3034 is an elongated patch, also called M 82. These are among the brightest galaxies in the sky. Observe them with low power on a moonless night. The other three galaxies shown on the charts, 2403, 2841, and 3310, are somewhat fainter than the Messier objects.

The double star ϕ UMa (or a 7545) is a very close visual double. The star a 6319, on the other hand, is readily resolved with a 1½-inch telescope, or even a binocular, into a pair of equal brightness. The double σ_2 UMa resolves into an unequal pair of magnitudes 5 and 8. The star fields in the head of UMa are not uninteresting for sweeping with a low-power binocular.

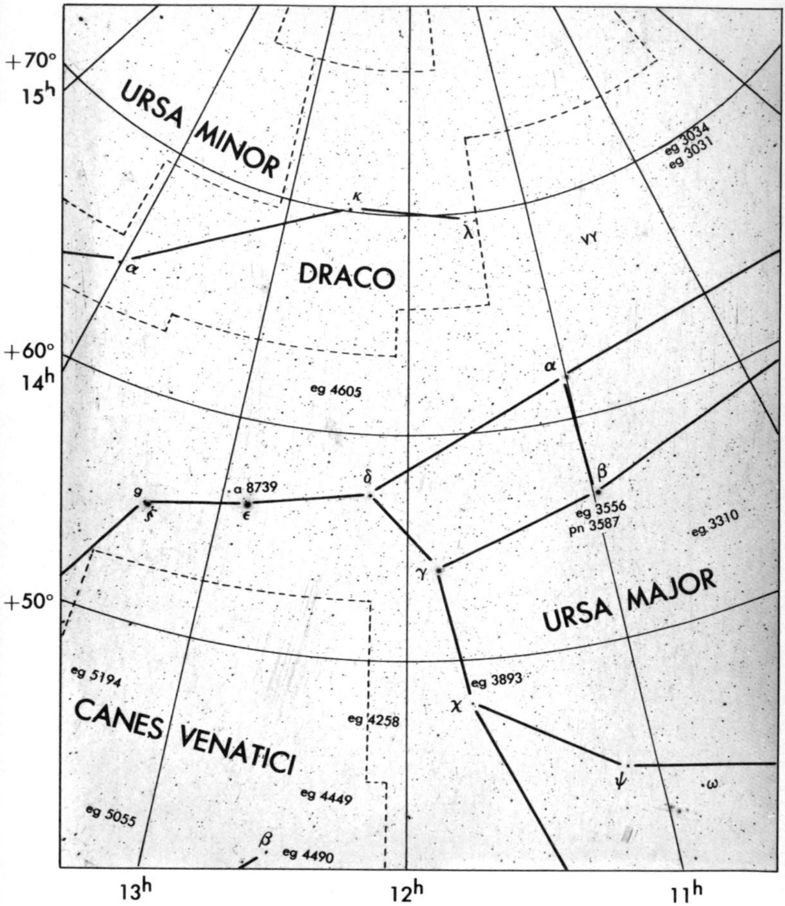

Atlas Chart 6. pn 3587 requires at least a 6-inch telescope and low magnification. Astronomers have called it the Owl, because two circular dark patches within the disk suggest the eyes of the nocturnal bird. This object also carries the designation M 97.

Mizar (ζ UMa), one of the most interesting doubles in the entire sky, also bears the alternate designation a 8891. In addition, Mizar possesses a faint naked-eye companion, g, commonly called Alcor. Ability to detect this companion tests one's visual acuity. The spectroscope shows that each component of Mizar is again

double, though no telescope is powerful enough to show them separately. The American Indians referred to Mizar and Alcor as the "Horse and Rider."

Of the seven Dipper stars, all but α and η (see also Chart 7) are moving essentially parallel to one another. These stars, with a few others, form a physical system, a very open cluster. As a result of the relative motion the familiar Dipper shape is slowly changing. In some 50,000 years the outline will have markedly altered.

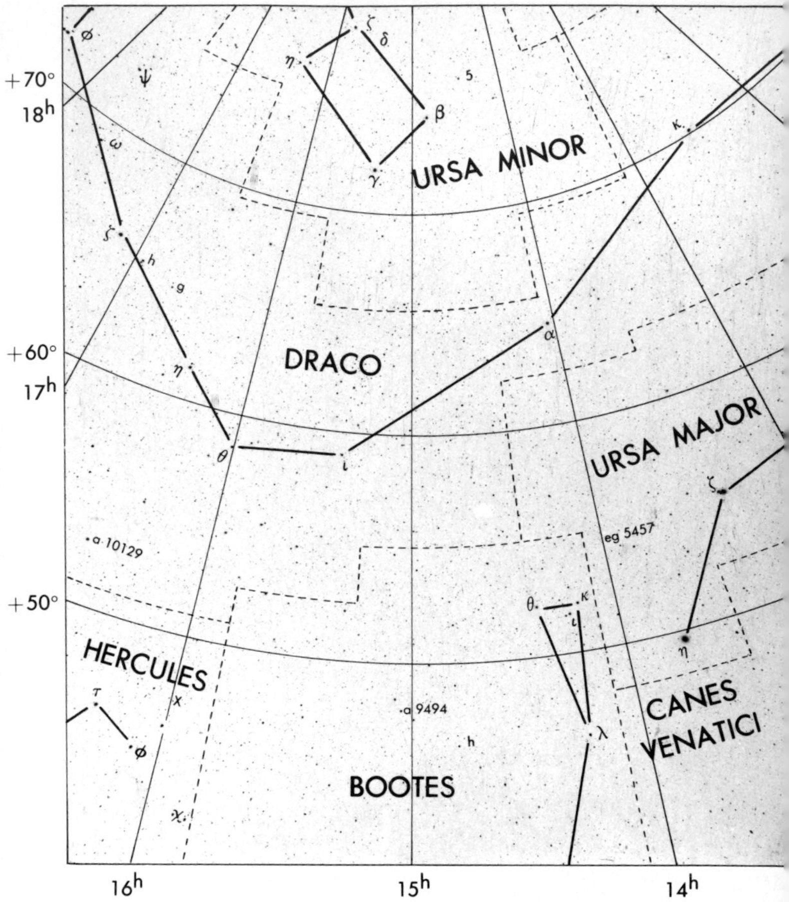

Atlas Chart 7. Few objects of special interest appear in this region. The double a 9494 in Boötes warrants attention. The star κ Boo is also double. eg 5457, also known as M 101, is an exceptionally beautiful spiral, one of the brightest in the sky. Note how Draco tends to encircle Ursa Minor, neatly separating it from its south-

erly neighbors, Ursa Major, Boötes, Hercules, and Lyra. α Dra, also known as Thuban, lies midway between the Guards (β and γ UMi) and Mizar (ζ UMa). Thuban was the Pole Star back in 3500 B.C.

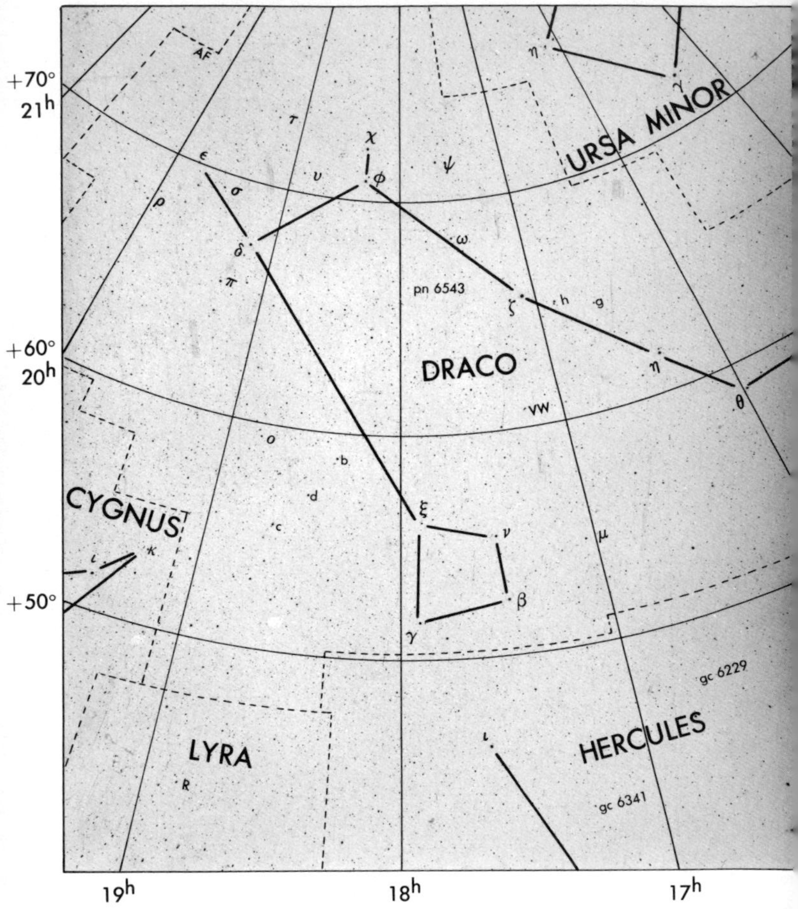

Atlas Chart 8. The head of Draco forms a conspicuous asterism, the Lozenge, not far from the bright star Vega and even closer to the foot of Hercules. Draco contains several interesting doubles. The star μ furnishes a test of resolution for a 3-inch telescope. The slightest optical aid, even opera glasses, will resolve the star ν.

pn 6543 is one of the brightest of these objects in the sky. A 3-inch telescope will show the disk, but somewhat larger instruments are necessary to reveal the internal structure, a bright irregular helix.

gc 6229 is faint compared with its neighbor, gc 6341 (also known as M 92). And this one, in turn, is much fainter than the spectacular gc 6205 (M 13), which lies to the south, beyond the boundary of this chart. Smaller telescopes show an enlarged hazy patch, with a condensation of brightness toward the center, instead of the individual stars.

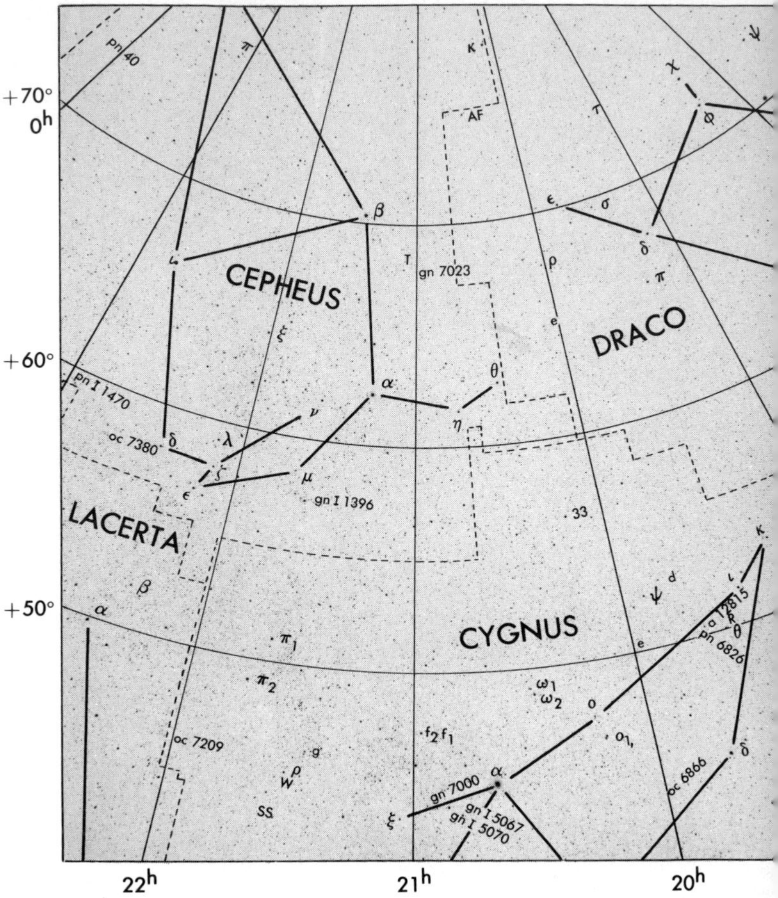

Atlas Chart 9. The area covered by this chart presents a star field more interesting than neighboring fields, because the Milky Way passes through Cepheus and Cygnus. The region contains a number of gaseous nebulae. Most famous is gn 7000, near α Cyg, commonly called the North American Nebula because of a resemblance to that continent. Although this object and its neighbors gn I 5070 (the Pelican Nebula) and gn I 5067 are extremely faint for telescopic observation, long-exposure photography with a large camera greatly enhances their interest.

In Cepheus, gn I 1396 is a large, faint structure just south of μ Cep, sometimes called the Garnet Star because of its deep red color. pn 6826 consists of bright patches in a faint oval of luminosity. The variable star SS Cyg rises sharply to a maximum about once every 50 days, on a somewhat irregular schedule. It is a favorite with many amateurs, who strive to observe it during one of its rapid, spectacular rises in brilliance. The brightness often changes markedly in two or three hours. See Figure 9 (p. 123).

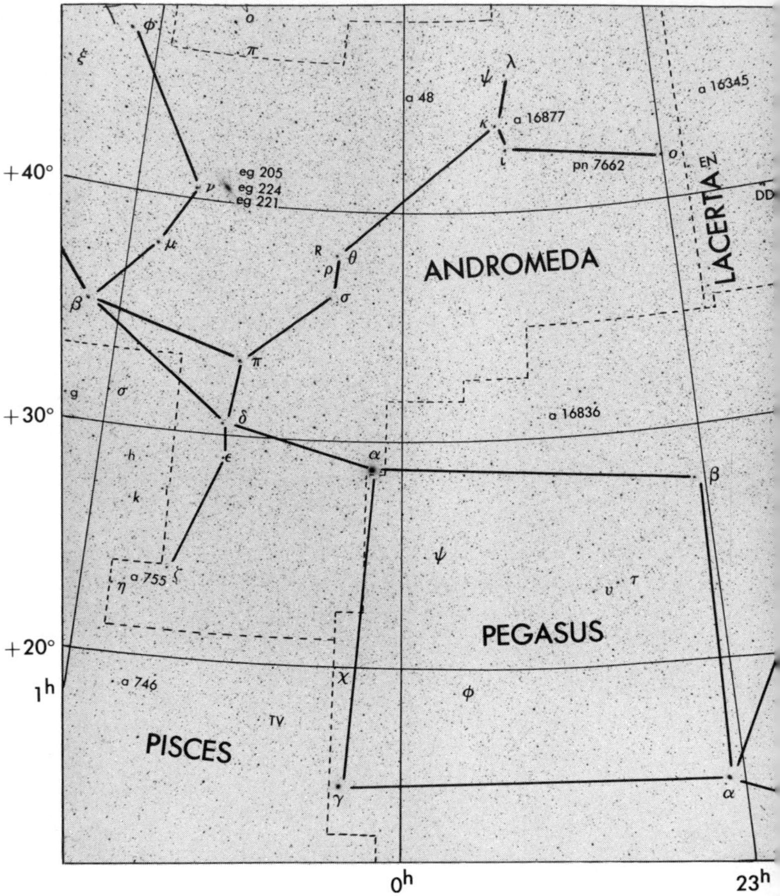

Atlas Chart 10. The region of Andromeda and Pegasus contains many interesting objects for telescopic observation. eg 224, also known as M 31, is the great Andromeda Nebula, clearly visible to the naked eye; a binocular shows it particularly well as an extended oval. This object is the most distant perceptible to the unaided eye, being approximately 1½ million light-years distant. This spiral has two faint elliptical companions, eg 221 (also known as M 32) and eg 205.

pn 7662 is one of the brighter members of the class. It has a bright inner ring surrounded by a faint outer oval approximately 28″ × 32″. The Andromeda stars ι, κ, ψ, and λ form a tiny Y-shaped asterism sometimes called Frederik's Glory.

Note that the line marking zero right ascension runs close to that side of the Square of Pegasus bordering Andromeda. This line, extended to the celestial equator, indicates the vernal equinox.

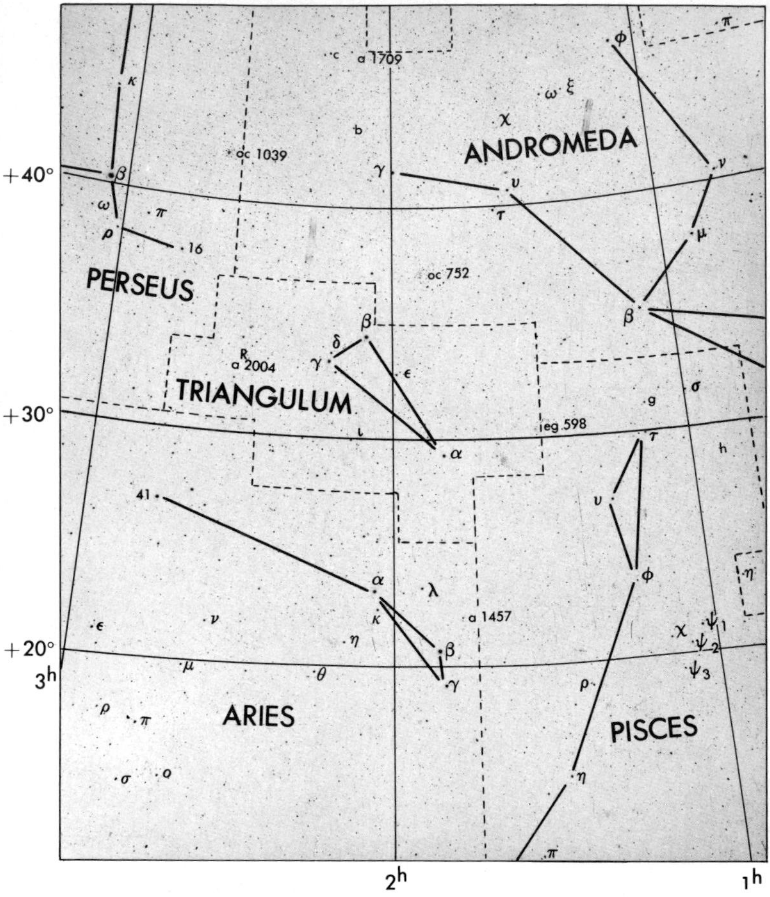

Atlas Chart 11. oc 1039 on the boundary of Perseus, also designated M 34, appears to the naked eye as a hazy object. Its distance is approximately 1500 light-years.

In Triangulum, the star ι, also a 1697, is a splendid double of contrasting colors, yellow and blue. A 2-inch telescope should readily resolve it. The double a 1457 also presents an interesting contrast of colors, white and green. The star γ And is one of the most beautiful doubles in the sky, its orange and green colors

offering an admirable contrast. ϵ Ari (a 2257) is a splendid double, an excellent test for a 3-inch telescope. ϕ And (a 940), a very close double, requires at least a 10-inch objective for resolution.

eg 598 (M 33) is the famous spiral in Triangulum. Since it is the brightest object of this class in the northern sky, with the exception of the Andromeda Nebula, the amateur will find it useful to test his telescope on this galaxy before trying to locate fainter objects.

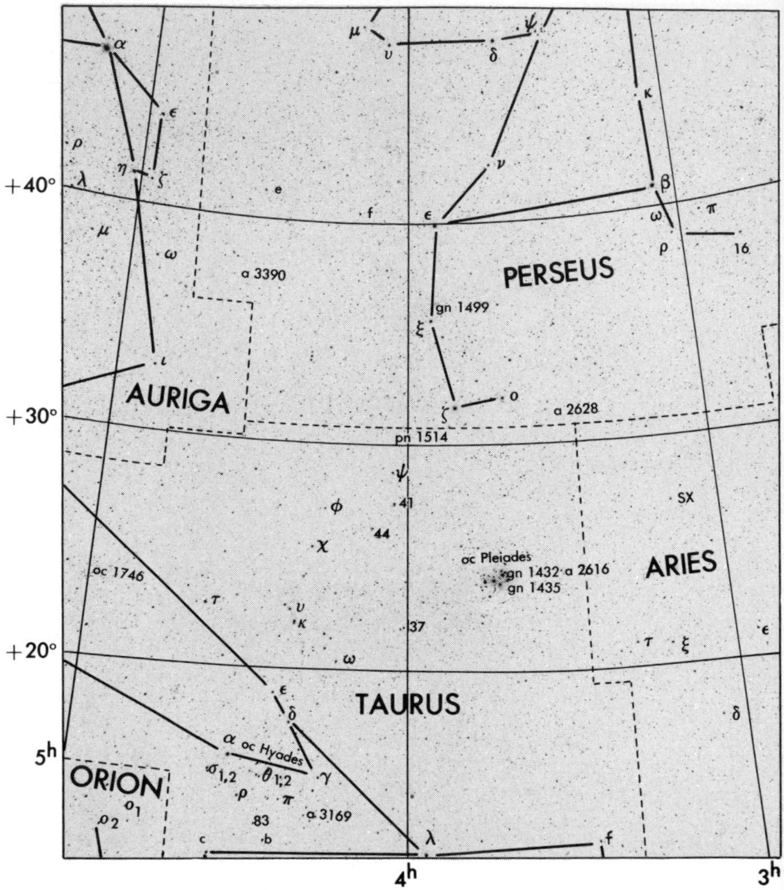

Atlas Chart 12. Two of the most spectacular open clusters in the sky, the Pleiades and the Hyades, appear on this chart. Although both of them are clearly visible to the naked eye, the revelation of their full beauty requires some telescopic aid. Since both clusters are large, be sure to use a low-power eyepiece. Binoculars will show clearly the numerous faint stars in these clusters. The naked eye can see 6, possibly 7, stars in Pleiades, arranged roughly in the form of a tiny dipper. Even a small telescope will greatly increase the number of stars visible, up to a limit of more than 100. Wisps of nebulosity cover the group, especially bright in the neighborhood of the more luminous stars. The nebulous back-

ground shows clearly on long-exposure photographs, but you will probably not be able to see it visually. The star cluster is 430 light-years away.

The Hyades contains fewer stars than the Pleiades and also is less compact. Use low power to see this extended V-shaped group, which forms the head of Taurus. The Hyades are approximately 130 light-years distant. Their white color contrasts interestingly with the red of nearby Aldebaran (α Tau). gn 1499 lies just north of ξ Per. pn 1514 consists of a 10th-magnitude star surrounded by a faint haze. Algol (β Per) is a variable star of the eclipsing type.

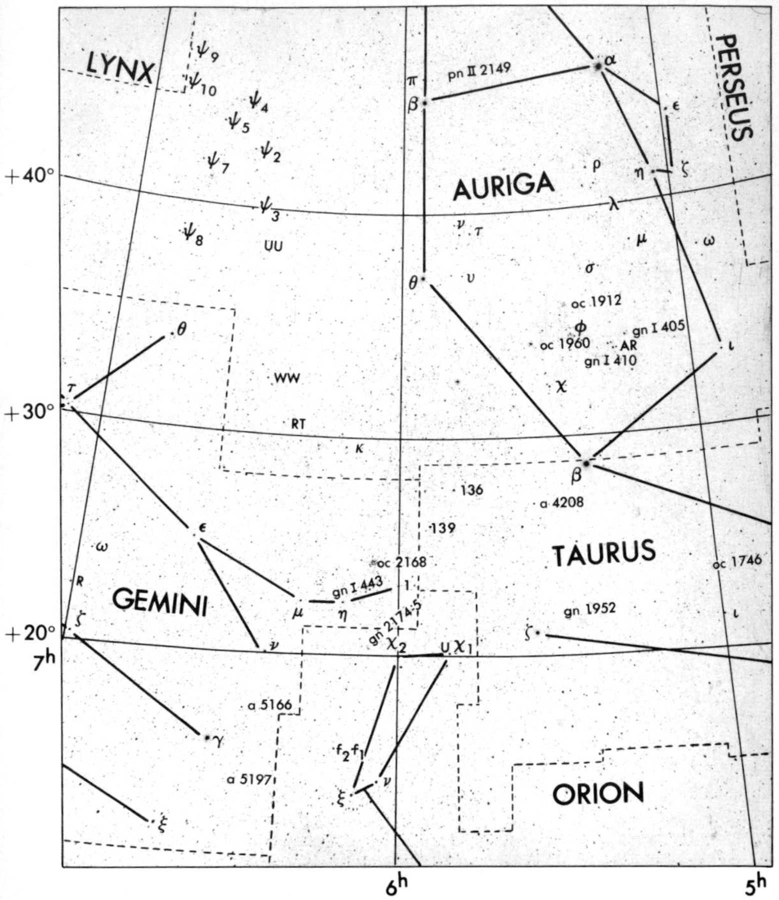

Atlas Chart 13. With binocular or a low-power telescope, sweep along the Milky Way from Cassiopeia and Perseus through Auriga and Gemini. Many rich star fields, not listed as actual clusters, will reward your search. In Auriga oc 1912, also known as M 38, is a beautiful oval group of stars at a distance of about 3000 light-years. Nearby is oc 1960 (M 36), a beautiful group 2600 light-years distant.

Several gaseous nebulae of interest occur in the region. gn I 405 is sometimes called the Flaming Star. gn I 410 is a combined

cluster and nebulosity. In Taurus, gn 1952 (or M 1), one of the most famous of all these objects, should probably be designated as a planetary. Termed the Crab Nebula, because of its peculiar structure, this object is evidently the debris from a nova, a star whose explosion the Chinese recorded in A.D. 1054. oc 2168 (M 35) is a fine naked-eye cluster. The star ϵ Aur is an eclipsing variable with one of the longest periods known, 27 years. Nearby is ζ Aur, also with a fairly long period, 2⅔ years.

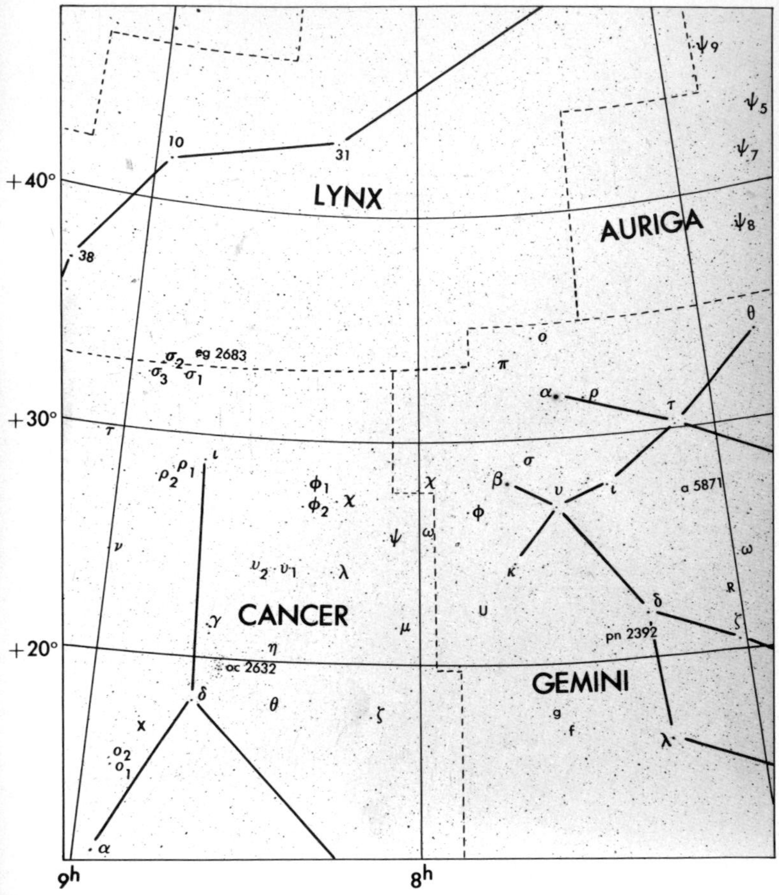

Atlas Chart 14. Castor (α Gem) is one of the most beautiful visual binaries. The two stars, magnitudes 2.0 and 2.8, slowly revolve about one another with a period of 477 years. It is a test for a 2-inch telescope, but a 3-inch will resolve it readily. The system is complex and has a number of faint *comes*, all members of the same physical system.

In Cancer, oc 2632 (M 44) is also termed Praesepe or the Bee-

hive. Although the naked eye sees the cluster as a hazy patch, a small telescope or binocular will resolve it into its component stars. Its estimated distance is 520 light-years.

pn 2392, one of the brightest examples of this class, consists of a bright ring set in a patchy disk 43″ × 47″.

The star ζ Cnc (a 6650) is an unusual triple. The period of the binary system is 60.1 years. ι Cnc (a 6988) is an orange-green pair.

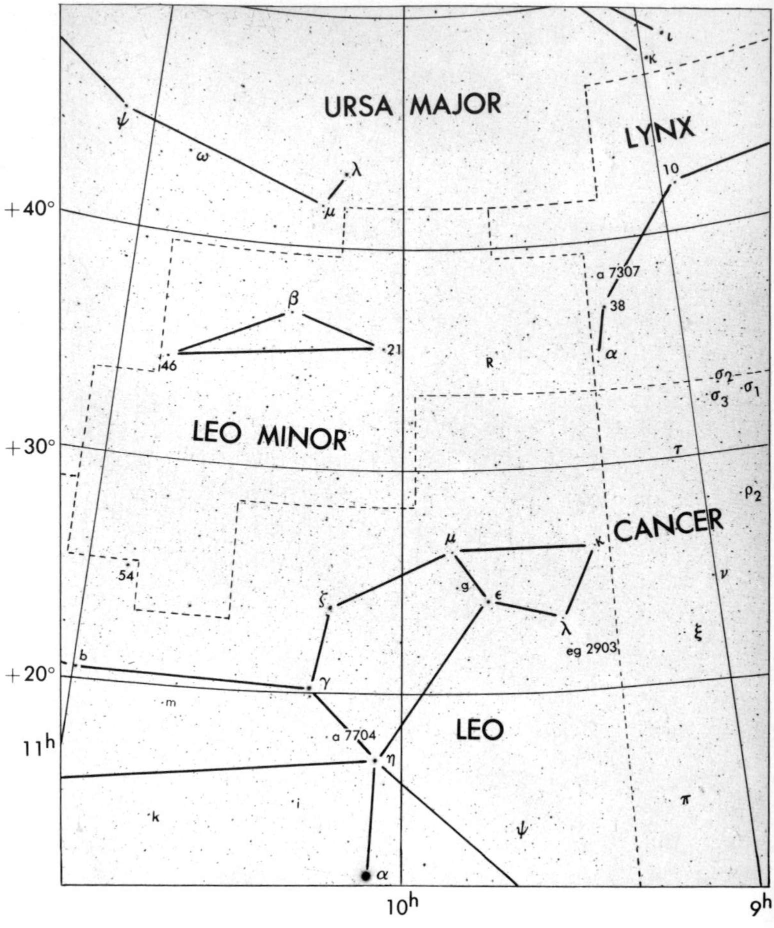

Atlas Chart 15. The constellation Leo Minor is a small triangle lying between the rear feet of Ursa Major and the back of Leo. The region contains a number of interesting doubles. One of the most striking is γ Leo, an easily resolvable pair of magnitudes 2.6 and 3.8. The star a 7979, designated as 54 Leo, is resolvable with very low power, splitting into a pair, of magnitudes 4.5 and 6.3. a 7704, on the other hand, requires at least a 4-inch telescope for separation of its two 7.5 magnitude stars. The region of ζ Leo is of some interest.

R LMi is a long-period variable, ranging in brightness from naked-eye visibility, 6.0, to magnitude 13.3 with a period of 372 days. Its deep red color, especially when near maximum, makes this star an interesting telescopic object. The area is singularly lacking in either gaseous nebulae or star clusters. The lone galaxy on the chart, eg 2903, shows as a faint, elliptical patch of luminosity.

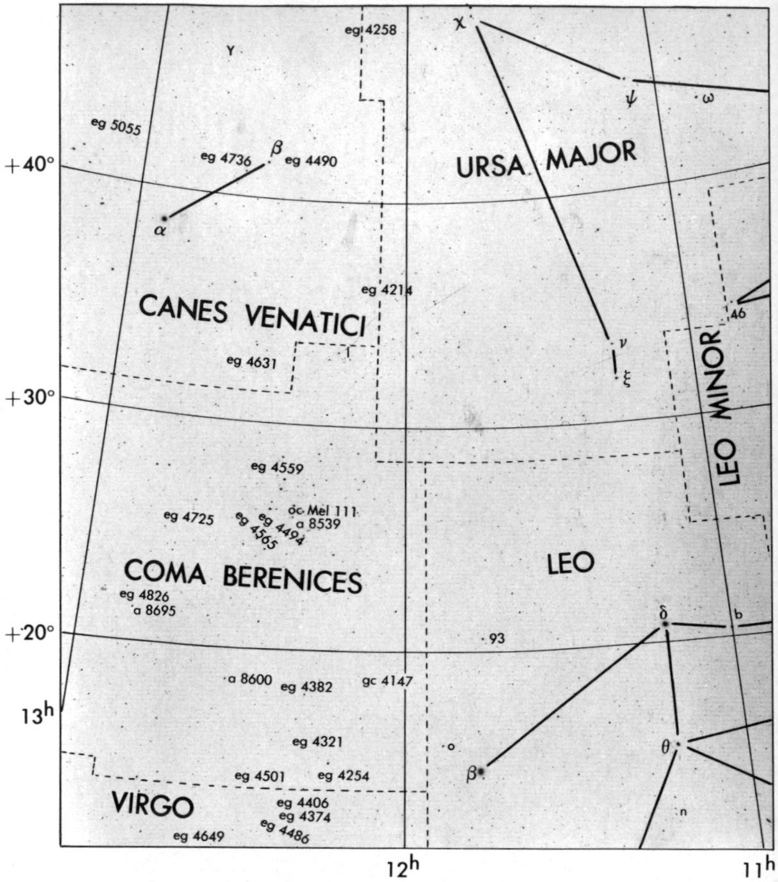

Atlas Chart 16. The star Denebola (β Leo) at the tip of the Lion's tail, is a beautiful blue-orange pair usually omitted from official lists of double stars because the system is not a true binary. Actually the stars are far apart in space even though they appear to be close together in the sky. An excellent object for a small telescope, ξ UMa (a 8119) should be resolvable in a 2-inch telescope. The stars are in relatively rapid motion for visual binaries, completing an orbit once every 60 years.

Most conspicuous is the naked-eye open cluster Mel 111, the official designation for the Coma Berenices cluster, often omitted from catalogs because of its large size. Nevertheless, this cluster

is a true stellar family. It appears best under extremely low power, preferably with field glass or binocular.

Characteristic of this part of the sky is the large number of galaxies. These also form a family, most of them millions of light-years distant. Those having Messier numbers are as follows: 4254 (M 99), 4321 (M 100), open spirals; 4374 (M 84), bright, amorphous; 4382 (M 85), 4406 (M 86), 4486 (M 87), bright oval patch; 4501 (M 88), beautiful spiral; 4649 (M 60), structureless; 4736 (M 94), beautiful bright spiral in CVn; 5055 (M 63), elongated spiral.

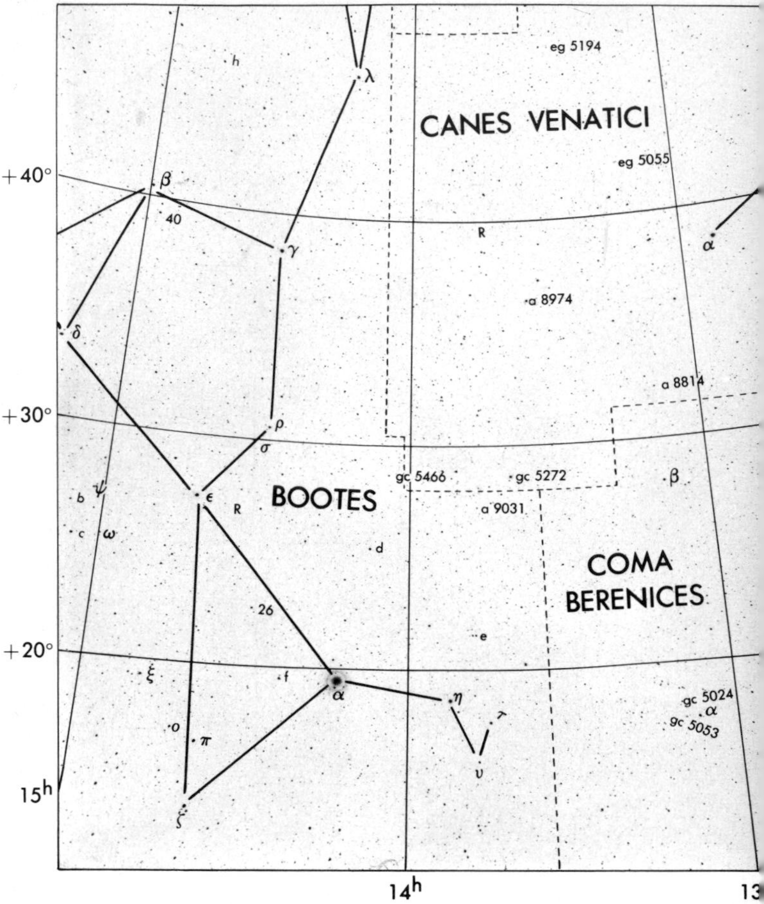

Atlas Chart 17. The most notable object in this region is eg 5194 (M 51), the famous Whirlpool Nebula in CVn, one of the brightest in its class.

Of the few globular clusters that appear in this region, 5272 (M 3) is perhaps the most interesting and the brightest. Its total magnitude, 4.5, makes it one of the brightest of these objects in the sky. gc 5024 (M 53) is considerably fainter, with a magnitude of 6.9. But these, in turn, are considerably brighter than gc 5053

and gc 5466, which are of the 10th magnitude and have no Messier number.

π Boo is a wide visual binary (a 9338). ζ Boo, a beautiful pair with magnitudes 4.4 and 4.8, requires at least a 4-inch telescope for resolution. The orbital period is 130 years.

ϵ Boo (a 9372) is an easily resolved pair of contrasting colors, orange and green. ϵ Boo is a wide pair with a period of 153 years.

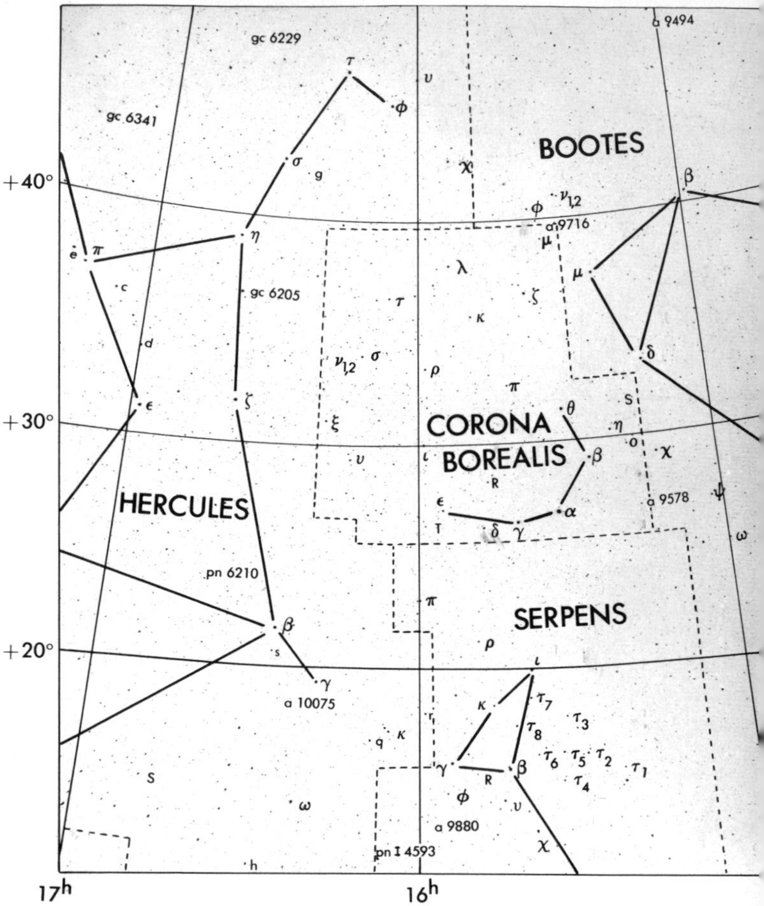

Atlas Chart 18. Of the three globular clusters that appear in this area, gc 6205 (M 13) is by far the most spectacular. This is the famous Hercules cluster, faintly visible to the naked eye as a fuzzy star of the 4th magnitude. It lies between ζ and η, about one-third of the way from the latter. gc 6341 (M 92) is about a magnitude fainter, though still a satisfactory object for the small telescope. gc 6229, on the other hand, is almost of the 10th magnitude and eight or ten times smaller than the other two. pn 6210 has a fairly bright inner ring with a faint outer ring 20″ × 43″.

η CrB (a 9617) is a close double, requiring about 42 years for a revolution. It has several faint *comes*. μ Boo (a 9626) is a wide white-orange pair, with *comes*. The principals revolve in a period of 224 years. a 9716 has a period of 56.7 years. ζ CrB is a wide pair of blue stars. γ CrB is a very close pair taking 101 years for revolution. σ CrB (a 9979) has several faint *comes* and a period of 215 years. a 10075 has a period of 317.5 years. ζ Her (a 10157) has a period of 34 years.

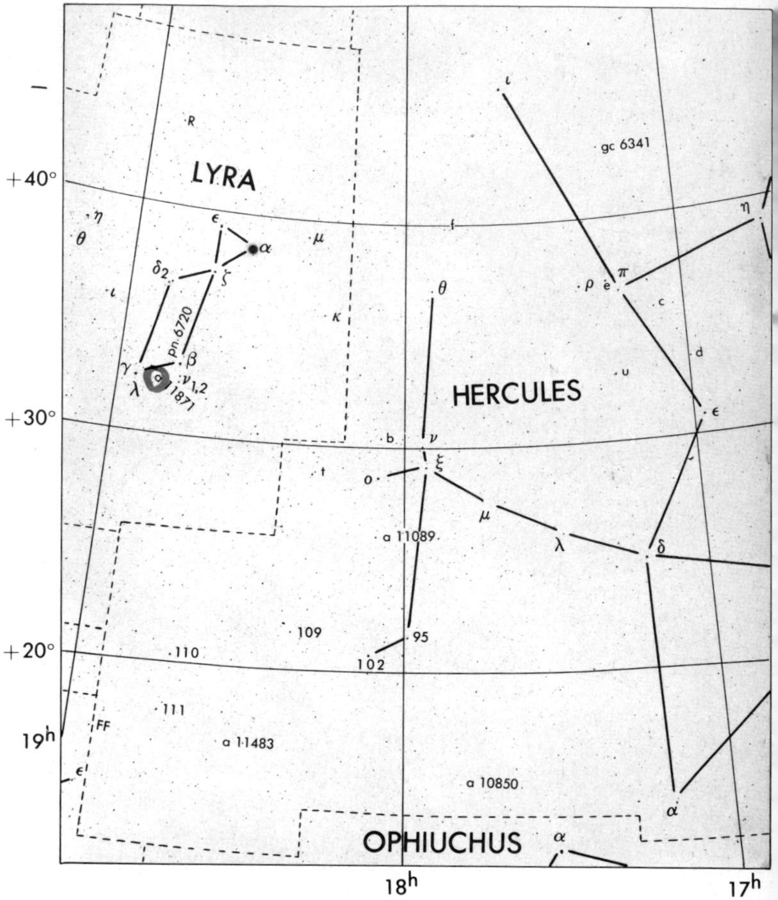

Atlas Chart 19. pn 6720, the famous Ring Nebula in Lyra, consists of a shell of luminous gas surrounding a very hot and very blue star. Although the star shows clearly on most photographs (see Fig. 12, p. 126), so much of its light falls in the ultraviolet that the central star is visible only with the largest telescopes. The atoms of the nebular gas absorb energy in the far ultraviolet and convert it into visible radiation, so that the gas becomes luminous. A 4-inch telescope will reveal the nebula clearly, though still larger ones are necessary to show structural details. This nebula, also known as M 57, lies approximately one-third of the way along the line from β to γ.

The star β Lyr is a spectroscopic double, a binary whose two

components are so close that they cannot be resolved in the telescope. The spectroscope reveals the double character by typical variations in the spectrum attributed to orbital revolution. The two stars so nearly touch one another that their mutual gravitational attraction distorts them into egg-shaped forms. Because they revolve in an orbit, the star is continually changing in brightness between the maximum of 3.4 and the minimum of 4.3, with a period of 12.91 days.

Vega (α Lyr) is a pure white 1st-magnitude star, of exceptional beauty in the telescope. With ϵ and ζ, α forms a small equilateral triangle. The star ϵ Lyr is one of the most interesting

(*continued on p. 248*)

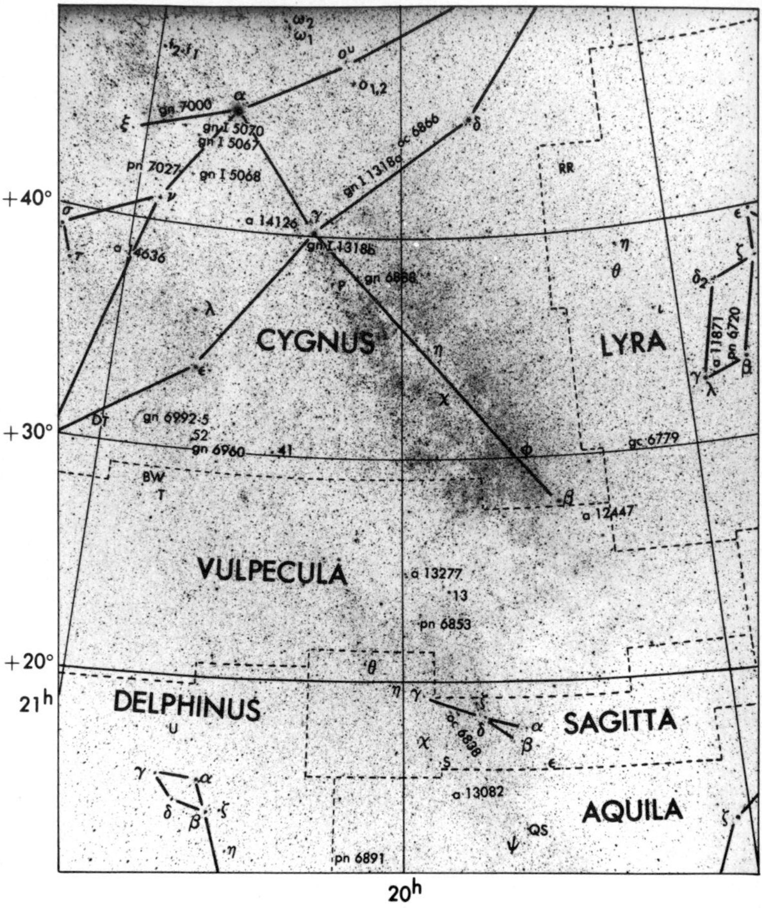

Atlas Chart 20. gc 6779 (M 56) is small and highly concentrated. pn 6853 is one of the largest in the sky, second only to 7293 in Aqr. The form vaguely suggests a dumbbell, which it is commonly termed. The total brightness is high, but the energy is spread over so large an area that the surface brightness is fairly low. Hence one needs a dark sky to view it well. Photographs reveal delicate internal structure. pn 7027 is of irregular shape, with four very bright condensations. oc 6838 (M 71) is rich in stars, despite its relatively small diameter.

Cygnus abounds with irregular gaseous nebulae. For a description of some of them see Atlas Chart 9. gn 6960 makes up the western half of a faint circular nebula variously termed the Wreath,

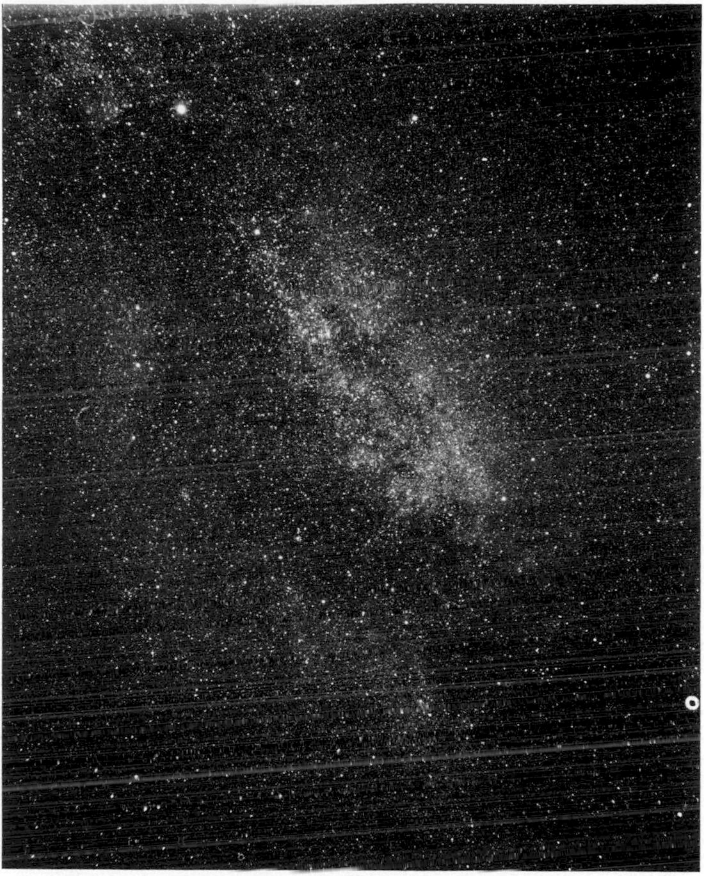

Loop, or Network. The pair gn 6992-5 represents the eastern half of the same loop. Photography is really necessary to show the form and structure of this interesting formation, which may in fact be a planetary or the debris from a star that once exploded.

A number of interesting double stars fall in the area. a 12447, a fairly close double, has a period of 355 years. β Cyg (a 12540) is one of the classic doubles because of the brightness of the stars, 3.2 and 5.4, and the interesting contrast of colors, orange and blue. This constitutes a true physical pair. δ Cyg (a 12880), with an orbital period of 300 years, shows a difference of five magnitudes between the stars.

(*continued on p. 248*)

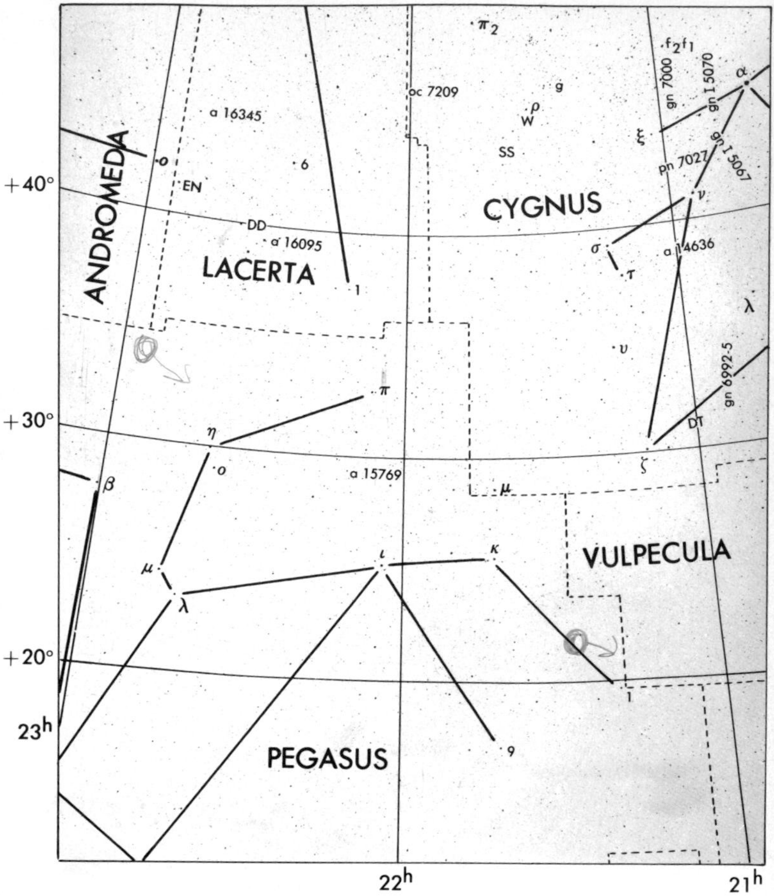

Atlas Chart 21. The star a 16095, a wide double resolvable in low-power binoculars or even opera glasses, consists of a pair of blue stars of magnitudes 5.8 and 6.6.

Some interesting star fields, for low-power telescopes, exist in the area between π_2 and g Cyg. oc 7209 is by no means an exceptional object, though a small telescope will resolve it.

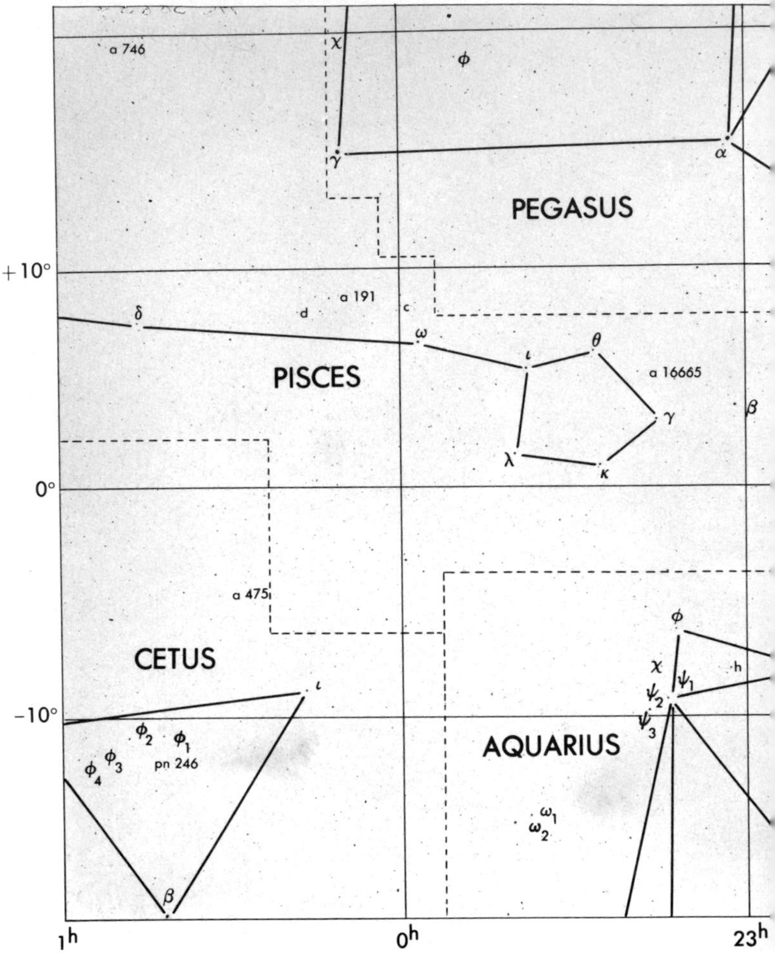

Atlas Chart 22. In Pisces, the stars γ, θ, ι, λ, and κ form a small, irregular pentagon that is distinctive in spite of the relative faintness of some of the component stars. They compose the asterism known as the Circlet, indicating the western fish. This asterism lies directly south of the Great Square of Pegasus. Two fainter stars, b and 19, are often included, making seven in all.

The faint double star a 16665, requiring at least a 5-inch telescope to resolve, has some faint *comes*. The two stars, with mag-

nitudes of 9.0 and 10.0, revolve around one another in a period of 85.7 years.

The double a 191 is a wide pair. a 475 is a triple, one component of which requires about a 10-inch telescope to resolve. a 746 requires similarly high resolving power. In Cetus, pn 246 is fairly bright, but the distribution of luminosity over the large area makes it a somewhat difficult object to observe.

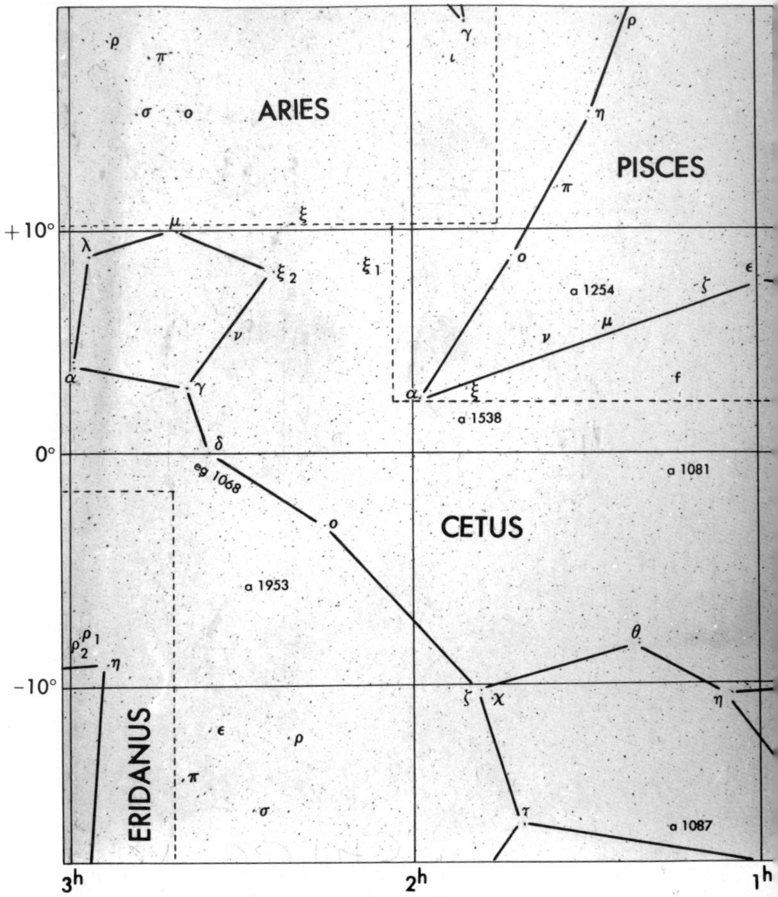

Atlas Chart 23. eg 1068 (M 77) is one of the brighter elliptical spirals. α Psc (a 1615) is a representative double for a 3-inch telescope, with components of magnitudes of 4.3 and 5.2. γ Cet (a 2080) should be resolvable in a 2-inch telescope, with magnitudes 3.7 and 6.2.

The so-called cords of Pisces converge to an acute angle at α Psc (El Rischa, the Knot). These point approximately toward

Mira (Wonderful), ο Cet, the long-period variable star about halfway down the Neck of Cetus. At maximum, Mira is the brightest star of the entire constellation, shining with the deep red hue. It slowly fades far below naked-eye visibility, however, and gradually returns to maximum, completing the cycle in about 330 days, on the average. Do not be surprised, therefore, if you fail to see it. See Figure 6 (p. 122) for special map.

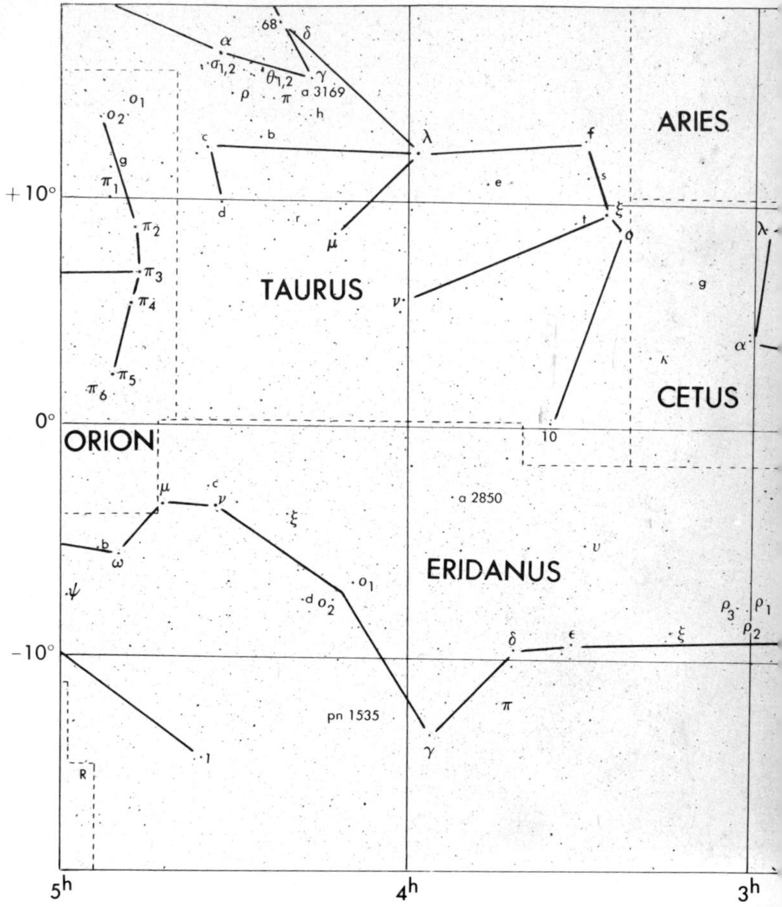

Atlas Chart 24. pn 1535 consists of a bright inner ring and a faint outer ring. The star fields become richer in Taurus and Eridanus. With binocular sweep southward from the Hyades through the shield of Orion.

a 2850 provides an interesting contrast, its stars being green and yellow, with magnitudes 5.0 and 6.3.

The star λ Tau, at the base of the Bull's neck, is an eclipsing variable with range 3.5 to 4.0 in approximately 4 days.

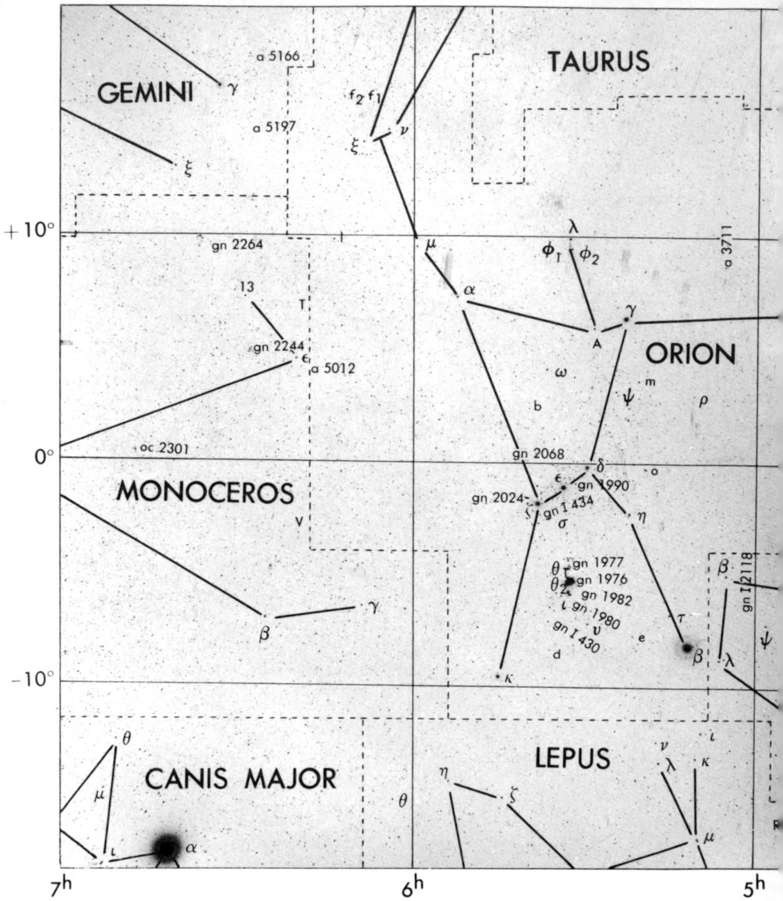

Atlas Chart 25. Although Orion does not strictly lie in the Milky Way, the hazy concentration of stars in this region is generally ascribed to the presence of one of the Galaxy's spiral arms. The entire region of Orion and Monoceros is interesting to sweep with low-power telescope or binocular. Gaseous nebulae abound in the region of Orion, most of them being part of the same underlying stratum of gas and dust, illuminated or excited by neighboring hot stars. gn 1976 (M 42), the Great Nebula in Orion, a magnificent luminous cloud surrounding θ_1 Ori, northern star in Orion's dagger, is plainly visible to the naked eye as a hazy object. Even the smallest optical aid reveals the wispy nebula structure (see Fig. 14, p. 128). A 2-inch telescope should readily resolve

θ_1 into four components, usually termed the Trapezium. The four stars, all very blue, have magnitudes 5.4, 6.8, 6.8, and 7.9. The distance of the stars and surrounding nebulosity is estimated at 550 light-years. This nebula is a magnificent object in larger telescopes, which show that some of the structure is due to clouds of obscuring dust as well as to luminous gas.

gn I 2118 is the Witch Head Nebula, 1.5° S of β Eri and 2° NW of β Ori. Sweep, with low power, the triangle formed by β and η Ori and β Eri. There are many faint irregular patches of nebulosity. γ Ori has some faint luminosity around it. Some luminosity lies in the head of Orion, between λ and ϕ_1. gn 1977

(*continued on p. 248*)

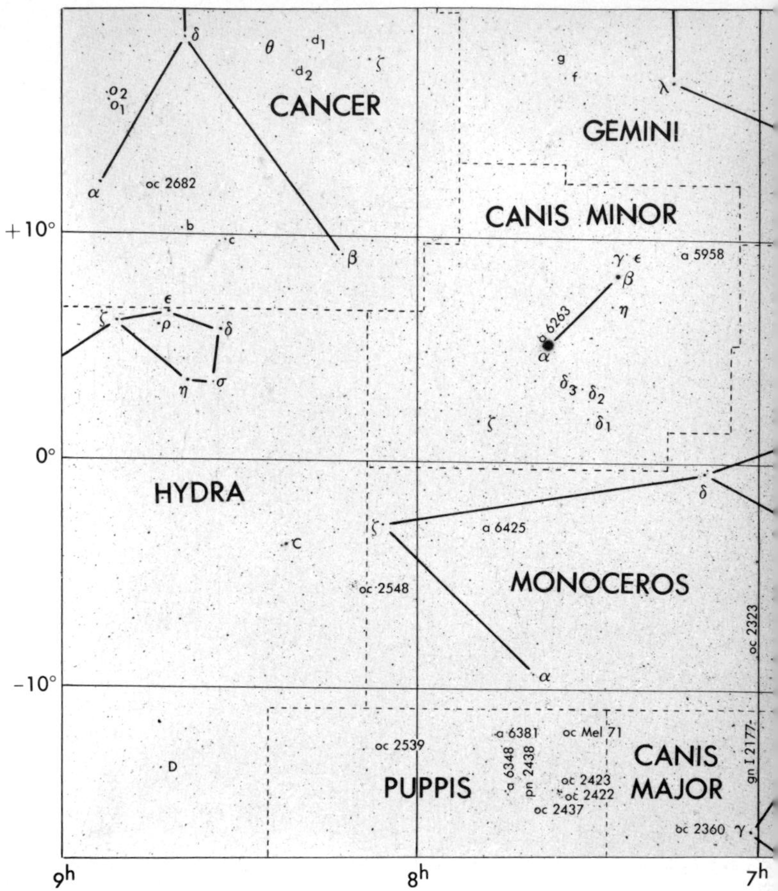

Atlas Chart 26. A number of open clusters appear in this region, which is worth sweeping with low power. The head of Hydra provides an interesting field for binoculars. oc 2323 (M 50) is rich. oc 2360 and oc 2422 provide splendid fields. oc 2437 (M 46) is somewhat fainter but contains more stars. oc 2548 is fairly

bright and rich. oc 2682 (M 67) requires a larger telescope. The estimated distance is 2900 light-years.

pn 2438 is an irregular patchy ring, on the northern edge of oc 2437. gn I 2177 is a nebulous patch. oc 2539 is rich in stars.

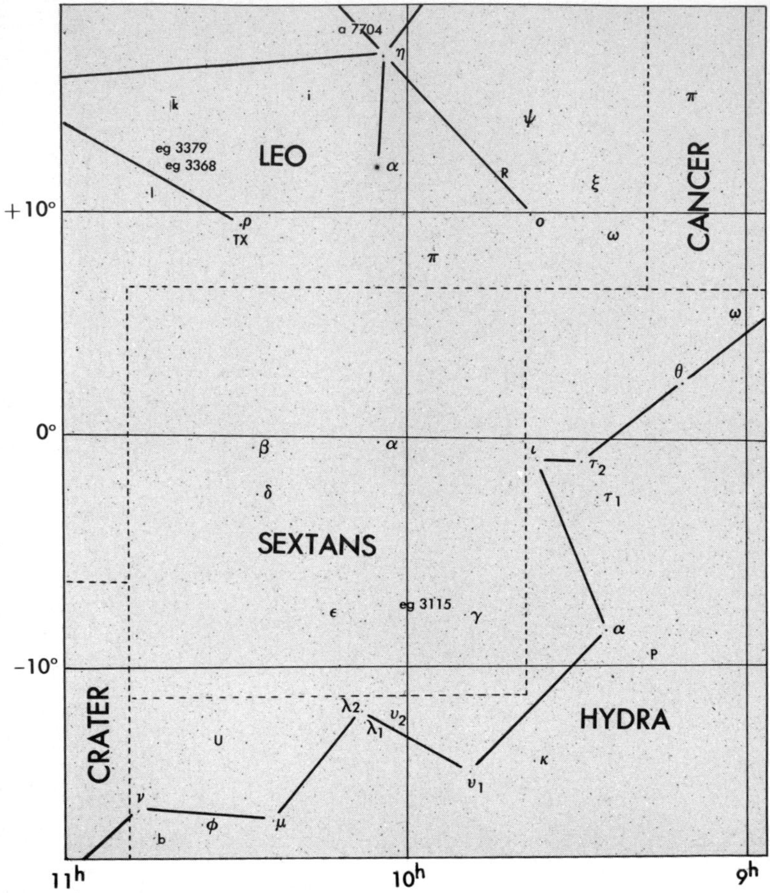

Atlas Chart 27. As we move away from the obscuring dust clouds of the Milky Way, the more distant galaxies now become visible. eg 3115 in Sextans is one of the brighter representatives of this class. The pair in Leo, eg 3368 and eg 3379, are considerably fainter.

R Leo, in the Lion's foreleg, is a long-period variable, ranging

from magnitude 4.4 to 11.6 in a period of 313 days.

Of the double stars, ω Leo (a 7390) is very difficult to resolve, requiring at least a 10-inch telescope for separation. The period is 117 years. Although Hydra is not particularly rich in stars, the region is not uninteresting for sweeping with a low-power binocular.

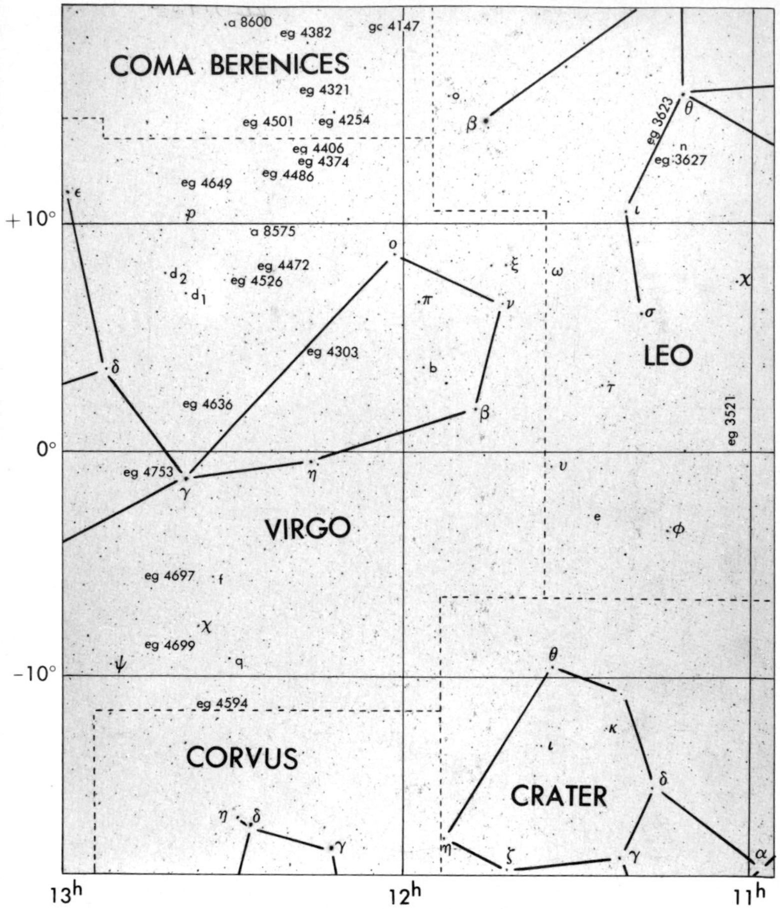

Atlas Chart 28. In Leo, eg 3623 (M 65) is one of the brighter spirals, highly elliptical in shape. eg 3627 (M 66) is also a bright elliptical spiral. eg 4254 (M 99) in Coma Berenices is a small spiral; eg 4321 (M 100) is an open spiral. eg 4382 (M 85) shows as a bright oval patch. eg 4501 (M 88) though faint shows beautiful spiral structure. eg 4649 (M 60) is without structure. In Virgo we find another clustering of galaxies, an extension of fields previously noted in Canes Venatici and Coma Berenices. eg 4303 (M 61) is a beautiful spiral, seen face on. eg 4374 (M 84) is a bright, amorphous nebula.

eg 4406 (M 86), eg 4472 (M 49), and eg 4486 (M 87) are relatively faint, difficult objects.

Of the double stars, a 8600 presents an interesting contrast of orange and green. γ Vir (a 8630) with its two almost equal components of magnitudes 3.6 and 3.7, is one of the finest double stars in the sky and a favorite with amateurs. A 2-inch telescope will separate it. The system is a true binary, with the stars revolving around one another in a period of 177.75 years.

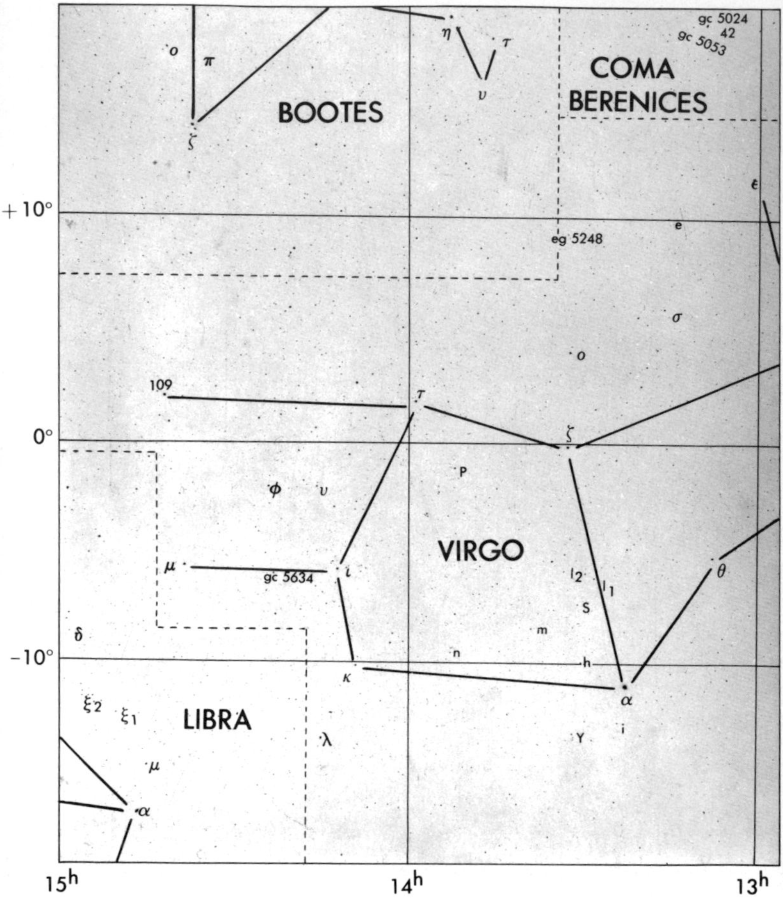

Atlas Chart 29. Few of the objects in this region are of special interest. As the accompanying photograph indicates, the region is singularly devoid of stars, a fact that makes Spica (α Vir), assume added brilliance. Neither eg 5248 nor gc 5634 is outstanding.

The region contains a few double stars, π and ζ Boo. The latter, having a period of revolution of 130 years, is the more notable.

The former is easily resolved, but ζ Boo requires a 5-inch telescope to show the component stars of magnitudes of 4.4 and 4.8.

S Vir is a long-period variable, whose outstanding red color near maximum of 6th magnitude makes it an interesting telescopic object. The average period is 377 days, during which time it falls to a minimum of about 13.0 magnitude.

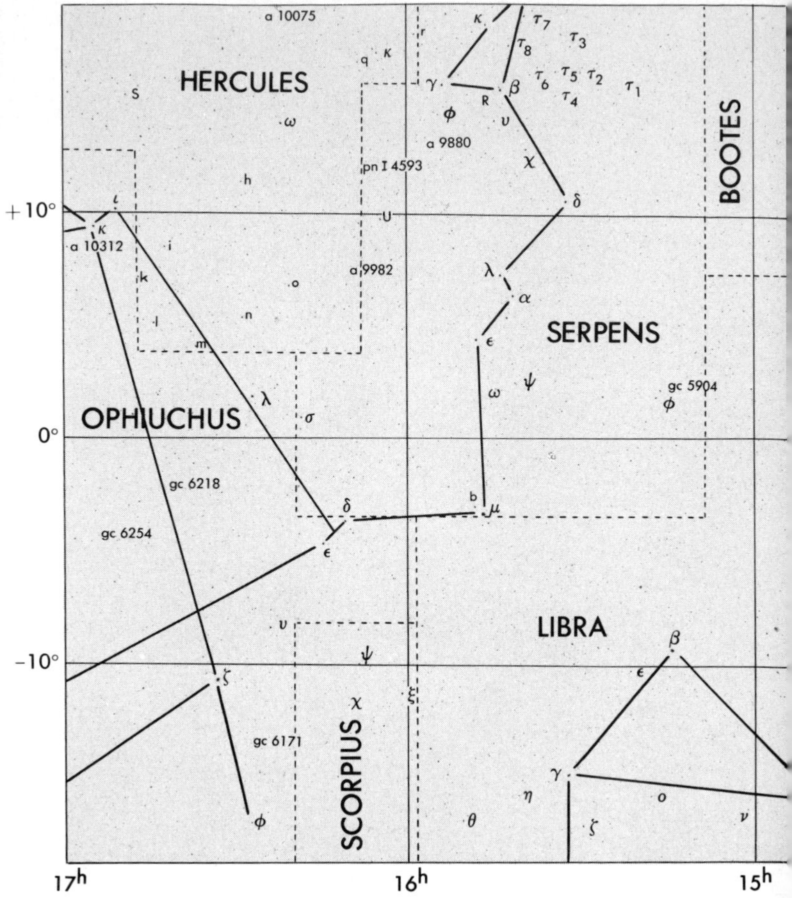

Atlas Chart 30. The head of Serpens will repay scrutiny with binocular or a low-power telescope. The richness of the sky background increases through Hercules to Ophiuchus. gc 5904 (M 5) is one of the finest of such objects in the sky, excelled only by 47 Tuc and ω Cen in the southern sky. gc 6218 (M 12) and gc 6254 (M 10) are also outstanding objects in the telescope. All three of these globular clusters are visible to the naked eye. gc 6171 is much fainter and smaller.

δ Ser (a 9701) is a fine double with magnitudes 4.2 and 5.2, easily resolved by a 2-inch glass. a 10075 has a period of 317 years. R Ser, a long-period variable, ranges from magnitude 5.6 to 14.0, with an average period of 357 days. The color near maximum is a startling crimson. S Her is another example of the same type of variable, ranging from 5.9 to 13.6 magnitude in 307 days.

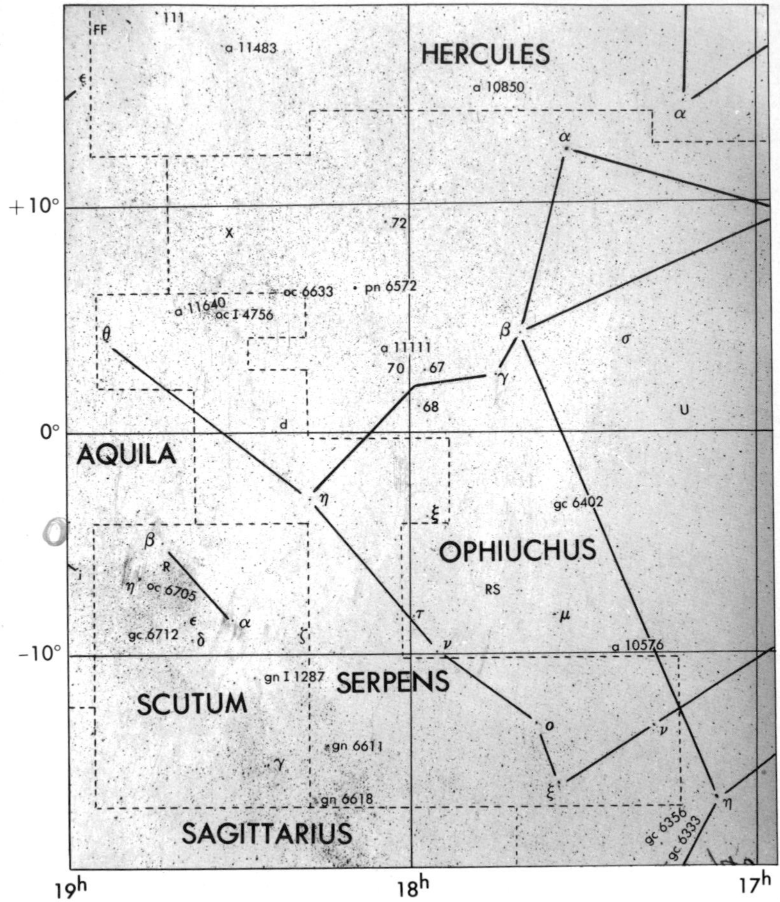

Atlas Chart 31. The star fields in Ophiuchus increase in brilliance toward the great concentrations in Aquila, Scutum, and Sagittarius. oc 6633 and oc I 4756 are easily resolved by relatively low power. The observer will also find a rich field a degree or so northeast of β Oph. With binocular or with a low-power telescope, sweep through the Milky Way in Scutum and the southeastern boundary of Serpens. oc 6705 (M 11) is magnificent in a large telescope. It lies at a distance of 4300 light-years.

pn 6572, a bright oval disk, is one of the most interesting objects of this class. Several outstanding gaseous nebulae lie in the region. gn 6611 (M 16) is actually a nebulous cluster. gn 6618 has been variously called the Omega, the Horseshoe, or the Swan

Nebula. The star background in this region is particularly rich. gn I 1287, 2° S of α Scu, requires slightly higher telescopic power. gc 6333 (M 9), gc 6356, and gc 6712 are all fairly faint.

68 Oph (a 10990) is a difficult double, consisting of stars of 4.4 and 9.2 magnitudes. The system is a complex one, possessing at least seven *comes*. τ Oph (a 11005) has a revolution period of 224 years. 70 Oph (a 11046) is one of the better-known binaries, with components of magnitudes 4.1 and 6.1 and a period of 87.7 years. It is readily resolvable in a 2-inch glass. a 11111 (73 Oph) is a difficult object to resolve, and requires a 10-inch telescope. The revolution period is 423 years. θ Ser (a 11853) is a wide pair, separable with opera glasses. The magnitudes are 4.5 and 5.4.

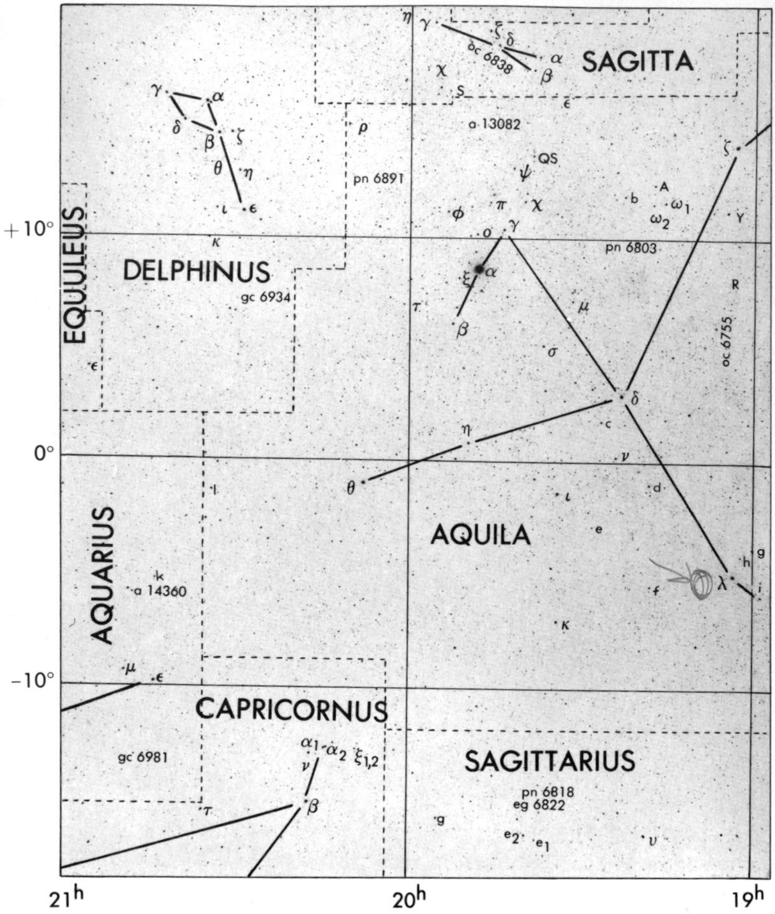

Atlas Chart 32. The Milky Way in Aquila presents many rich star fields, particularly beautiful in binoculars and low-power telescopes. Through the constellations Cygnus, Sagitta, Aquila, Scutum, and Ophiuchus, the Milky Way appears to be divided lengthwise. Enormous clouds of dust between us and the more distant stars obscure the brilliant Milky Way background, forming what astronomers generally call the Rift.

oc 6838 (M 71) is a magnificent object. pn 6803 exhibits a small bright disk. pn 6818 appears as an irregular oval ring; pn 6891 is a bright disk surrounded by a fainter ring, about 15″

diameter. gc 6981 (M 72) is small and a difficult object without large telescopes. eg 6822 is particularly remarkable because it lies so close to the Milky Way. Evidently we view it through a "window" in the obscuring dust that fills this region. Although it is one of the nearer and apparently larger galaxies, the intervening dust cloud obscures it so that it seems relatively faint.

π Aql (a 12962) is an interesting double, but to resolve it requires at least a 3-inch telescope. γ Del (a 14279) is a wide double, resolvable in binoculars, with components of magnitudes 4.5 and 5.5. a 14360 has a period of 135.6 years.

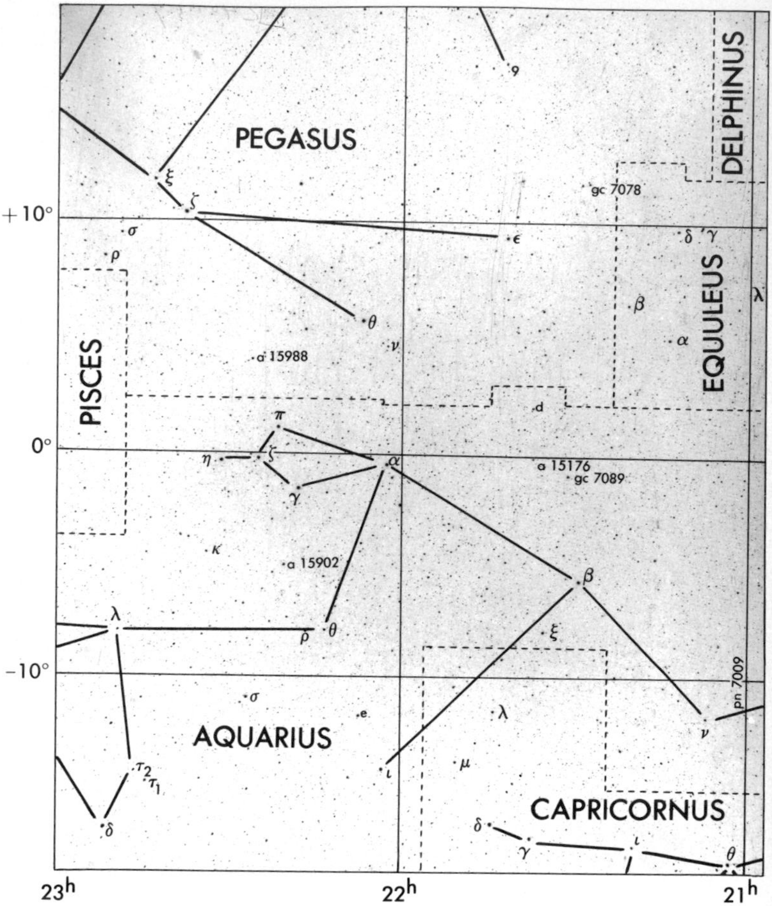

Atlas Chart 33. pn 7009 is one of the brightest members of this class. It consists of a bright inner ring surrounded by a patchy outer disk 26″ × 30″, with faint extensions to 44″. It appears to possess a faint external ring and hence is sometimes called the Saturn Nebula. The two globular clusters in this region, 7078 (M 15) and 7089 (M 2) are clearly visible to the naked eye. Even the smallest optical aid shows them as hazy patches, condensed toward the center. They are to be counted among the brightest objects of this class.

The double star a 15176 requires an 8-inch telescope for resolution. The pair is close and needs only 71 years for complete revolution. The four stars γ, η, ζ, and π Aqr delineate a small Y-shaped asterism, a characteristic feature of this constellation. The star ζ at the center of this Y is one of the finest doubles in the sky, with components of 4.4 and 4.6 magnitude. It requires a 3-inch telescope for resolution. a 15988 has a period of revolution of 150 years.

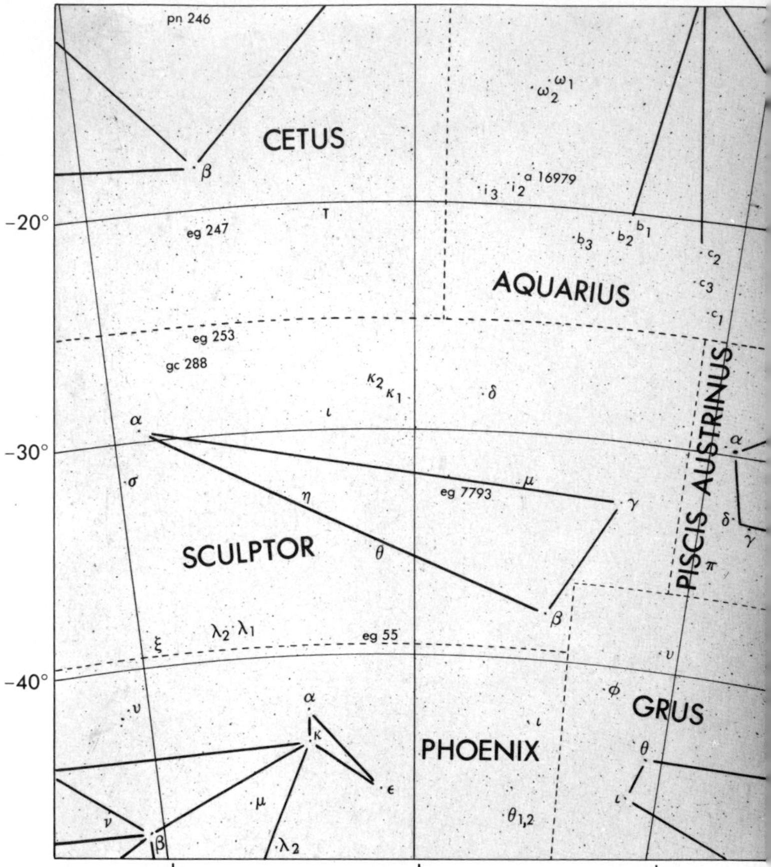

Atlas Chart 34. With the obvious thinning of the stellar background in this region, the galaxies begin to appear. eg 253 is a large and beautiful spiral, the finest and brightest in the sky with the exception of the great Andromeda Nebula. eg 247, though

nearly as large, is about four magnitudes fainter. It, too, is a fine spiral. pn 246 is large but quite faint. Observe it with low power. θ Gru (b 54) is a double, readily observable in a 3-inch telescope.

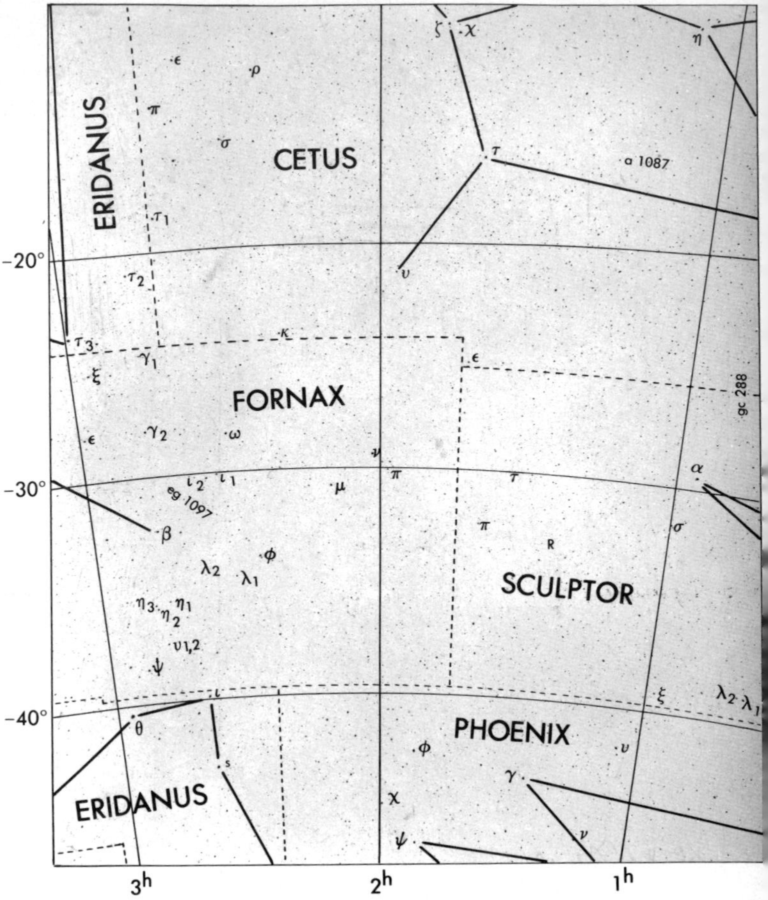

Atlas Chart 35. The region is particularly barren of stars and even galaxies are rare. eg 1097 is faint and gc 288 is relatively uninteresting. a 1087 is an interesting double, and a good test for a

3-inch telescope. θ Eri (b 5) is a wide double, resolvable with a binocular. The components are bright, of magnitudes 3.4 and 4.4.

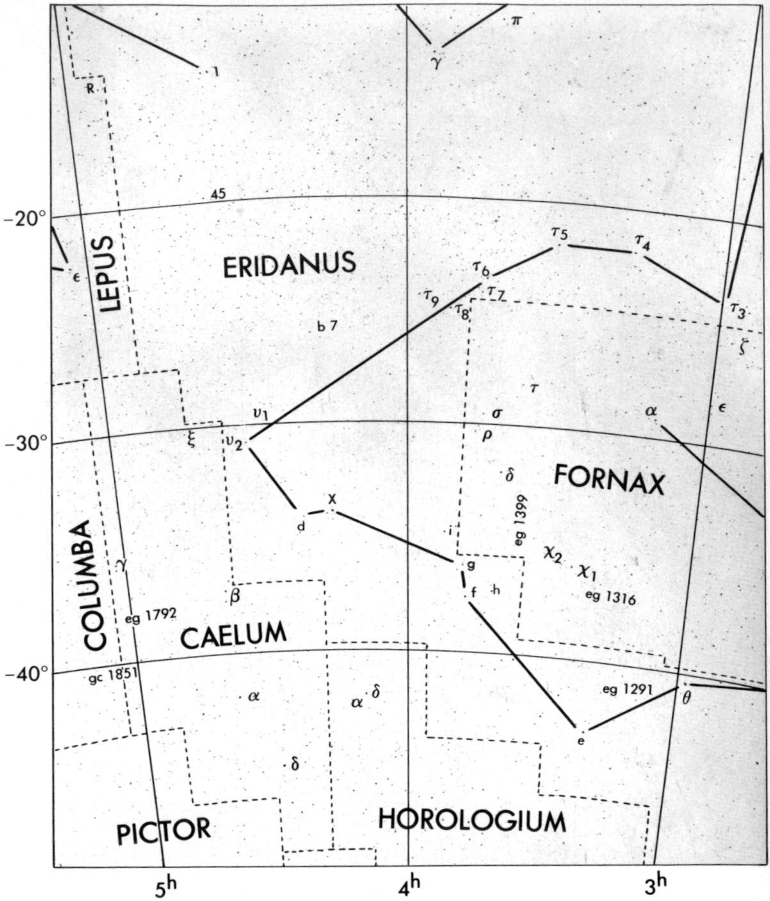

Atlas Chart 36. This area of the sky begins to improve as far as richness in background is concerned, especially in the northern part of Eridanus. None of the galaxies is conspicuous. gc 1851, though barely visible to the naked eye, is an interesting object in

the telescope. f Eri (b 6) is a wide double, with components 4.9 and 5.4 in magnitude. b 7, a double difficult even for a 5-inch telescope, has a revolution period of 65 years.

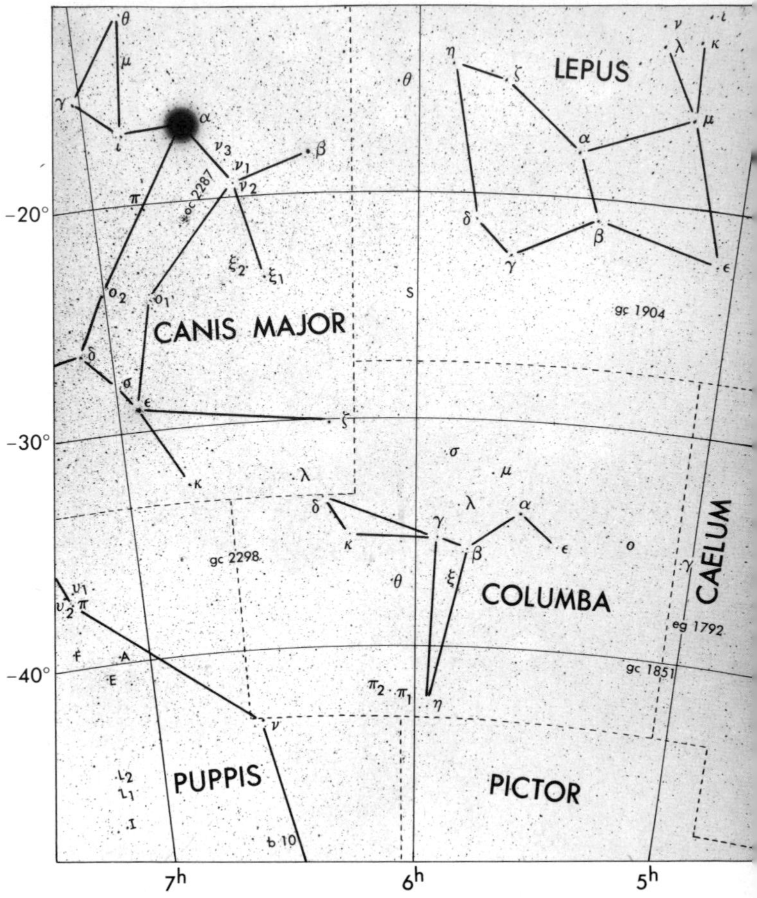

Atlas Chart 37. This region offers many beautiful star fields. Canis Major and Puppis in particular are well worth sweeping with low-power binocular or telescope. oc 2287 (M 41) is one of the finest in the sky and readily resolvable with medium optical power.

Sirius (α CMa), the brightest star in the sky, has an 8th-magnitude companion at a distance of approximately 10″. Despite the large separation, however, the object is difficult to see in a small telescope because of the dazzling brilliance of the primary. Under exceptional conditions a 6-inch telescope will show it. Ordinarily, an 8-inch telescope is the minimum for a clear view. The separation varies as the faint star swings about its companion in a period

of 50 years or so. The minimum distance is about 2″. The companion is notable for several reasons. More than a century ago Bessel discovered that Sirius did not remain exactly fixed in the sky but seemed to be pursuing a small orbit. From this fact he inferred the existence of a then unseen companion, both stars revolving around their common center of gravity. Alvan Clark discovered this companion star in 1862. Since it has approximately the same surface temperature as Sirius, the difference in brightness must be attributed to difference in size. The companion, Sirius B, proves to be representative of an unusual class of stars called white dwarfs. Such stars are highly compressed, and B has a

(*continued on p. 248*)

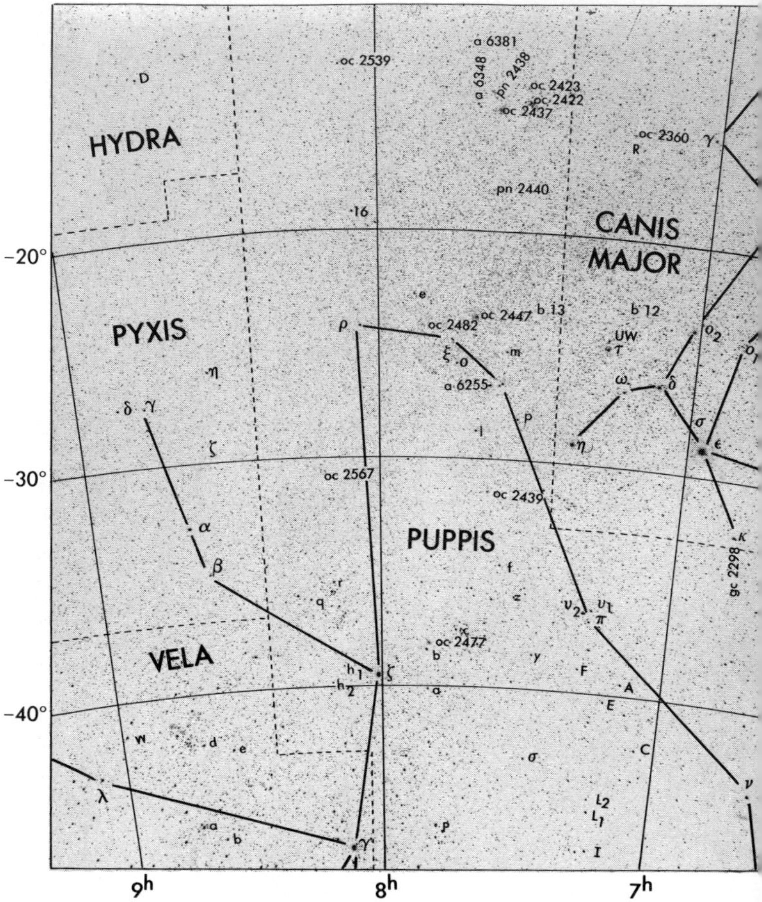

Atlas Chart 38. The Milky Way through Canis Major, Puppis, and Vela contains, in addition to many open clusters, rich star fields not specifically recognized as clusters but nonetheless beautiful. oc 2360 and oc 2422 present splendid fields observable with relatively low power. oc 2437 (M 46) is especially notable. oc 2447 (M 93) also has a rich field. oc 2477 is one of the most beautiful clusters in the area, with its 300 component stars lying at a distance of 3300 light-years. pn 2438 is an irregular patchy ring on the north edge of the cluster M 46. pn 2440 consists of bright

condensations of nebulosity with faint extensions. It lies in a region rich in stars. a 6255, also known as k Pup, is a bright and interesting double, easily resolvable with very small telescopes. The stars are almost of equal magnitude, 4.5 and 4.6, and extremely blue in color. The star b 12 has a wide separation, an orange-white pair of magnitudes of 4.8 and 6.8. The star b 13, also known as n Pup, is a wide double with almost equal 6th-magnitude components.

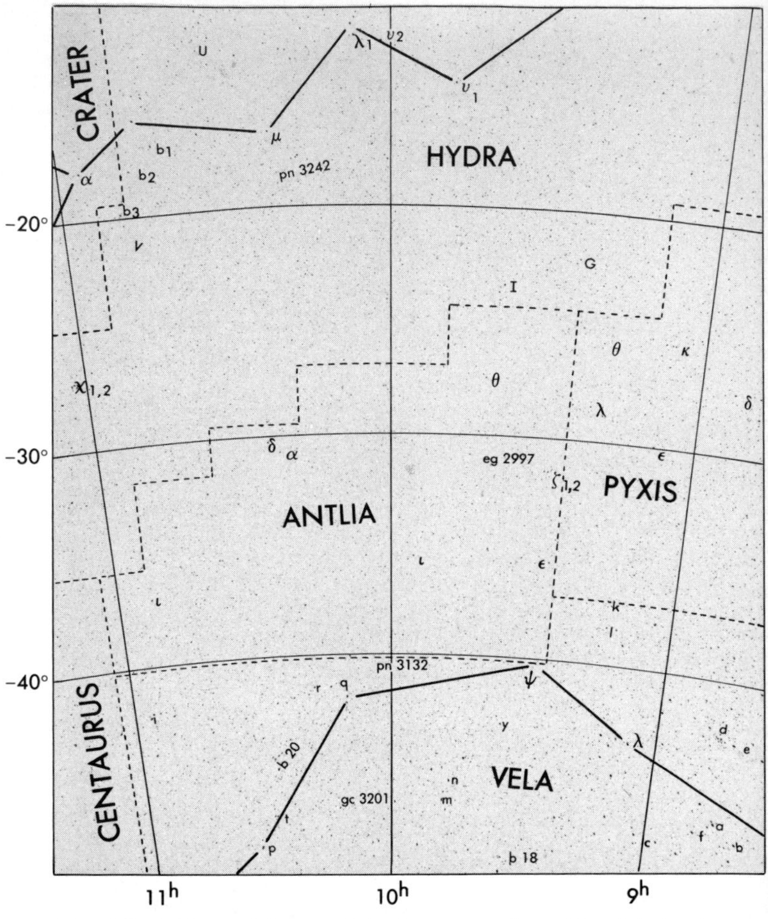

Atlas Chart 39. pn 3242 consists of an inner ring with a faint outer disk, 35″ × 40″. The number of stars rapidly diminishes in this area with increasing distance from the Milky Way. Although

Vela has some interesting star fields, Antlia is one of the least interesting regions of the entire sky, to the telescope as well as to the naked eye.

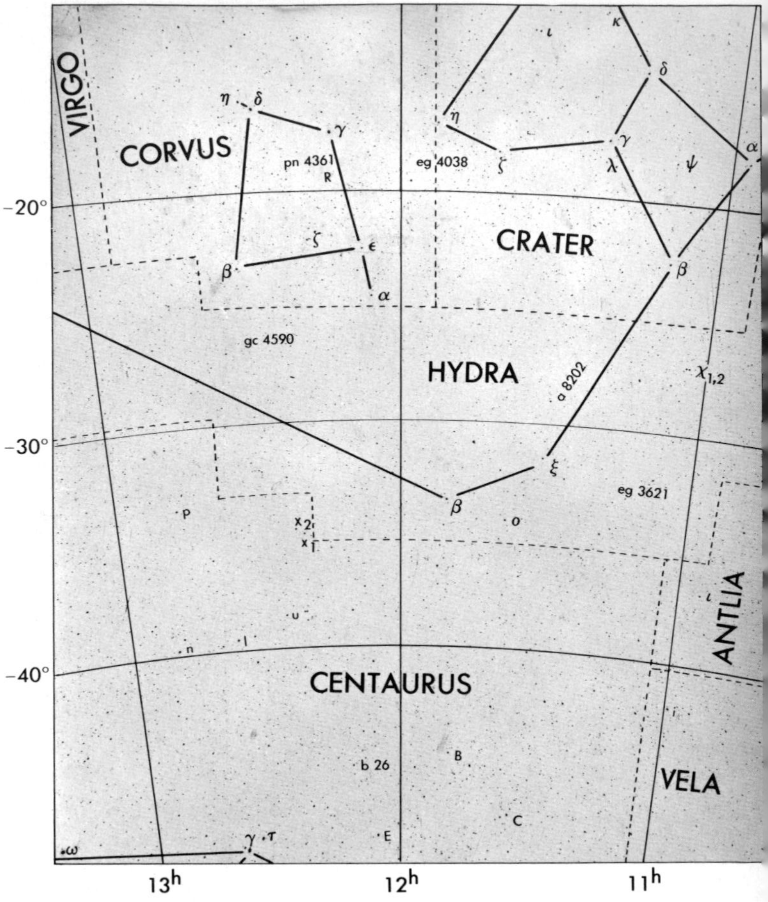

Atlas Chart 40. The region of Corvus and Crater is interesting, visually and telescopically, but Hydra and northern Centaurus contain few stars. pn 4361 is a large, irregular oval mass, faint and difficult to see except with a low-power eyepiece. gc 4590 (M 68) is relatively bright, though small and highly concentrated. a 8202 is a wide pair of yellow stars, of almost equal magnitude,

5.8 and 5.9, which seems to be moving together in space and probably forms a physical system. It also appears in Table 11 as b 23. β Hya (b 24) is a fine pair, of magnitudes 5.0 and 5.4, requiring a 5-inch telescope for resolution. b 26 is also known as D Cen. R Crv, a long-period variable, ranges from 5.9 to 14.4 in 317 days.

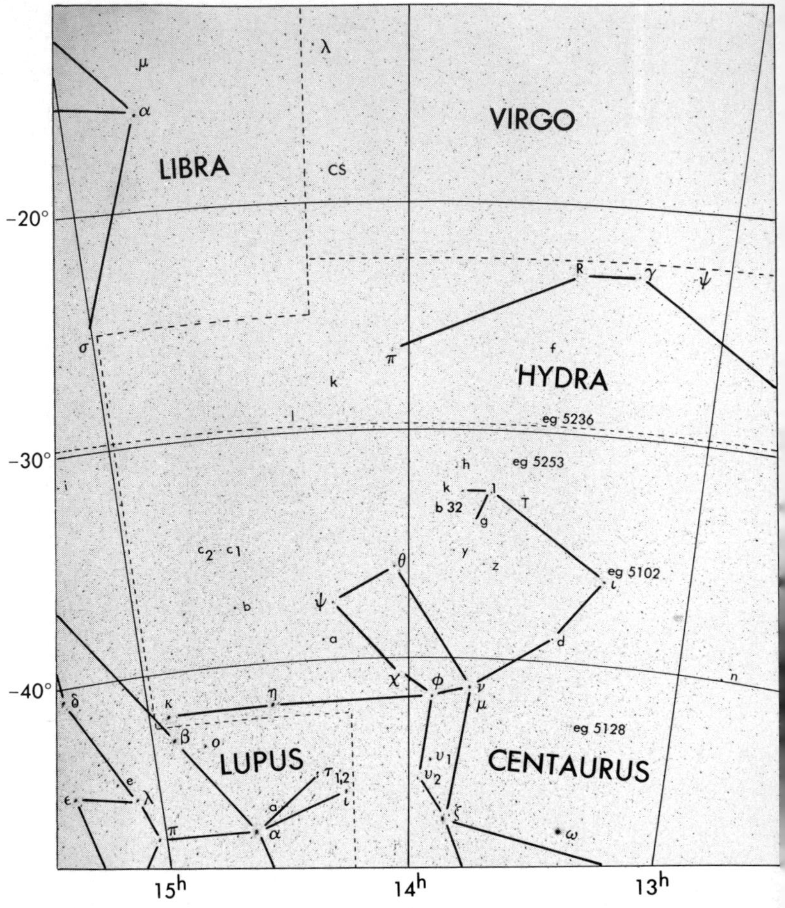

Atlas Chart 41. The sky is growing somewhat richer in this area, but the obscuring dust clouds of the Milky Way are still sufficiently transparent to reveal the faint galaxies in the background. Of these, eg 5253 (M 83) is exceptional. It is certainly to be counted among the brightest of the spiral nebulae.

R Hya, a long-period variable, ranges from magnitude 3.6 to 10.9 with a period of 387 days. The star is notable for its deep red color. b 32 (3 Cen) is a blue pair of magnitudes 4.7 and 6.2 and wide separation. ε Lup (b 38) requires a 4- or 5-inch telescope for resolution.

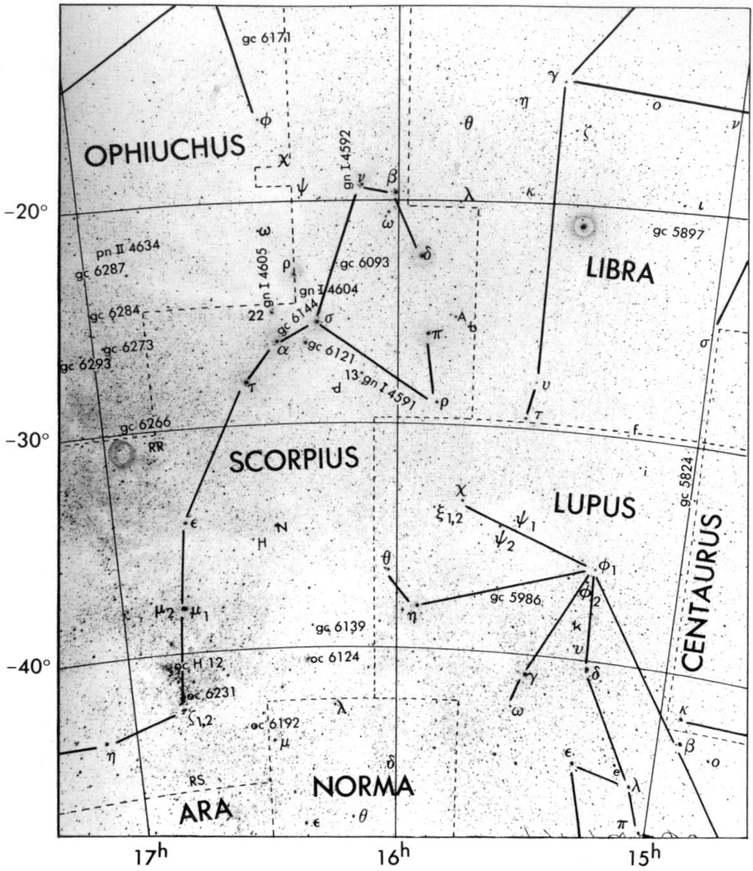

Atlas Chart 42. This area is one of the most interesting and exciting regions of the entire heavens. Clusters of all sorts, both open and globular, abound. The Milky Way presents many rich fields, rewarding the observer who sweeps them with binocular or a low-power telescope.

gc 5897 and gc 5986 are only slightly less brilliant than gc 6093 (M 80). gc 6121 (M 4), gc 6266 (M 62), and gc 6273 (M 19) are excellent objects for telescopic observation.

Of the open clusters, 6124 is quite rich. oc 6231 and oc H 12 make an interesting pair, almost a double cluster, though they do not resemble stellar clusters. gn I 4591 is a faint haze around 13 Sco. σ Sco also has a faint nebular appendage. gn I 4604 is an interesting nebular patch with a rich stellar background and many

dark lanes around the star ρ Oph. Nearby, gn I 4605 is a nebulous envelope around the star 22 Sco. The star π Lup (b 34) is a fine double, of magnitudes 4.7 and 4.8, both blue, requiring a 3-inch telescope for resolution. λ Lup (b 35) consists of a pair in "rapid" motion, a difficult object even with a 5-inch telescope. κ Lup (b 36) is a very wide pair with magnitudes 4.1 and 6.0. γ Lup (b 40) is a bright double of magnitudes 3.7 and 4.0. The stars are very close, however, and difficult to separate even in an 8-inch telescope. The period of revolution is 104 years. ξ 1 Lup (b 41) is a very wide double, with stars of 5.4 and 5.7 magnitude. β Sco (b 43) consists of a pair of magnitudes 2.9 and 5.1. The stars are well separated and the pair should easily be resolvable with low

(*continued on p. 248*)

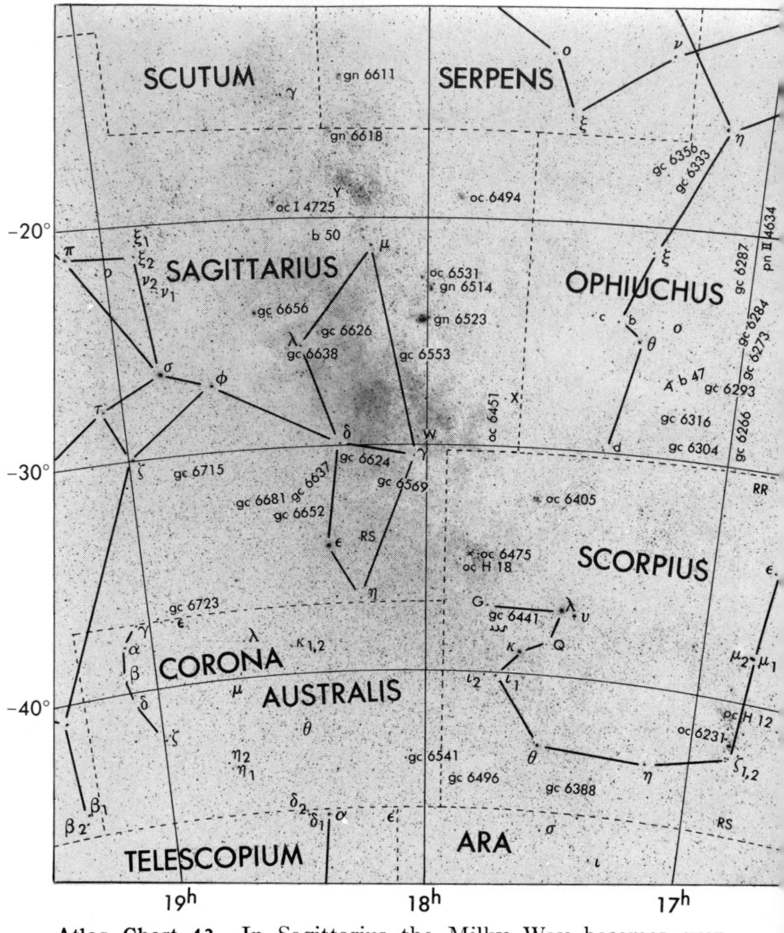

Atlas Chart 43. In Sagittarius the Milky Way becomes even richer. The amateur will find it profitable to spend hours sweeping these fields with a binocular, noting and recording the more interesting areas as seen in his own instrument. Globular star clusters dominate the region of this chart. gc 6333 (M 9) and gc 6356 make an interesting pair, with the former more conspicuous. gc 6541 is one of the brightest clusters, even though it does not possess a Messier number. gc 6626 (M 28) and gc 6637 (M 69) are slightly fainter but interesting objects nevertheless. gc 6656 (M 22) shares the honor with gc 5904 (M 5) of being the brightest globular clusters visible to observers in the Northern Hemisphere (see Atlas Chart 30). 6681 (M 70), 6715 (M 54), and 6723 com-

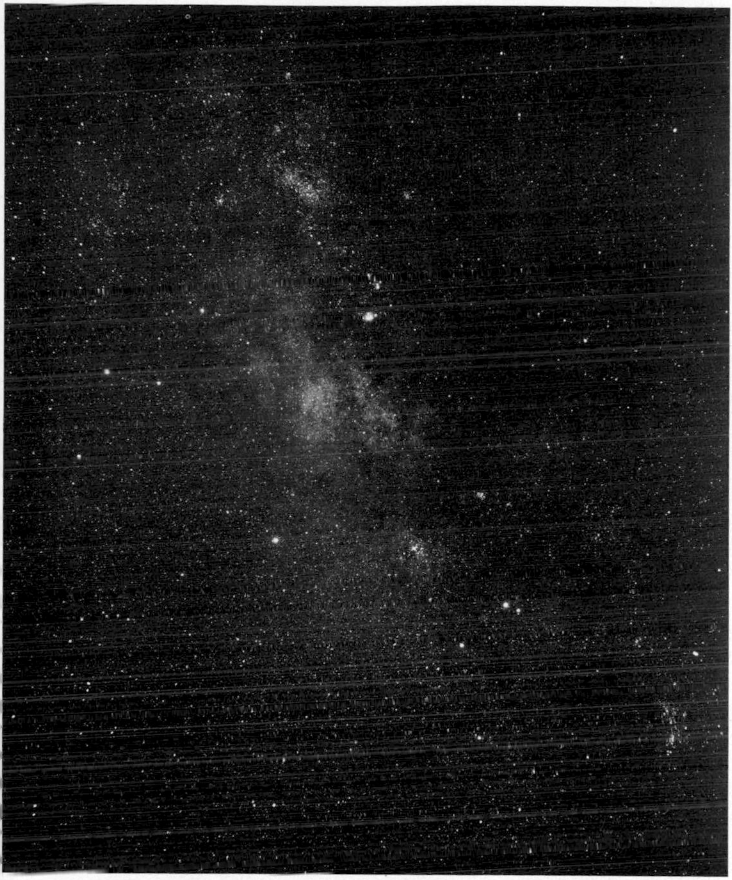

plete the list of the more interesting globular clusters in the area. oc 6405 (M 6) and oc 6475 (M 7) are beautiful objects in a small telescope. Neither is as rich as oc 6494 (M 23) or oc 6531 (M 21). The latter lies at an estimated distance of 2900 light-years. oc I 4725 (M 25) falls in a region of exceptional beauty. Here, the Milky Way background is almost as rich as some clusters.

gn 6514 (M 20) is the famous Trifid in Sagittarius. Note the dark lanes forming a triple fork. Nearby, gn 6523 (M 8) is even more spectacular, itself being visible to the naked eye. This is the Lagoon Nebula, which has an especially rich field with a cluster involved. A spectacular object in the telescope, gn 6611

(*continued on p. 249*)

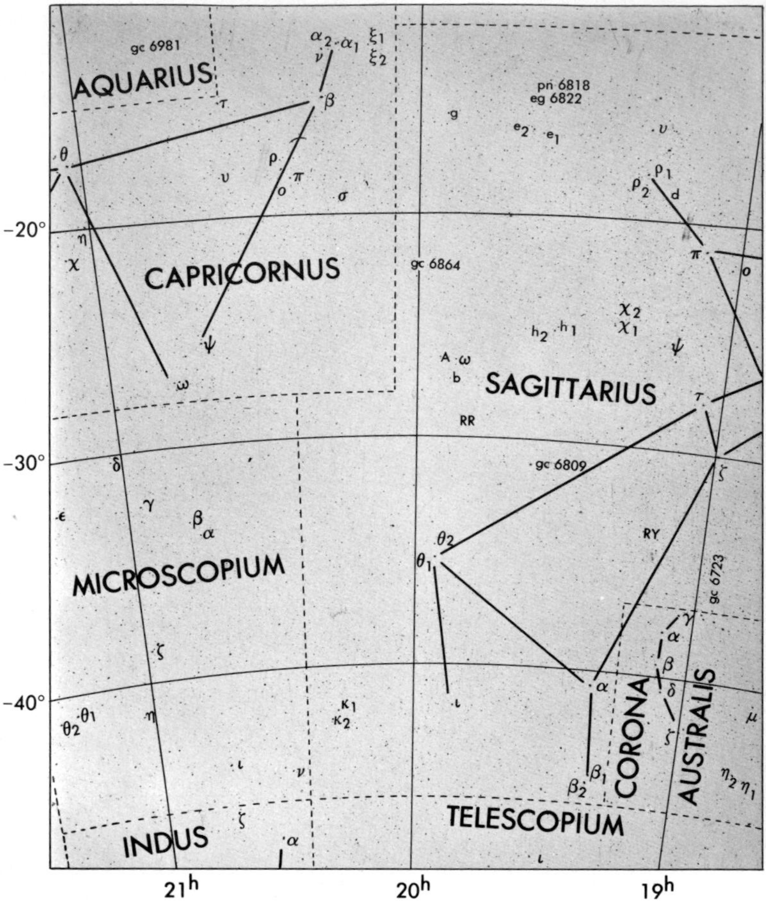

Atlas Chart 44. The fringes of Sagittarius will still repay low-power sweeping, though the richness of the field diminishes rapidly toward the east. pn 6818 is an irregular oval ring. gc 6809 (M 55) is an outstanding object, readily visible to the naked eye and interesting

in a small telescope. In a telescope large enough to resolve the component stars, the cluster is magnificent. gc 6864 (M 75) and gc 6981 (M 72) are very much fainter. γ CrA (b 52) is a readily resolvable 5th-magnitude double with a period of 120 years.

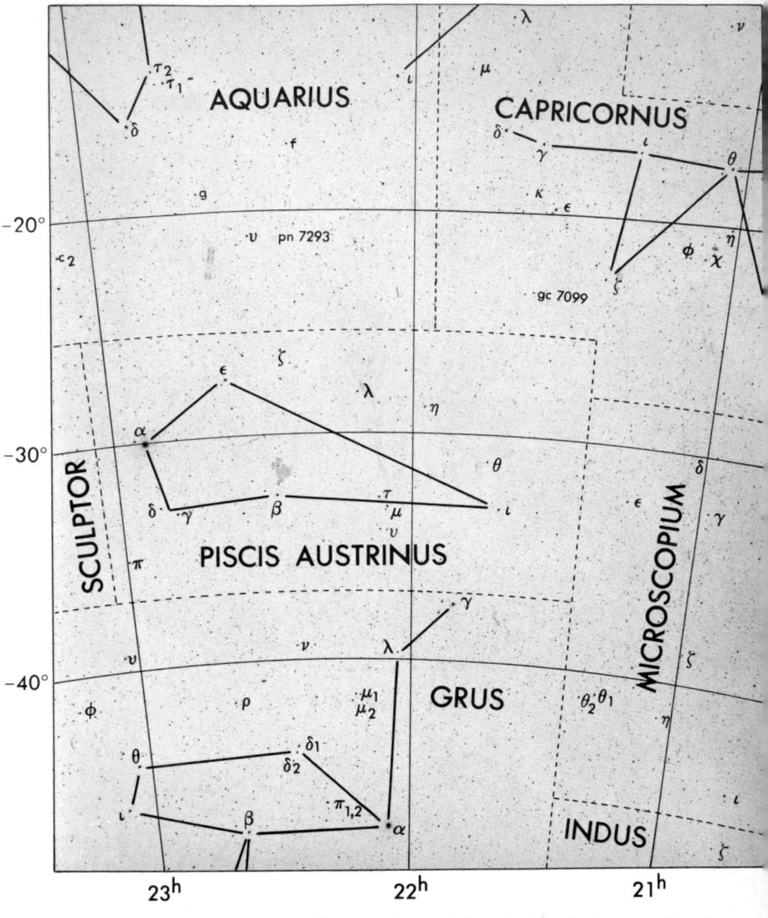

Atlas Chart 45. gc 7099 (M 30) is readily visible in a small telescope, though difficult to resolve into its star components. pn 7293, in Aquarius, is bright as far as total magnitude is concerned. Its great size, however, makes it difficult to see except with very low power. This is the famous giant Helical Nebula.

θ Gru (b 54) is an easy double in a 2½-inch telescope, with the components of magnitudes 4.5 and 7.0.

The region is uninteresting except for the bright star α PsA (Fomalhaut), whose brilliance appears intensified by the comparative darkness of the starry background.

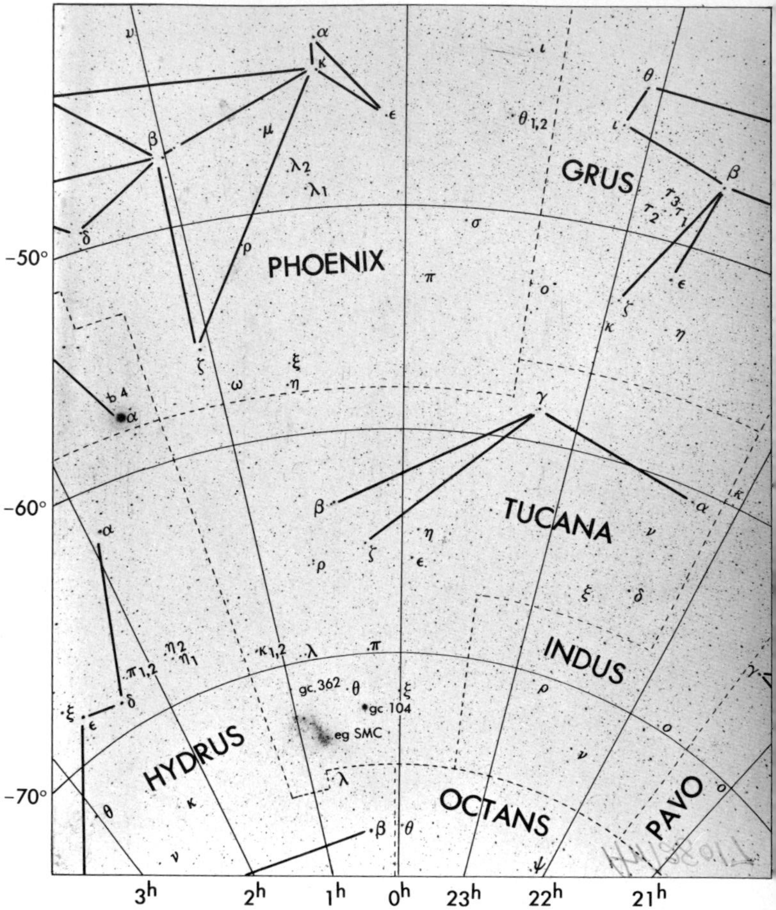

Atlas Chart 46. By far the most interesting objects in this area are the galaxy, the Small Magellanic Cloud, and the unusually large and bright globular cluster 104 (also known as 47 Tuc). The Small Magellanic Cloud, which looks like a minute section of the Milky Way, is an irregular group of stars constituting a separate universe. It will repay observation with all varieties of telescopic power from low to high. And even with the eye the hazy outline is interesting because of the relative darkness of the sky background.

β Tuc (b 1) is a wide double, resolvable with low power into a beautiful pair of 4.5 magnitude stars. Actually the system is sextuple, though the other components are relatively faint. β Phe is a test object for a 3-inch telescope. The pair, of equal magnitude 4.1 and distinct orange color, makes one of the more beautiful doubles in the sky. κ Tuc (b 3), a white-blue pair, resolvable in a 1½-inch telescope, has at least two fainter components to make the star quadruple.

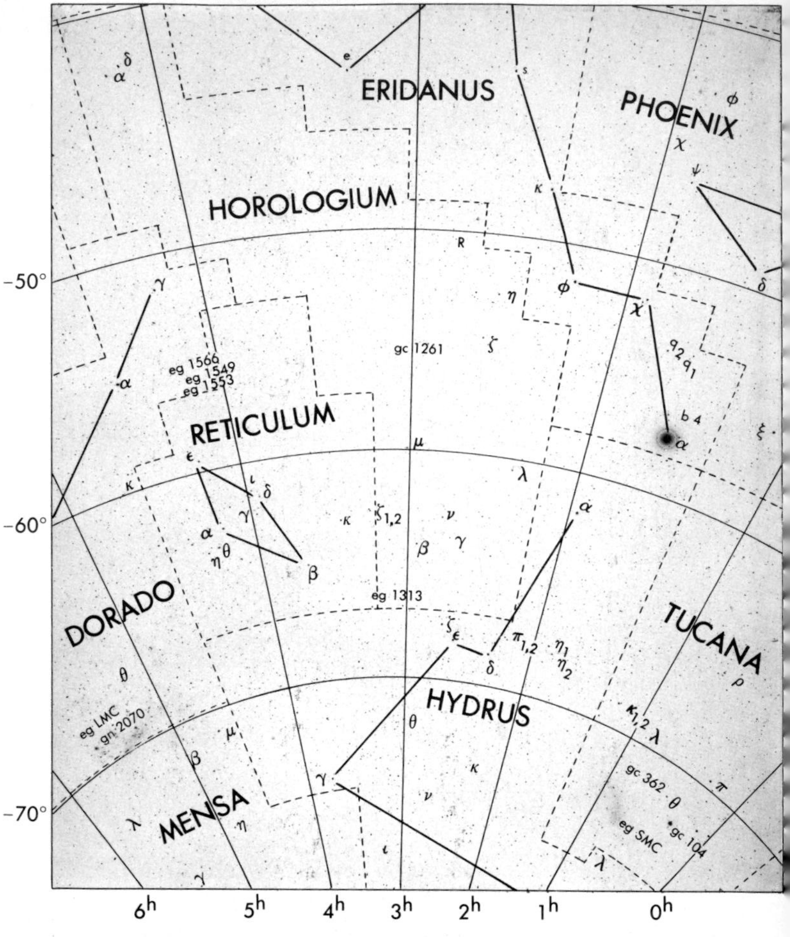

Atlas Chart 47. The area is devoid of bright stars, with the exception of α Eri (Achernar). eg 1549, eg 1553, and eg 1566 show

nothing particularly notable. b 4 (p Eri), a wide orange pair of magnitude 6.0, lies just north of Achernar.

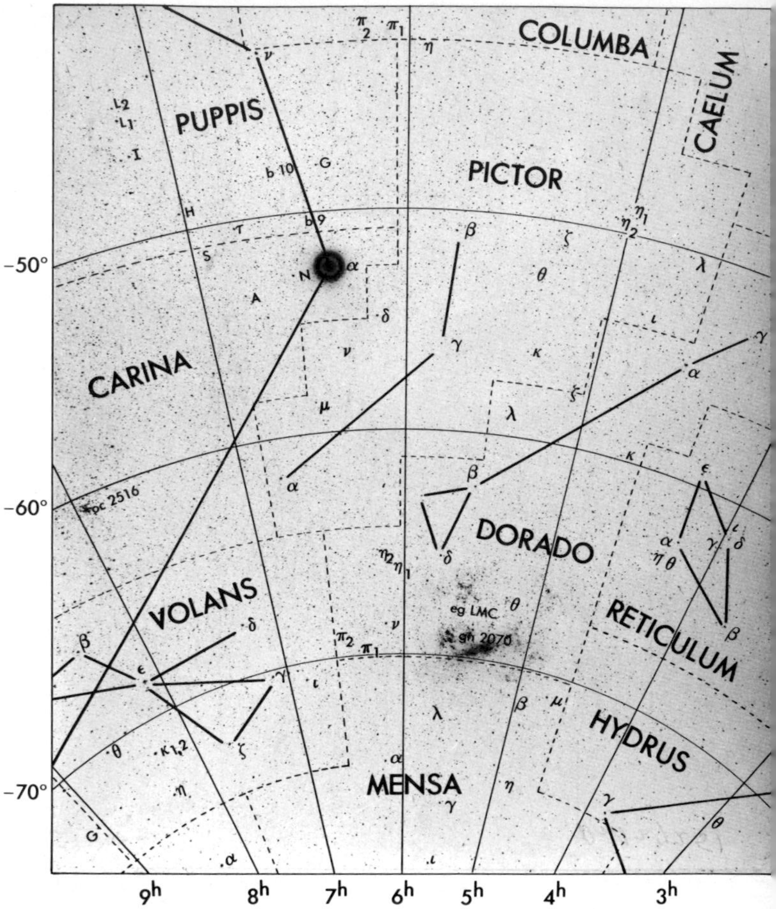

Atlas Chart 48. This region is particularly notable for the large irregular galaxy, the Large Magellanic Cloud, and for the second brightest star of the sky, α Car (Canopus). The Large Magellanic Cloud is one of the finest objects in the sky, resolvable even with relatively small optical aid. It contains gn 2070 (30 Dor), a great looped nebulosity in a large, bright complex field. The nebulosity is visible to the naked eye, and with the exception of the Orion Nebula probably is the most outstanding example of gaseous nebulosity in the visible heavens. The Large Magellanic Cloud lies at a distance of about 100,000 light-years. oc 2516, which forms an isosceles triangle with Canopus and the Large Magellanic

Cloud, is a fine telescopic object, with a rich field. ι Pic (b 8) is a wide double. b 9, a multiple system whose fainter pairs are in rapid motion, is an interesting object in a 5-inch telescope, though a 10-inch is necessary in order to split the fainter pair. b 10, a wide pair of magnitudes 5.1 and 7.4 with contrasting orange and green colors, is a beautiful object in a small telescope. γ Vol (b 11) is a wide orange-yellow pair of magnitudes 3.9 and 5.8. The Cepheid variable, β Dor, ranges between magnitudes 4.5 and 5.7, with a period of 9.8 days. L_2 Pup is a long-period variable with relatively small range, between 3.4 and 6.2 magnitudes.

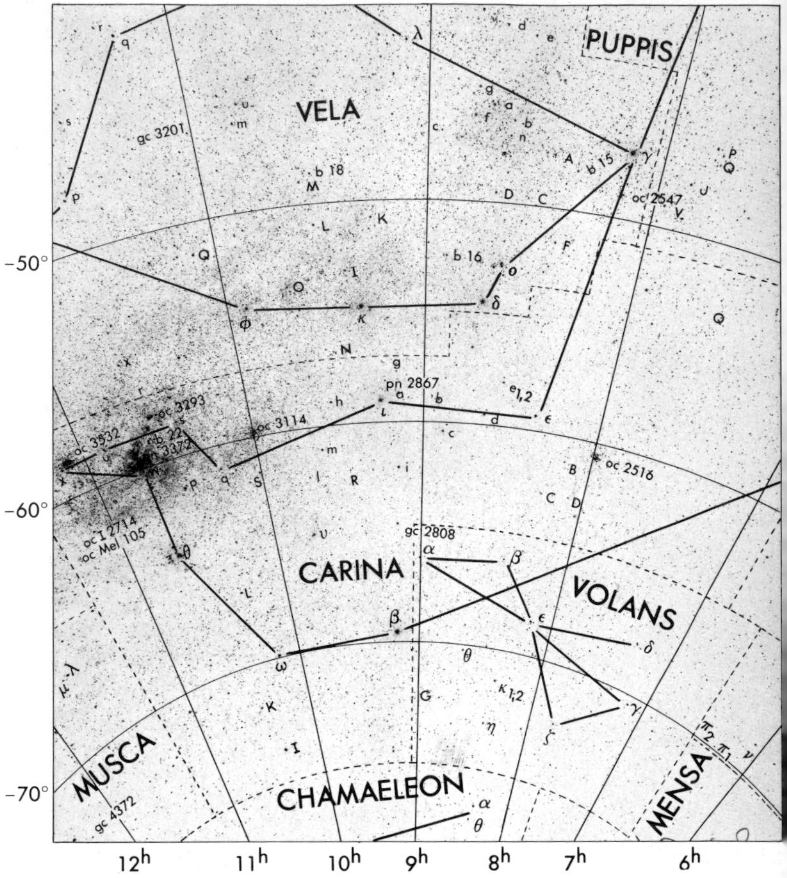

Atlas Chart 49. The Milky Way in Vela and Carina is brilliant and will repay the amateur for many hours of random scanning with binocular or small telescope. The star fields are extremely rich.

The numerous open clusters all merit optical inspection: oc 2516, oc 2547, oc 3114, oc 3293, oc 3532 (at an estimated distance of 1300 light-years), oc I 2714, and oc Mel 105. It is sometimes difficult to tell which of these clusters is a true physical system and which is an apparent enhancement of the background through accidental concentration of the stars. gc 2808 is a faint naked-eye object.

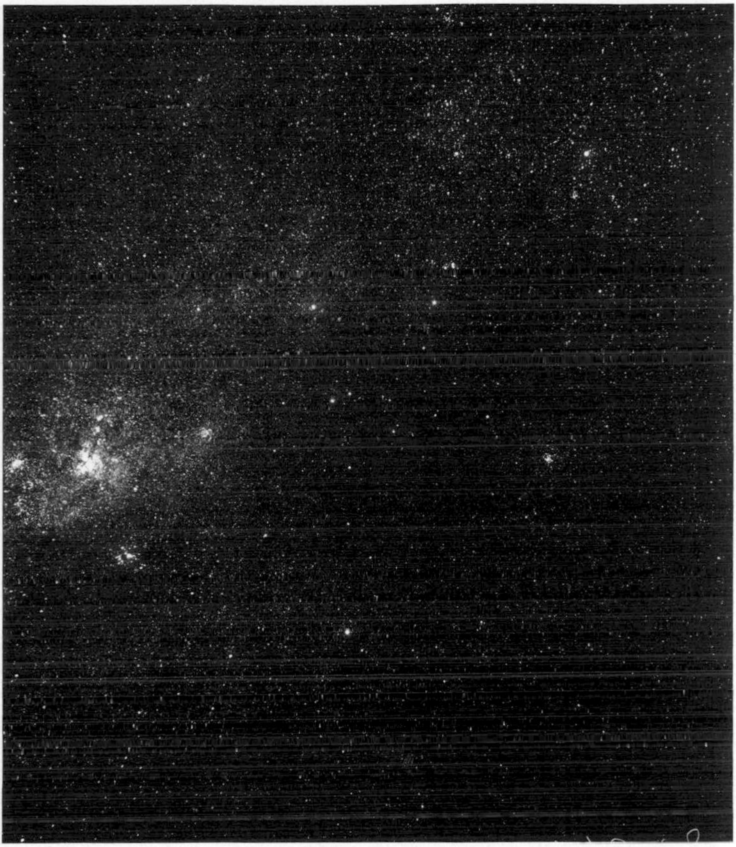

γ Vel is a very wide pair, magnitudes 2.2 and 4.8, intensely blue in color. b 16, resolvable in a 3-inch glass, is a blue-orange pair of magnitudes 4.9 and 7.7. ϵ Car (b 19) is an easy object for a 2-inch telescope, with magnitudes 3.2 and 6.0. p Vel (b 21) requires the largest telescopes for resolution. The object consists of a white-green pair, magnitudes 4.6 and 5.1, in rapid motion, completing the orbit in the exceptionally short time of 16 years. b 22 (t_2 Car) is a wide orange-green pair.

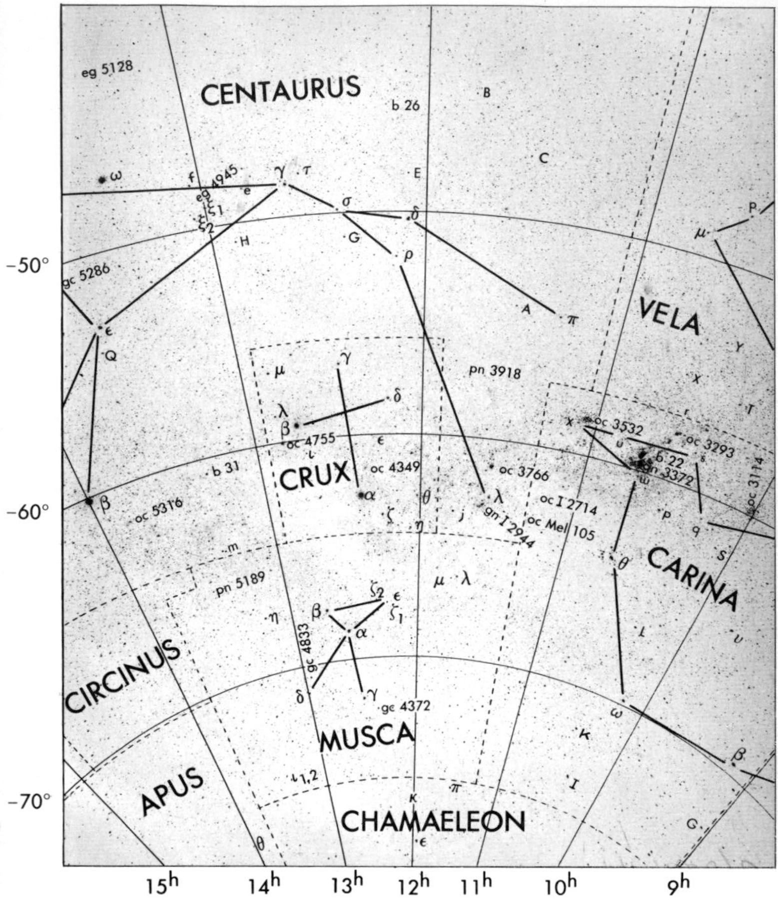

Atlas Chart 50. The Milky Way through Carina, Crux, and Centaurus continues to be brilliant. Among the open clusters, oc 4755 (κ Cru) is particularly notable. This naked-eye object lies at an estimated distance of 3300 light-years. The globular clusters are interesting, though not exceptionally bright. gn 3372 (η Car) warrants special attention. This object is part of a bright and extended nebulous region, with intermingled bright and dark patches. The region is particularly notable because a nova once appeared here. The new star was variable in brightness during the 19th century, finally reaching 1st magnitude in 1827. After some more fluctuations, the star reached magnitude 0 in 1838, and −1 in 1843, with a minimum in between. The star has now

faded to 7th magnitude, but it still bears watching because of its unusual past history. gn I 2944 is a faint halo around the star λ Cen. East and slightly south of the Cross is a dark area, caused by a vast dust cloud that obscures the bright Milky Way background. This is the famous Coalsack. The few stars that the telescope shows in this region are undoubtedly "foreground" objects, lying between us and the obscuring matter. To the naked eye, the area appears completely black.

eg 5128, an irregular object resembling the Magellanic Clouds, is one of the brighter examples of this class.

b 26 (D Cen) is an easy double for a 2-inch telescope, with
(*continued on p. 249*)

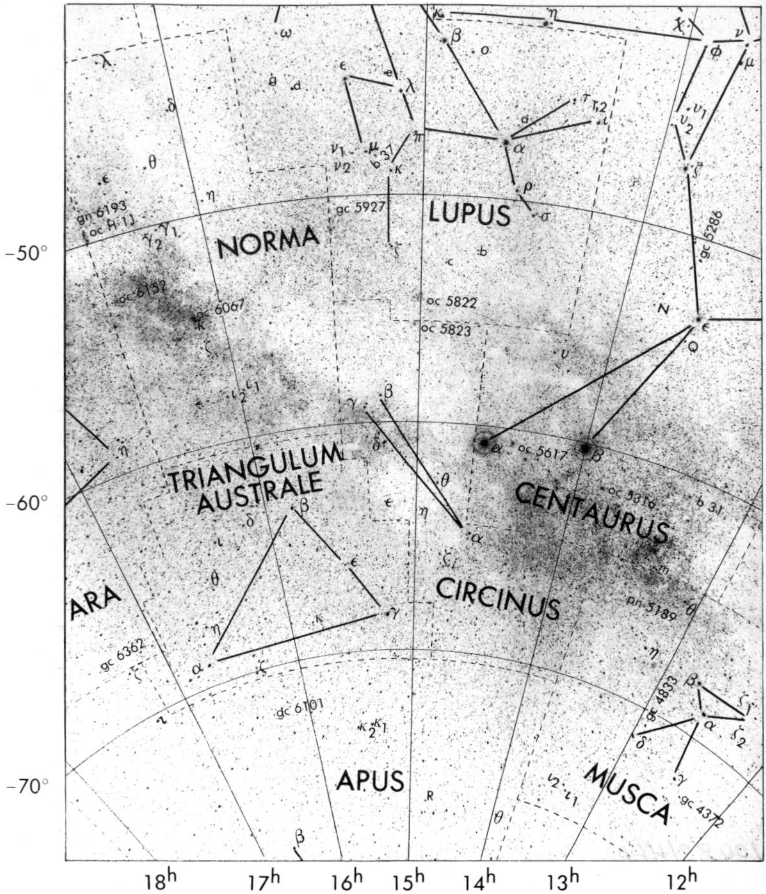

Atlas Chart 51. Through Centaurus and Norma the Milky Way continues brilliant, divided here and there by extended clouds of obscuring dust. The most distinctive feature of this region is the pair of 1st-magnitude stars, α Cen and β Cen, the former known in navigational circles as Rigil Kent. The latter has no definite designation, though Wazn was one name ascribed to β.

oc 5822 and oc 5823 are interesting, especially because of the difference in size, the former being much the larger. α Cen (b 33) is a fine yellow-orange pair, separable with a 1½-inch telescope into components of magnitudes 0.3 and 1.7. The pair displays orbital motion with a period of 80 years. α Cen lies at a distance of 4.3 light-years, the nearest of all stars, with the exception of

a nearby faint companion with common proper motion, evidently a physical member of the same system, often referred to as Proxima Centauri.

π Lup (b 34) is a fine blue pair of magnitudes 4.7 and 4.8, readily resolvable in a 3-inch telescope but somewhat difficult in a 2-inch glass. λ Lup (b 35) is a close pair in rapid motion, an object difficult to see with anything less than a 5-inch or 6-inch glass. The components have magnitudes 5.0 and 5.4. κ Lup (b 36) is an extremely wide double. b 37 is a good test double for a 3-inch telescope. It has a 7th-magnitude companion. ϵ Lup (b 38) needs a 5-inch glass to separate the 3.9 and 5.6 components. γ Cir,
(*continued on p. 249*)

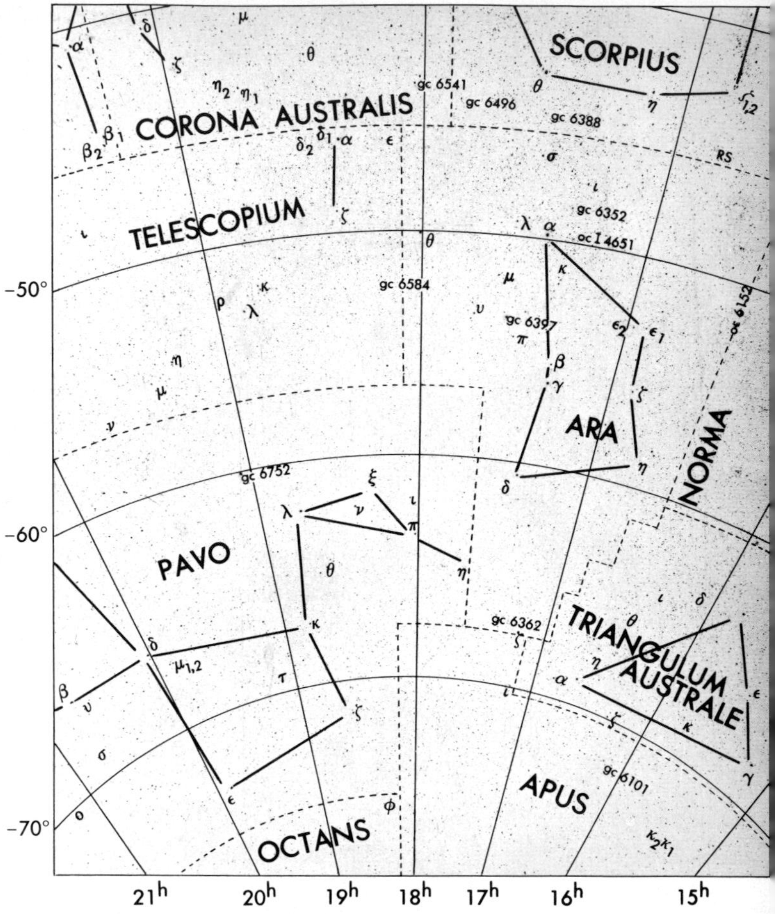

Atlas Chart 52. The Milky Way diminishes rapidly in intensity toward the southeast. oc I 4651 is rich and interesting. gc 6397, gc 6541, and gc 6752 are naked-eye globular clusters of exceptional

brilliance; the first is much the largest. ξ Pav (b 49) is an orange-green pair with considerable difference in magnitude between the primary and secondary, which have magnitudes 4.3 and 8.1.

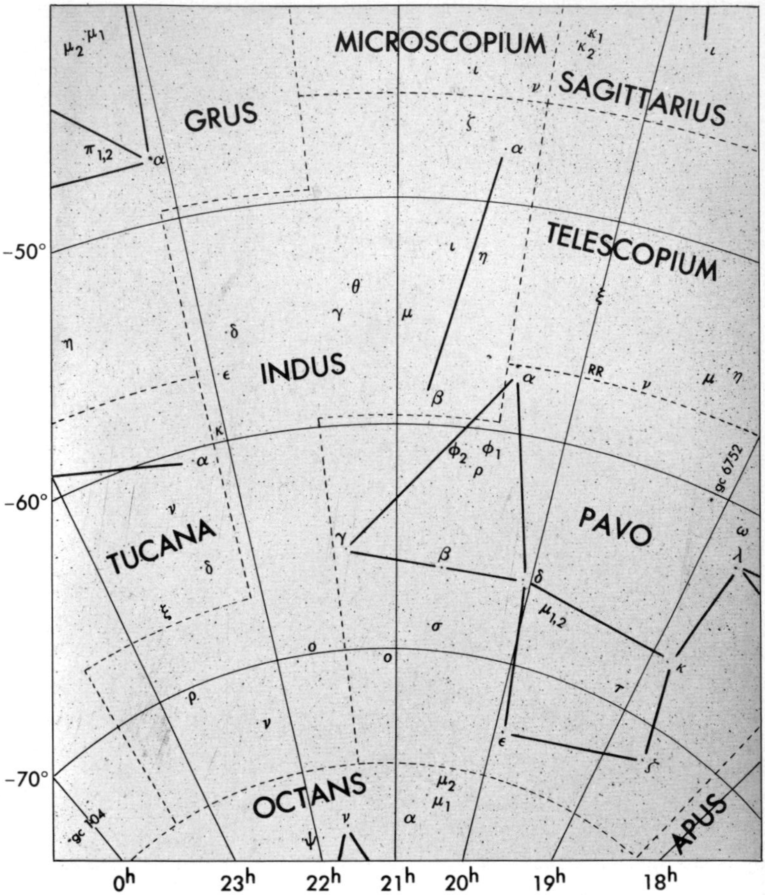

Atlas Chart 53. The constellations within this area are relatively poor in stars. There are only two bright stars in the region, α Pav (Peacock) and α Gru (Al Na'ir).

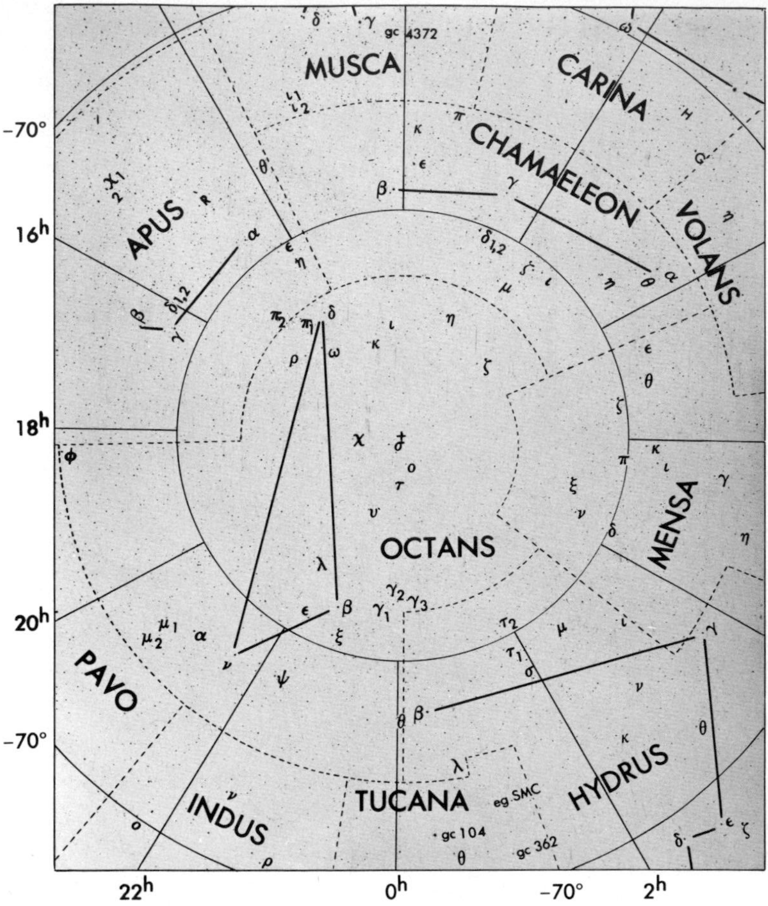

Atlas Chart 54. The south celestial pole is barren. The background does not differ greatly from that of the northern pole. The latter, however, has a number of reasonably bright stars that at least make the basic constellations interesting. The 5.5 magnitude star σ Oct is the southern pole star, very inconspicuous compared with Polaris.

ε Cha (b 25) is a double with magnitudes 5.4 and 6.2, separable only with a 5-inch telescope. λ Oct (b 53) is a double, easily observed in a 2-inch scope, with contrasting orange and green components of respective magnitudes 5.5 and 7.7.

Atlas Chart captions, continued

Atlas Chart 19 continued

double stars in the heavens. A person with good eyesight can just detect the duplicity. But a telescope of 2 inches or greater aperture will separate the naked-eye pair and split each of the stars into a second pair, making four stars in all — a double double. Use as high magnification as the seeing conditions will permit.

α Her, itself a variable ranging between magnitudes 3.1 and 3.9, is readily resolved in a small telescope into a red-yellow pair (a 10418). ρ Her (a 10526) is another easily resolved pair. 95 Her (a 10993) is a wide green-orange pair.

a 11871 requires a 5-inch telescope for resolution. Its orbital period is 62 years.

Atlas Chart 20 continued

a 14636 (61 Cyg), famous for being the first star whose distance was actually measured, is an interesting orange pair 10.9 light-years distant.

Within the angle formed by the stars α, γ, and ϵ Cyg, there is a dark patch of nebulosity, obscuring the Milky Way behind. Astronomers refer to this as the Northern Coalsack.

Atlas Chart 25 continued

refers to nebulosity around the northern star of Orion's dagger. gn 1980 is a faint extension of the great nebula in the vicinity of ι Ori. gn 1982 (M 43) is a bright nebulosity with tail forming part of the Great Nebula. gn 1990 is a faint nebula around ϵ Ori. gn I 430 is a fan-shaped appendage to d Ori. gn I 434 contains the famous Horsehead Nebula, a bright streak extending south from ζ Ori. gn 2024 is a fan to the east of ζ Ori. gn 2068 (M 78) is a wispy cloud north of the Belt. gn 2244 is a nebulous cluster 2° E of ϵ Mon. gn 2264 is another nebulous cluster. oc 2301 consists of several overlapping groups.

In Orion, η, θ_2 and ι are all interesting doubles. β Mon is an easily resolved triple.

Atlas Chart 37 continued

density about 50,000 times greater than that of water. A teaspoonful of its matter would weigh almost a ton. The star contains almost as much matter as our sun, but it is compressed into a sphere whose diameter is only slightly more than three times that of the earth.

Atlas Chart 42 continued

telescopic power. Note that the brighter component has a close companion of magnitude 8.5. The nearby star ν Sco (b 44) consists of a very close pair of magnitudes 4.5 and 6.0 with a distant companion of magnitude 6.5. ρ Oph (b 45), previously referred to in relation to its involvement in gaseous nebulosity and cosmic dust, is also a fine double, of magnitudes 5.2 and 5.9. α Sco (b 46), the bright red star Antares, possesses a faint blue companion of magnitude 7. At least a 6-inch telescope will probably be necessary to reveal the companion. The color contrast is particularly interesting.

Atlas Chart 43 continued

(M 16) is another nebulous cluster. gn 6618 carries several alternate designations, descriptive of the peculiar shape. This is the Omega, Horseshoe, or Swan Nebula. It presents a rich background.

μ_1 and μ_2 Sco form a naked-eye double, an interesting object in field glasses or binoculars. b 47 (36 Oph) possesses equal components of magnitude 5.3. A 2-inch telescope will separate it. o Oph (b 48) is a wide double of magnitudes 5.4 and 6.9, presenting a nice orange-green contrast. The components of b 50 (21 Sag) have a somewhat greater magnitude difference, 5.0 and 8.3. Because of this difference a 3-inch, or even a 4-inch, telescope is necessary to show the interesting orange-green pair. ζ Sgr (b 51) is an extremely close double, resolvable only in the largest telescopes. It has a period of 21 years. There are several interesting variable stars in the region. X and W Sgr are Cepheid variables of range more than a magnitude.

Atlas Chart 50 continued

magnitudes 5.6 and 6.8. α Cru (b 27) is one of the finest doubles in the sky, a blue pair easily resolved in a 2-inch or even smaller telescope, with magnitudes 1.6 and 2.1. γ Cen (b 28) requires at least a 12-inch telescope for resolution of the 3.1 and 3.2 magnitude pair, which revolve around one another with a period of 203 years. β Mus (b 29) is a good test object for a 3-inch telescope, with magnitudes 3.9 and 4.2. μ Cru (b 30) is a very wide blue pair, with magnitudes 4.3 and 5.5, easily resolvable in field glasses or opera glasses.

Atlas Chart 51 continued

which requires similar optical aid to split the 5.2 and 5.4 magnitude pair, will show an interesting color contrast of blue and yellow. ι_1 Nor (b 42) is a multiple star. The close pair, of magnitudes 5.4 and 5.8, needs at least an 8-inch telescope for resolution. The components revolve around one another in the short period of 26 years. The bright star has a 7.5 companion at a distance of 11″.

VII

The Moon

We are accustomed to regard the moon as the earth's satellite. And yet, if there should be any astronomers on Mars or Venus who look up into the sky and impartially view all the planets (except possibly their own), they almost certainly would not think of the moon as a satellite. They would undoubtedly classify the earth-moon combination as a double planet, unique in the solar system. Martian or Venusian amateur astronomers would no doubt find it a favorite object for telescopic observation. Six planets of the solar system possess satellites, but these moons, except for our own, are insignificant compared with their primaries.

You all have seen how our sister planet appears to the naked eye. You have seen it as a narrow crescent, low in the western sky just after sunset. You have seen it wax to full and wane again to crescent form. You may have noted the vague shadows that cross its disk, shadows in which the imaginative see various figures: the man in the moon, the Madonna and Child, the crab, the beetle (Fig. 21), the lady reading a book (Fig. 22), the rabbit (Fig. 23), the donkey, and dozens of others.

MOON LORE

Fig. 21. The beetle. Fig. 22. The lady reading a book. Fig. 23. The rabbit

Even the slightest optical aid destroys the illusion. The dark markings stand out as rolling plains; the brighter areas consist mainly of mountainous regions, revealed like those on a relief map.

With a 2-inch or larger telescope and an eyepiece magnifying about 50 times you can easily see individual mountain peaks and vast mountain ranges. The most conspicuous features are the thousands of craters, ranging in size from gigantic holes 150 miles in diameter down to minute crater pits a few thousand feet across, depressions barely visible with the largest telescopes.

The changing shape of the moon results from the sun's illumination of the sphere (Fig. 24). When the moon lies in the general direction of the sun, most of the illuminated hemisphere is turned away from us. When the moon rises at sunset, we see the entire illuminated face, the full moon. When the moon is a narrow crescent, light from the earth faintly illuminates the lunar surface. With this "earth shine" on the moon, we faintly see the darkened disk edged on the sunward side by a bright crescent — the so-called "old moon in the new moon's arms."

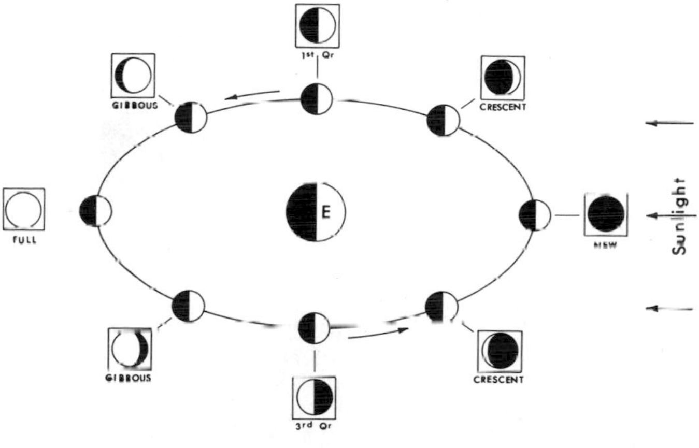

Fig. 24. Phases of the moon
Framed sketches show the moon as it appears in the sky

The best telescopic views of the moon occur between the crescent and quarter phases, when the indirect illumination makes the elevations stand out in relief, because they are edged with shadows.

Certain lunar features should be understood before the Moon Maps are studied. *Maria* (sing., *mare*) are rolling plains, somewhat darker than their surroundings. The term signifies "seas," although the areas are completely devoid of water. The largest is the Oceanus Procellarum. *Palus* (marsh) and *lacus* (lake) designate small dark areas whose surfaces somewhat resemble those of the maria. *Sinus* (bay) designates strong indentations on the borders of the maria.

Ring formations, which abound on the moon, include a variety of types, ranging from the *walled plains*, whose central floors are relatively smooth, to the true *craters*, whose internal structure is generally rough and uneven. Many of the craters possess one or more *crater cones* in their interiors. Some of these cones have a minute crater or depression at their summit. *Crater pits* are an entirely different class of object, being circular depressions ranging in size from a few hundred yards up to 8 or 10 miles in diameter. Although the craters have arisen from collisions with giant meteors, many of the crater pits appear to have originated from some sort of secondary volcanism. The chain of crater pits near Copernicus could not possibly have formed from random meteoric impacts. A few genuine extinct volcanoes appear, resembling terrestrial forms: massive cones of lava or cinders, surmounted by a relatively small but definite crater.

Individual mountain *peaks* appear. And numerous mountain *ranges*, some forming the boundaries of the greater maria, add variety to the lunar scene. Here and there we see *cliffs*, some of them occurring where one side of a crack has risen or fallen, a process that geologists term "faulting."

Miscellaneous *depressions* abound on the lunar surface. A few sharp scars we call *valleys*. When the illumination is right, sets of jagged black lines emerge in certain regions. These are *rills* or *clefts*, where the surface has apparently cracked to produce a sharp ravine. Sinuous rills, which resemble meandering river valleys, are quite common.

The largest and presumably newest craters show long bright streaks radiating away from them. They are most conspicuous near full moon (Fig. 25). These *rays*, as we call them, are believed to be rock dust splashed from the crater when the giant meteor struck the lunar surface. The rays from Tycho are the most extensive. One of them clearly crosses the middle of Mare Serenitatis, from south to north. Detailed descriptions of some of the more interesting lunar features will be found beneath the 12 sectional maps of the moon.

The ever-changing illumination of the lunar surface produces an almost infinite variety of lunar landscapes. The *terminator*, as we call the shadowy boundary between night and day, slowly creeps across the face of the moon to create the gradual sequence of phases from new, to crescent (Fig. 26), to first quarter, to gibbous, to full, to gibbous, to last quarter, to crescent and back to new moon again — all in about 29 days. Even in the course of several hours, however, the terminator moves fast enough to reveal the sun rising or setting on rugged lunar peaks. The long shadows exaggerate the ruggedness of the lunar features and some-

Fig. 25. Full moon, with ray structure

times lead even trained observers to wrong conclusions. The various successful landings of Apollo spacecraft on the moon have greatly changed our ideas of the lunar surface. Meteors, crashing into the moon, have plowed the surface and broken up the rocks into fine powder. In the vacuum the fragments adhere to one another, forming a compact soil, capable of supporting man or vehicle. Numerous tiny spheres, apparently of volcanic glass, are present. The rock is basic in character, resembling terrestrial basalt, a solidified lava.

The lunar surface is extremely dark, almost black, with perhaps

Fig. 26. Crescent moon, age 4.6 days

a tinge of brown. The rocks apparently contain no water of crystallization. Without air or water, the surface of the moon undergoes extreme changes of temperature, from about that of boiling water at midday to about that of liquid air just prior to dawn. The moon's force of gravity is about one-sixth that of earth.

The question of what changes, if any, occur on the surface of the moon has been argued for years. The basic problem is to distinguish between real changes and apparent ones resulting merely from different angles of illumination. The only sound solution involves a comparison of photographs taken at exactly the same phases. To date, this procedure has revealed no convincing evidence of change. The most controversial problem is that of the crater Linné, mentioned in the caption for Moon Map 10. Did a giant meteor, crashing into the moon, obliterate the earlier crater without forming a new one? The question is still open.

Observers have from time to time reported apparent changes in color or visibility of the floor of various craters. Such impressions, which depend on memory, do not constitute the best scientific evidence. However, certain records have occasionally tended to confirm the reality of such variations, notably D. Alter's photographs of the crater Alphonsus, which in blue light appeared to be washed out. The Russian astronomer Kozyrev thought that the effect might be due to an effusion of gas from the crater and watched for a recurrence of the phenomenon. On November 2, 1958, he saw a reddish glow followed by an unusual brilliance of the central peak. A spectrogram taken at the time showed evidence of a gas that contained carbon molecules, presumably glowing as does the aurora borealis, because of the excitation effects of solar corpuscular radiation. Since the crater Alphonsus did not seem to suffer a permanent alteration, the eruption, if real, must be described as gaseous emanation rather than any earthlike form of volcanic activity. Until fully confirmed by other records, this report cannot be regarded as beyond question. But such observations suggest that the moon may not be quite the dead, unalterable world usually pictured by astronomers. And amateurs the world over, in Operation Moonblink, have redoubled their efforts to detect further evidence of other eruptions.

The moon circles around the earth once every 29½ days, turning on its axis once during the same interval. As a result, astronomers see only one side of the moon. Actually, astronomers can observe about 5/9 of the lunar surface because the moon moves more rapidly when it is near the earth and less rapidly when it is farther away. But the rotation on its axis is always uniform. In consequence we see some distance first around one edge of the moon and then some distance around the other. But most of the far side of the moon is inaccessible to observation except from a rocket orbiting the moon.

Photographs obtained by U.S. and Soviet rockets have enabled us to map the entire lunar surface in great detail. The U.S. Orbiter series, especially, has revealed how rough and irregular the surface actually is, down to the smallest scale. The records prove that the moon's far side is even more rugged and pockmarked than the visible face. The far side has few areas resembling the large, dark markings known as maria.

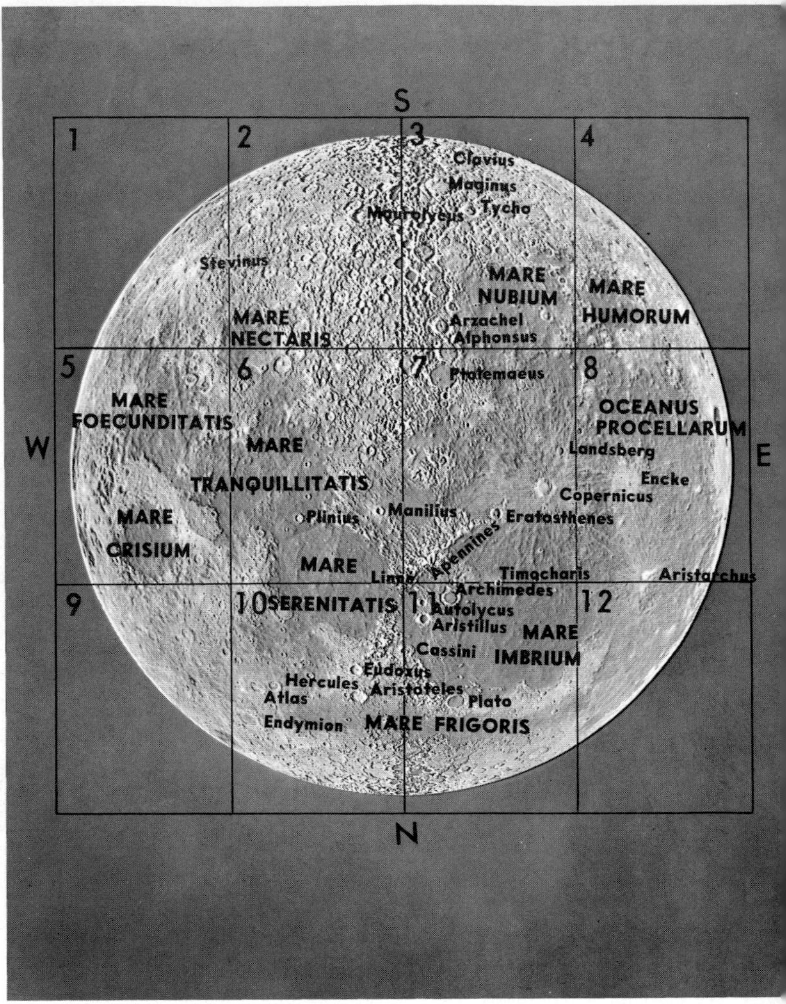

Moon Map A provides the key to the 12 larger-scale maps that together portray the entire visible surface of the moon. This map is a composite of the moon's left half taken near first quarter and of the right half taken near last quarter, the basic photographs for the 12 sectional maps. This procedure provides better shadows and sets forth the craters in bold relief. Some degree of overlap occurs at the borders of contiguous maps. Segments of some of the larger lunar features, such as the maria, necessarily appear on more than one map. Lick Observatory provided copies of their

original photographs. The photographs show the moon as seen in the telescope and not as it would appear to the naked eye. South is always at the top of the page, north at the bottom.

Note that the names attached to the various objects do not derive from an orderly system of mythology, as do many of those of the constellations. Chosen arbitrarily by various observers and at various times, the names are hodgepodge in origin. They range from those of classical heroes to those of modern astronomers.

Moon Map 1 shows the moon's southwestern limb. At the extreme edge the dark Mare Australe appears, greatly foreshortened by projection. The Rheita Valley is one of the more remarkable lunar features, one whose existence is difficult to explain. Broken here and there with craters, the valley extends southward with diminishing width a distance of about 100 miles. The crater Rheita juts into the valley at its extreme northwestern boundary. Rheita, about 35 miles in diameter and ringed with peaks that rise to 14,000 feet, contains a small central peak. Fabricius, near-

by, is a fine crater 55 miles in diameter. Its terraced interior gives the impression of a double wall, most conspicuous on the northwest. Stevinus is a remarkable crater about 50 miles in diameter with a central peak and a system of rays. The region from Rheita to Fracastorius, on the southern boundary of Mare Nectaris, is rugged indeed, pock-marked with hundreds of craters, of which Neander (34 miles in diameter), Steavenson (22 miles across), and Santbech (44 miles in diameter), are perhaps the most remarkable.

(continued on p. 282)

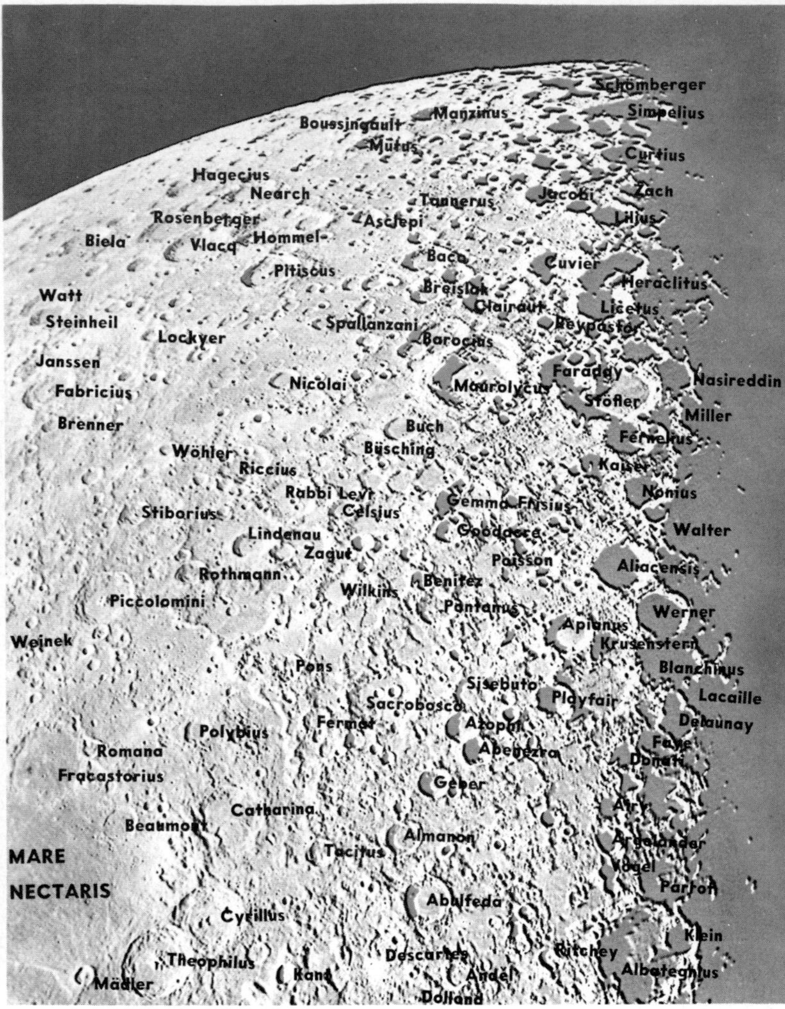

Moon Map 2 presents, in combination with the one to the east, Moon Map 3, the most rugged regions to be found on the surface of the moon, which lie in the neighborhood of the lunar south pole. It is hard to conceive how wild and cut up this area actually is. Toward the southern limb, Rosenberger and Vlacq constitute a double crater, with respective diameters of 50 and 57 miles. The walls of the latter rise to altitudes of 9000 or 10,000 feet. It also possesses a fine complex central twin peak, with a small depression between the peaks. Maurolycus is a magnificent walled plain

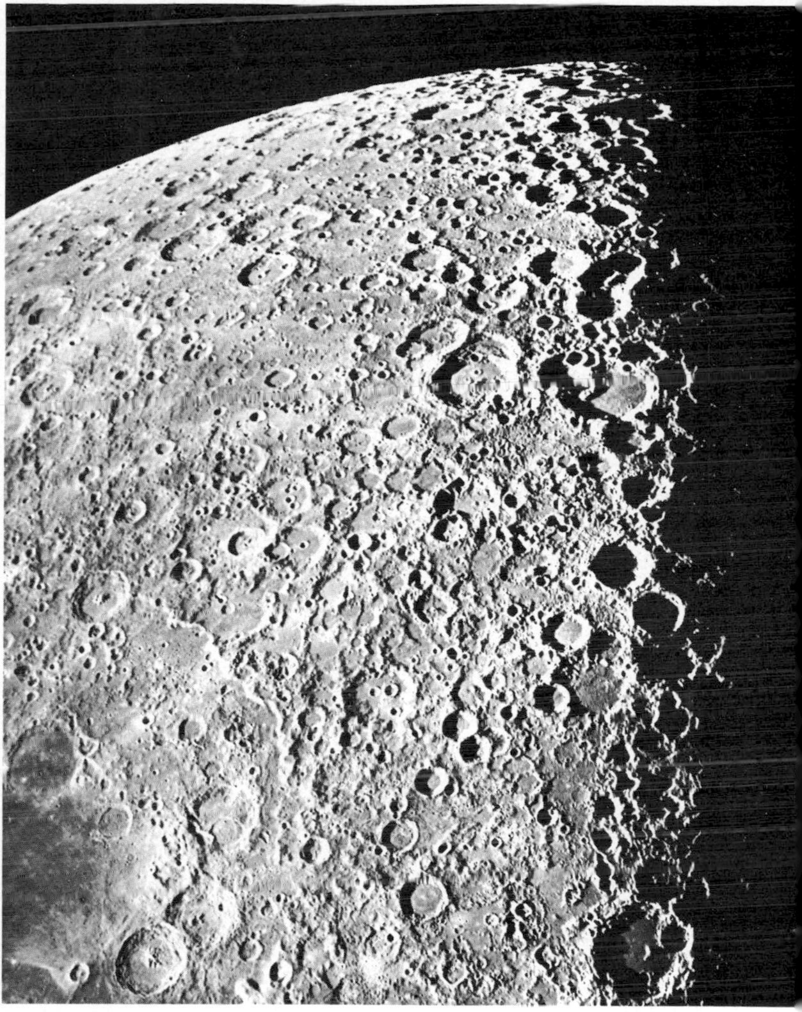

having a complex terraced wall ruined here and there by single craters or groups of craters. The diameter is about 70 miles and the surrounding ring rises to 14,000 feet in places. Stöfler, a fine walled plain, was originally even larger than Maurolycus, but the original walls have been ruined, chiefly on the west, by a series of craters between these two objects. The partially ruined crater Nonius, about 20 miles across, is noted for its rugged internal structure.

(*continued on p. 282*)

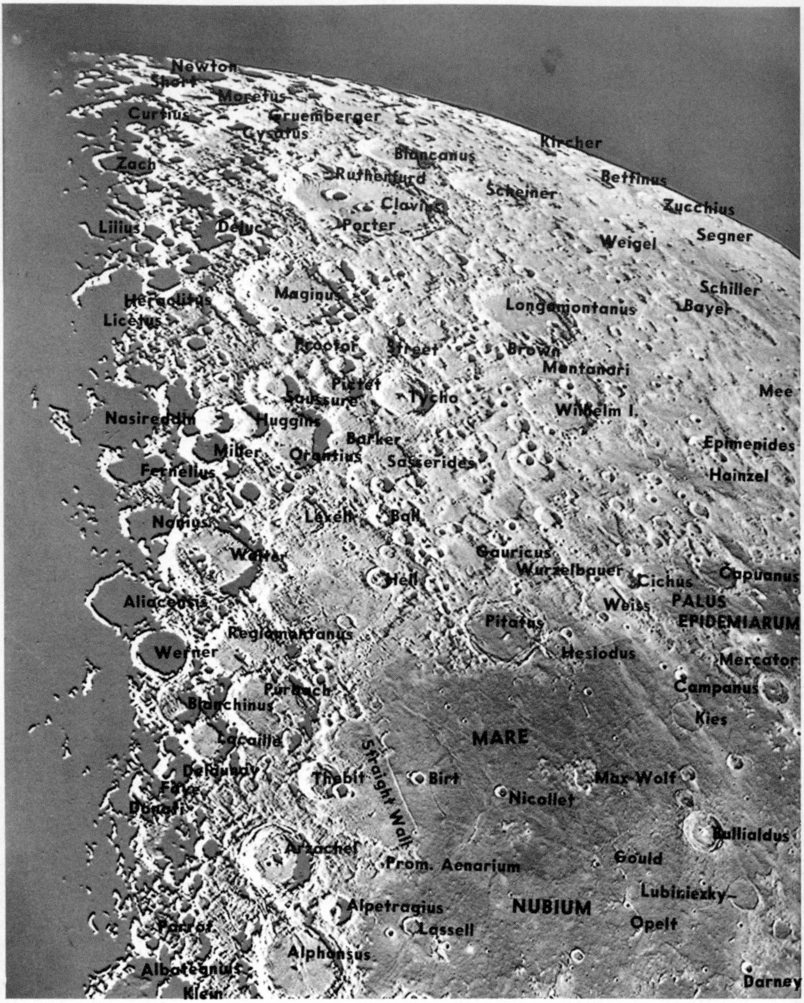

Moon Map 3 shows wild, rugged country similar to that of the previous map. Clavius, largest of the lunar craters, covers an area about equal to that of Massachusetts and New Hampshire combined. This crater, 140 miles across, is ringed by mountains rising to 12,000 feet, with occasional peaks to 17,000 feet. The floor presents a complex structure, particularly striking when the sun is rising or setting. Maginus, 110 miles in diameter, is notable for the profusion of craters along its rim. Longomontanus, 90 miles in diameter, has partially obliterated a large crater that lay

to the west and has, in turn, itself been partially ruined by newer craters.

Tycho is one of the most perfect of the lunar craters and probably one of the most recent formations. A sharp central peak stands in the middle of the 54-mile crater, whose rugged walls, terraced both inside and out, rise to heights of 16,000 feet. Tycho is the center of the greatest ray system possessed by any lunar crater. It is particularly prominent at the time of full moon. The

(*continued on p. 282*)

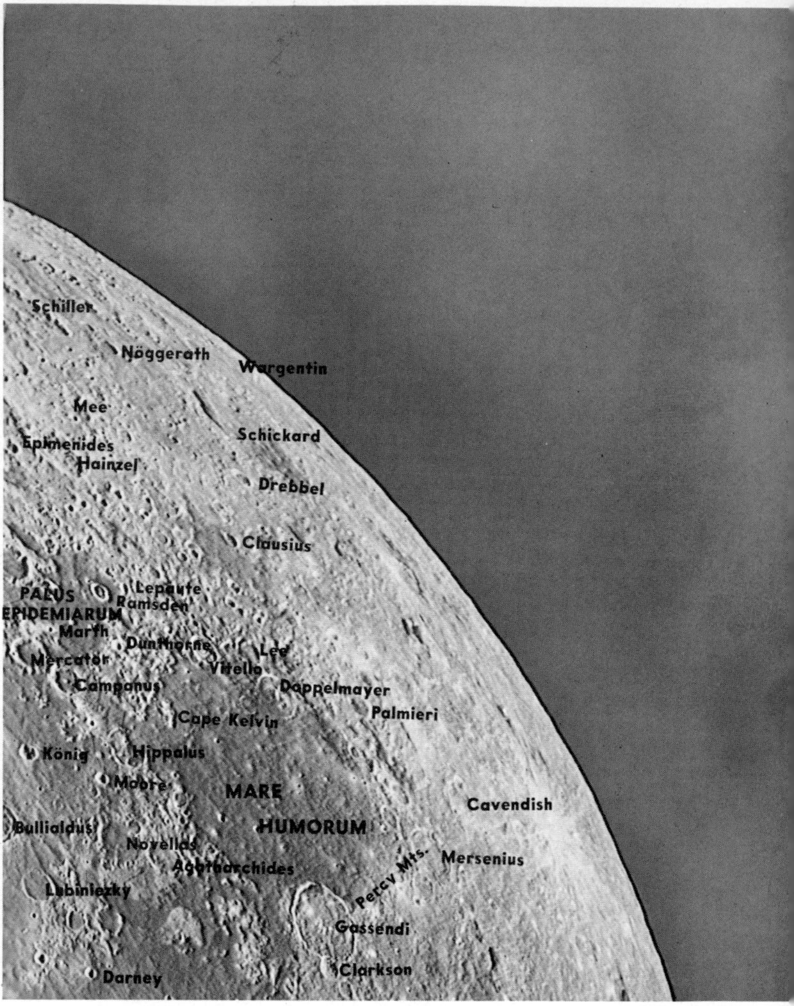

Moon Map 4 depicts the southeastern limb of the moon. Schickard is a magnificent walled plain 134 miles across, one of the largest on the moon. It is particularly notable for the variation of shading on the crater floor. The surrounding walls are relatively low. Hence Schickard is best seen when the old moon is a thin crescent, just before sunrise. Wargentin, on the southern boundary of Schickard, is a unique object on the moon, a crater apparently filled to the brim by a subsequent flow of lava. It resembles some of the mesas or table mountains on the earth. One observer

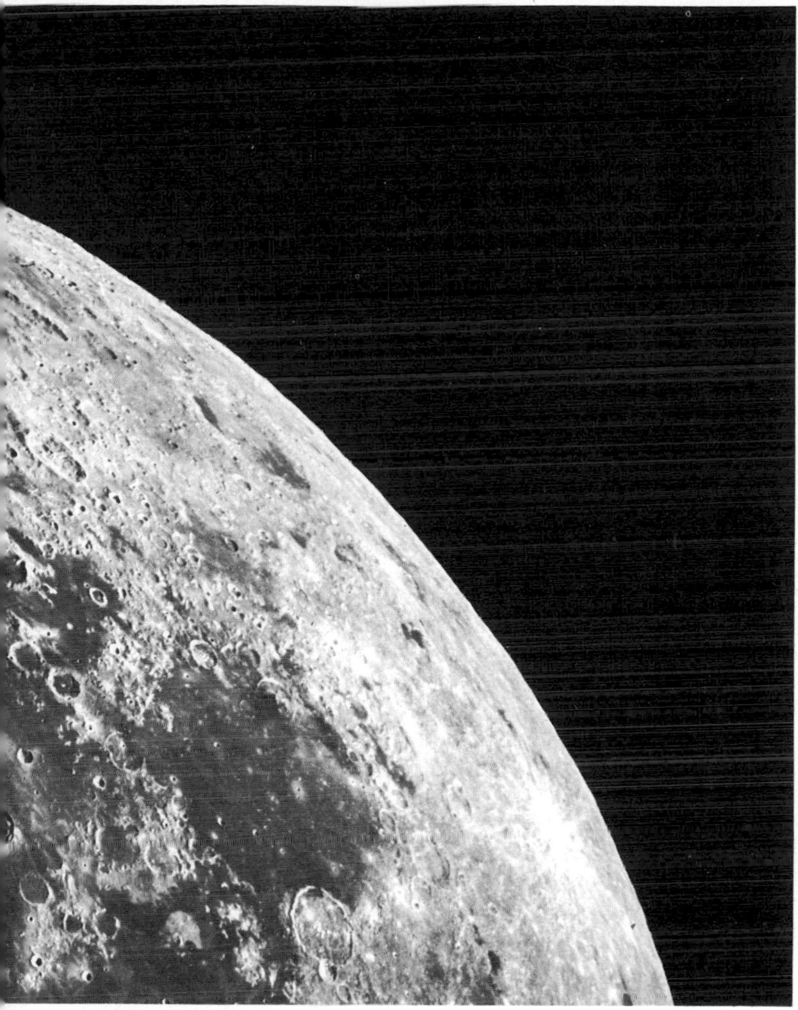

graphically described it as resembling a round slab of cheese (presumably green). The Palus Epidemiarum is a dark area connecting Mare Nubium and Mare Humorum. It contains the fine crater Ramsden, 12 miles in diameter, which has a system of rills extending to the west. Campanus and Mercator, both 30 miles in diameter, form an interesting pair. The floor of Mercator is quite level, whereas that of Campanus is more broken.

Note the many ruined and ghost craters around the boundaries

(*continued on p. 283*)

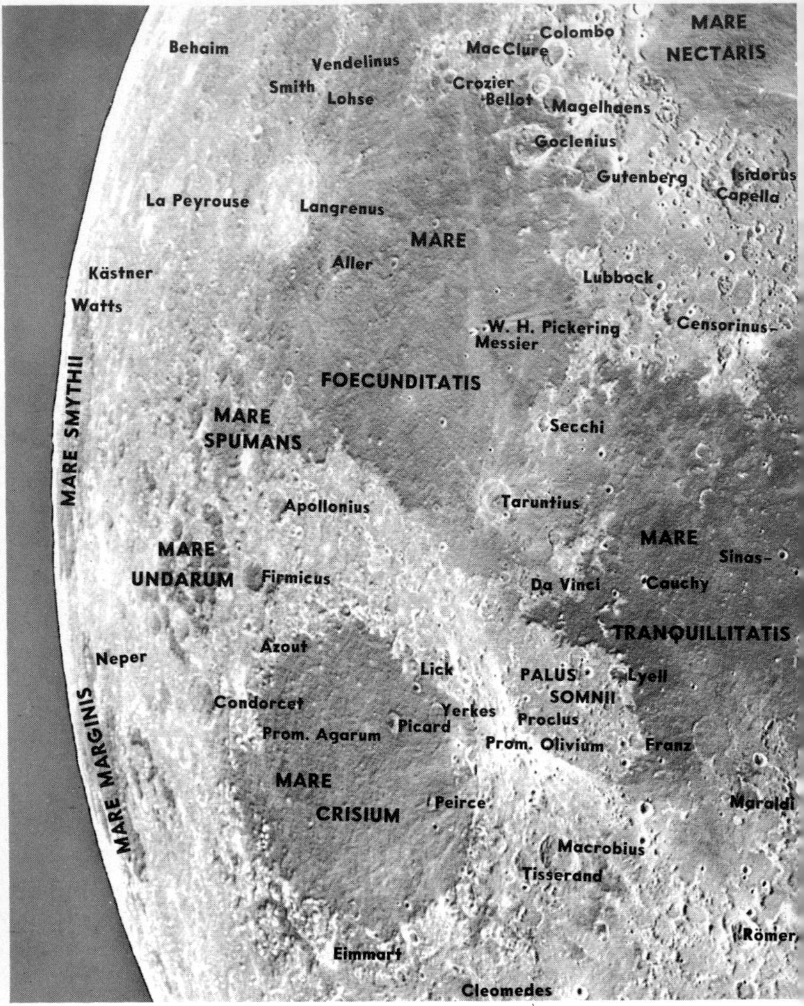

Moon Map 5 shows the moon's western limb. Here maria abound, each having a distinctive character. Foecunditatis and Tranquillitatis have irregular boundaries, whereas Mare Crisium is more nearly circular. Spumans and Undarum scarcely deserve the designation "mare," so broken are they with other formations. Marginis and Smythii are small, highly foreshortened dark areas at the limb, always difficult to see.

Langrenus, a magnificent crater 85 miles in diameter, surrounded by walls that rise to 9600 feet, is visible at all illu-

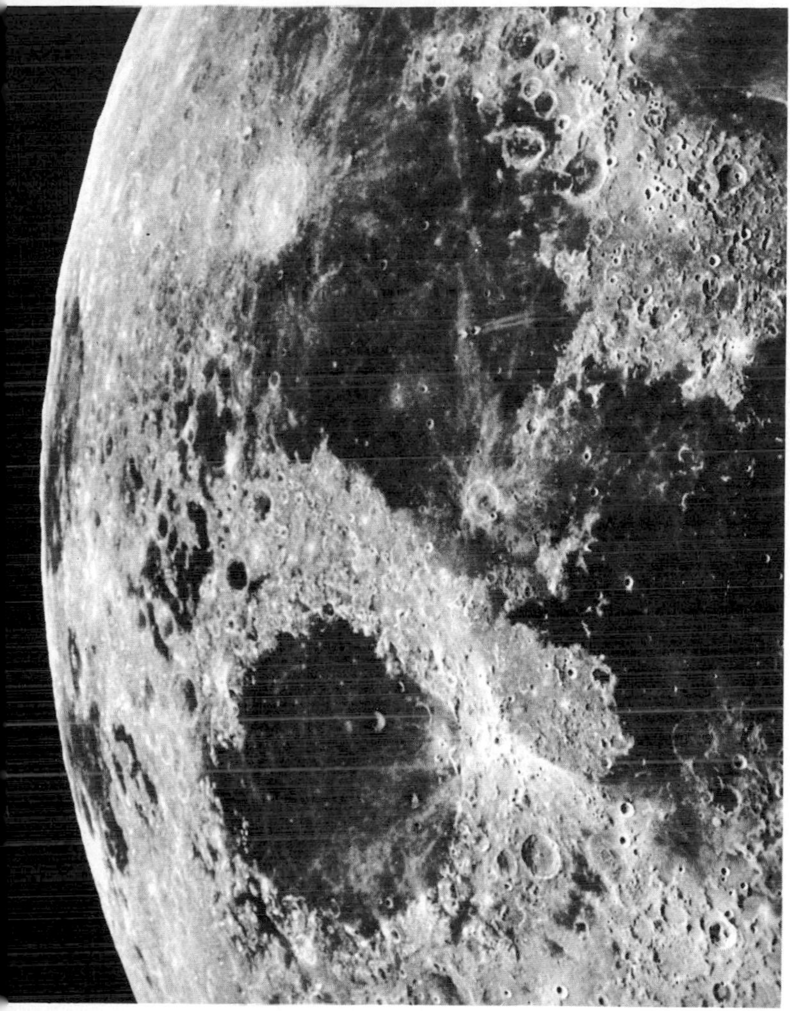

minations. It is the center of a fine ray system.

The crater pair Messier and W. H. Pickering are especially notable because of the peculiar ray system extending eastward from the latter, giving the appearance of a comet's tail. The crater was probably formed by a meteor that came in at a sharp angle, splashing matter mainly in one direction. Apparently Pickering, the older of the two, partially shielded the eastward region, so that the rays developed into two streaks of material, in

(*continued on p. 283*)

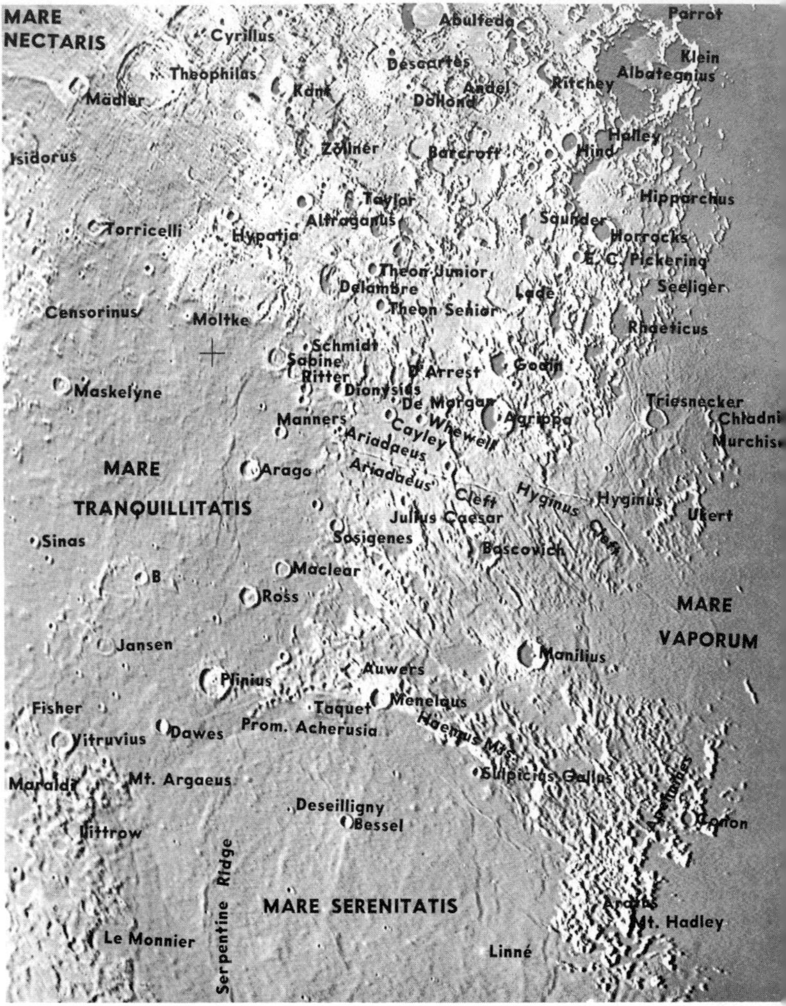

Moon Map 6 presents the interesting contrast of the relatively flat Mare Tranquillitatis toward the west and Mare Serenitatis toward the north and the rugged mountains toward the southeast. Mare Vaporum lies in the northeastern portion, separated from the other mares by the Haemus Mountains and the Apennines.

Albategnius is a plain 80 miles in diameter, surrounded by walls rising from 9000 to 14,000 feet. The slightly larger Hipparchus, just to the north, is closer to ruin. In many places the surrounding walls have crumbled into heaps of debris. Delambre is a nearly

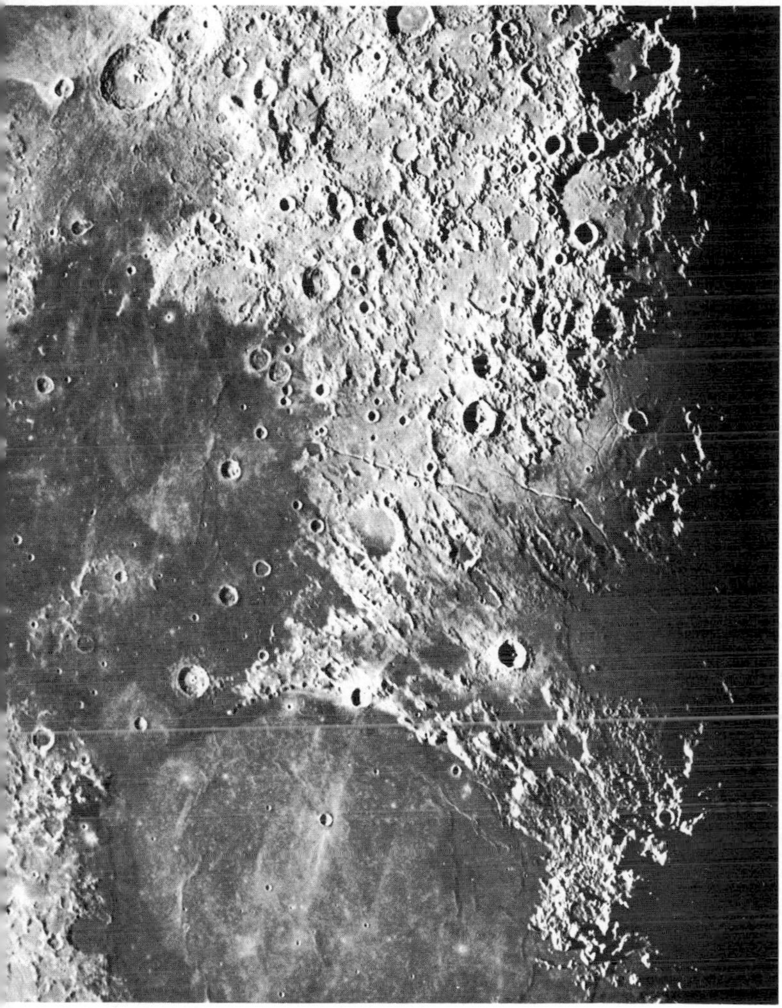

perfect crater, with a double peak and one or more craterlets in the interior. Triesnecker, only 14 miles in diameter, is most noted for the series of clefts and crevices emanating from its western wall, extending south toward Rhaeticus and northward toward the tiny crater Hyginus, with its remarkable cleft system which in turn interlinks with the Ariadaeus cleft. A 2-inch telescope readily reveals most of these clefts.

Sosigenes, named for the astronomer who revised the calendar

(continued on p. 283)

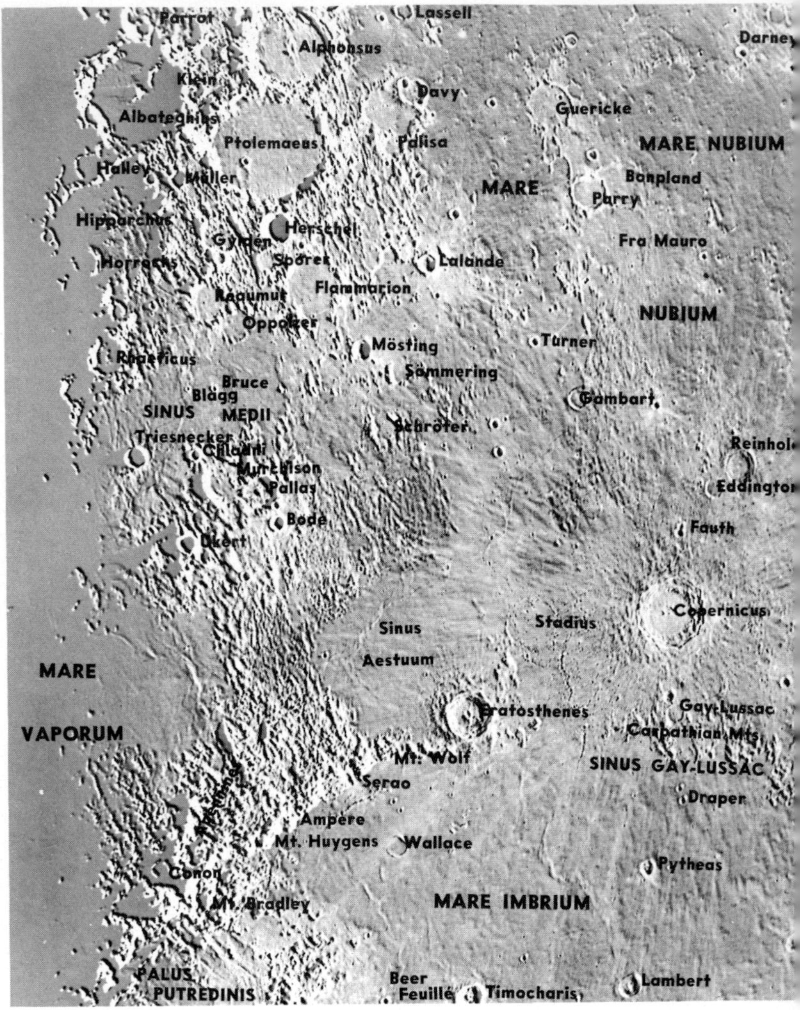

Moon Map 7 is dominated on the south by Mare Nubium and on the north by Mare Imbrium. Ptolemaeus, a walled plain 115 miles across, is part of a triple group including Arzachel and Alphonsus (see also Moon Map 3). It is not unlike a miniature mare. Notice the striking valley or cleft extending north from the northeastern wall of Alphonsus toward the crater just west of Lalande. The partially ruined craters Guericke and Parry and the ghost Bonpland are striking objects in Mare Nubium.

The two great craters Copernicus (56 miles across) and Eratos-

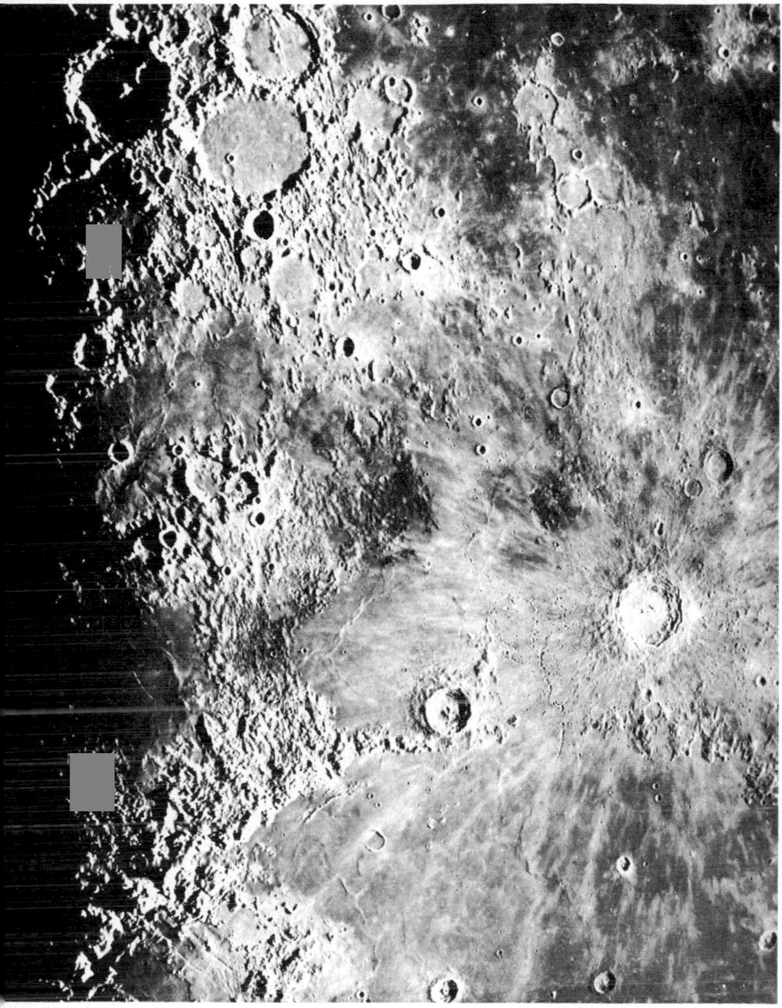

thenes (37 miles across) are among the most remarkable objects on the lunar surface. They both possess rough, terraced walls. Their ray systems, however, are not as extensive as that of Tycho (see Moon Map 3). Eratosthenes seems to lie on a peninsular extension of the Apennines. Between these two craters, extending north from the ghost crater Stadius, lie several chains of crater pits, clearly the result of some form of volcanism. The ruined crater Wallace, with its distinctive horseshoe shape, is an in-

(continued on p. 284)

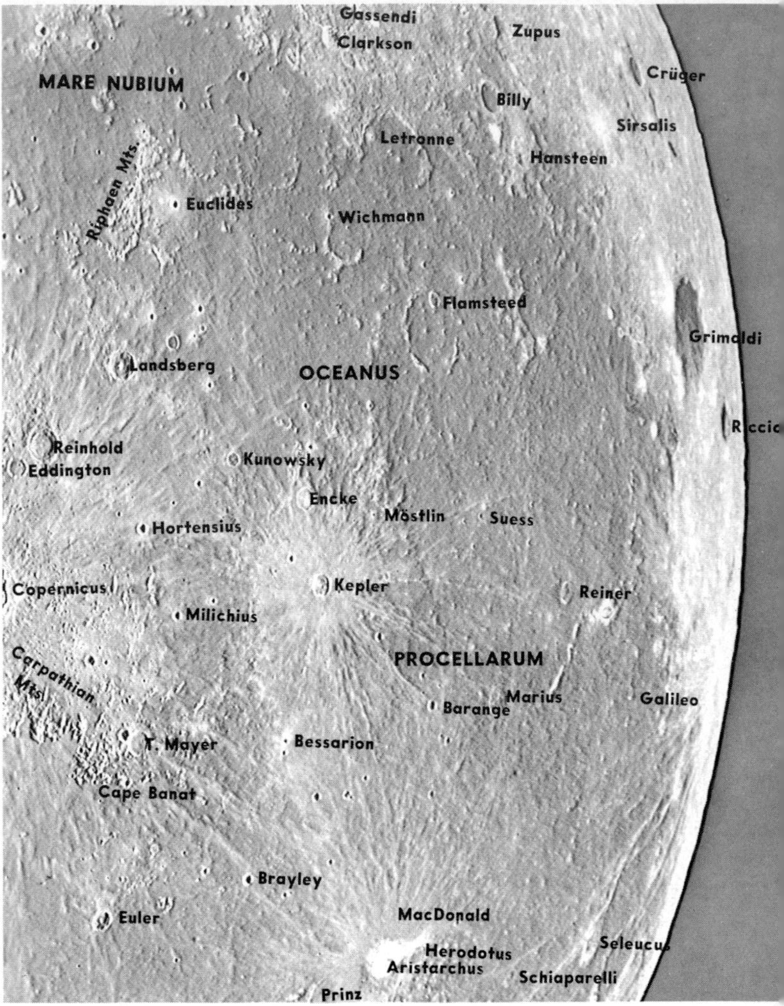

Moon Map 8 displays chiefly rolling plains spotted here and there with craters, most of them of the ruined or ghost variety. On the south, Mare Nubium opens into the greatest of all the seas, the Oceanus Procellarum. Toward the eastern limb, Grimaldi, a walled plain 150 miles in diameter, and Riccioli, 100 miles across, are more like maria than craters. They might have been so designated had they been nearer the center of the moon. Flamsteed, a

well-defined crater 9 miles in diameter, lies on the southern edge of a great broken mountain ring about 60 miles across. Kepler is the finest crater on this map, with a central peak and a well-defined ray system. Euler, 19 miles across, possesses an exceptionally deep interior, some 6000 feet. The Carpathian Mountains extend northeastward from Copernicus to Cape Banat.

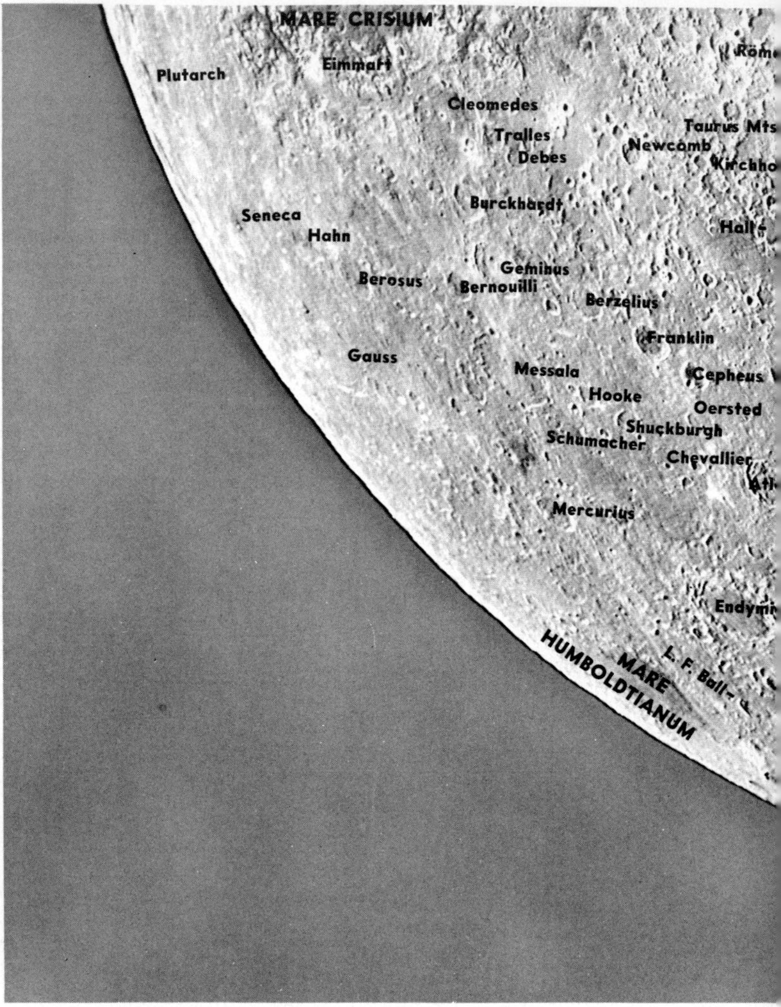

Moon Map 9 shows the moon's northwestern limb. The region contains many interesting craters, best seen near new moon, when the sun is rising in this area. Cleomedes, a large walled plain 78 miles in diameter, shows many internal irregularities, including numerous small craters. Geminus is a striking ringed plain, 54 miles across, with steep terraced walls rising 12,000 feet above the exterior and 16,000 above the interior of the deep crater. Franklin,

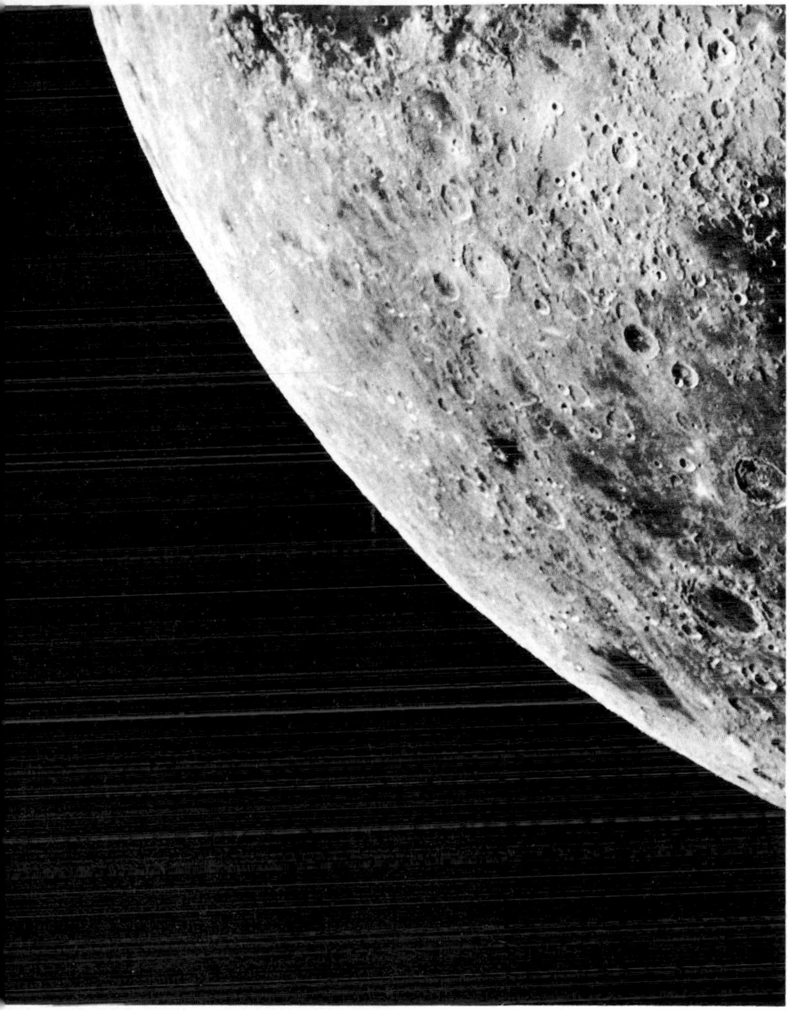

33 miles across, was named for the famous American statesman-writer-scientist. Note the bright triangular patch just to the west of the great crater Atlas (55 miles in diameter). With Hercules (see also Moon Map 10) it forms a double crater. Endymion is a fine walled plain 80 miles in diameter, with broken walls that extend in places from 10,000 to 15,000 feet above the interior. The small dark area near the limb is Mare Humboldtianum.

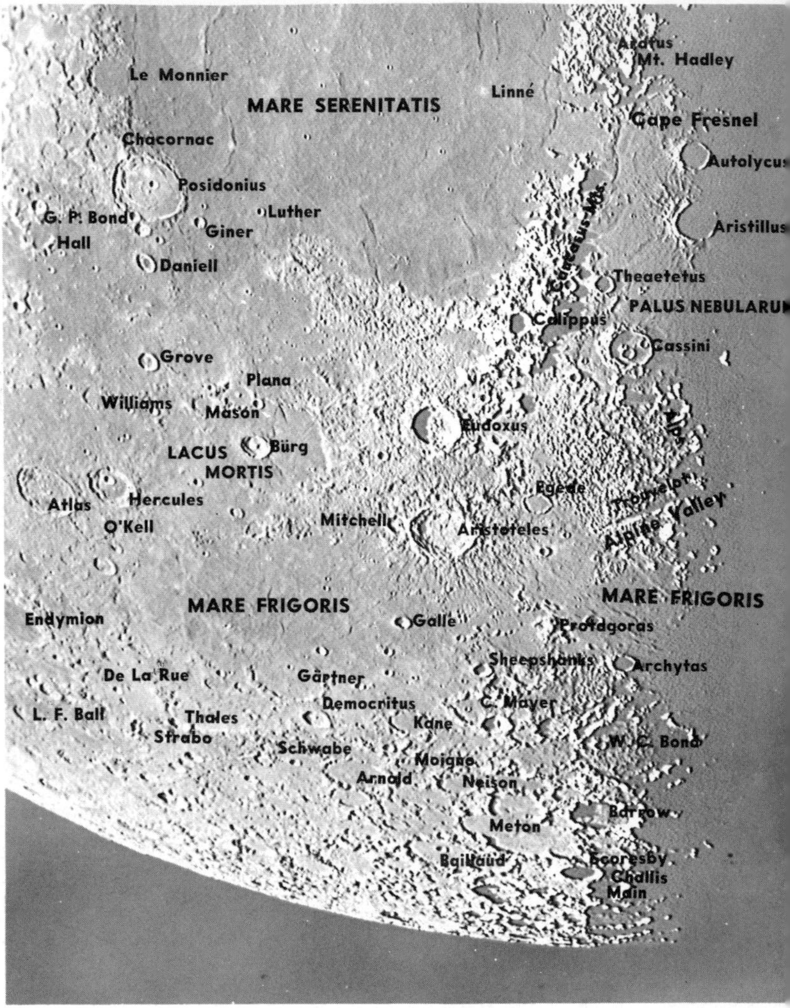

Moon Map 10 is dominated on the south by Mare Serenitatis near whose eastern border we find a bright spot marked Linné. This region is probably the most controversial on the entire surface of the moon. Early observers described Linné as a deep crater 5 or 6 miles in diameter. But in 1866 Schmidt announced that he could find no crater at all and that it must have disappeared. Since that time, only a bright spot has remained in the position of the initial crater. The question, still unresolved, is whether some major lunar catastrophe actually destroyed an important

crater or whether the early observers were merely mistaken.

Posidonius, 62 miles in diameter, on the northwestern border of Mare Serenitatis, is a remarkable walled plain in that the larger crater appears to contain the remnants of a smaller ruined crater, a phenomenon difficult to explain on any hypothesis concerning crater origin.

The Caucasus Mountains on the northeastern border of Mare Serenitatis have two magnificent craters in the northern bound-

(*continued on p. 284*)

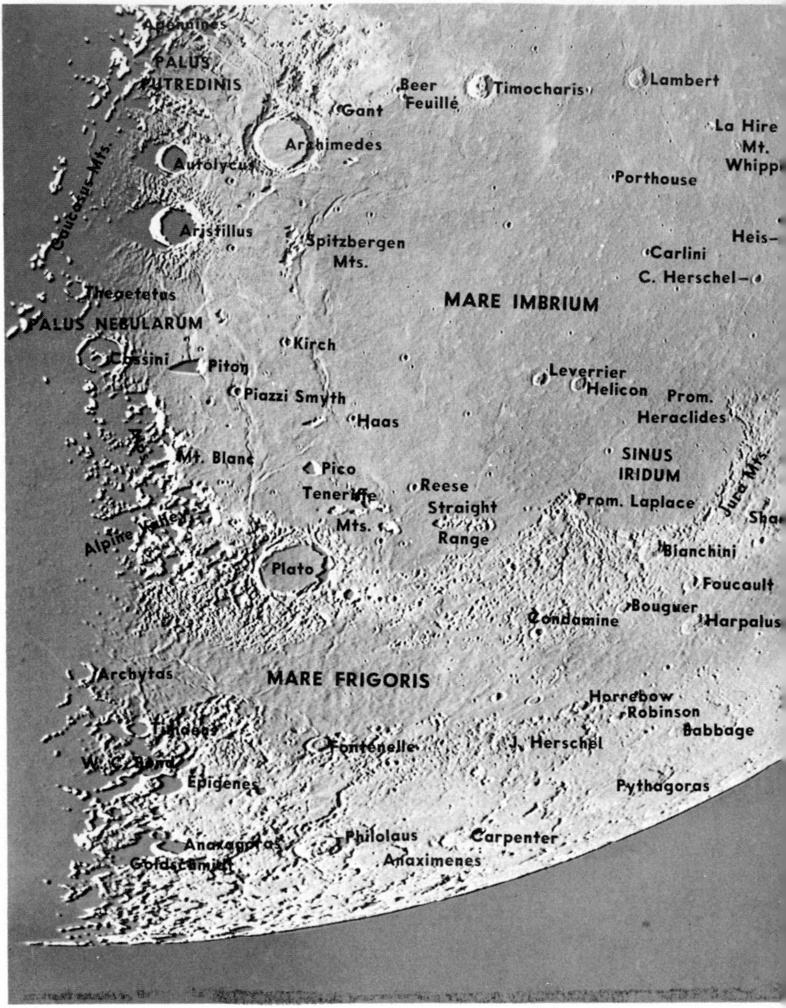

Moon Map 11 contains an unusual variety of distinctive craters and other lunar formations. The great walled plain Plato is one of the most striking objects. The floor, some 60 miles in diameter, is relatively smooth, though you may see a few crater pits under favorable illumination. The jagged profile of the shadows indicates the ruggedness of the surrounding mountains, which rise to heights greater than 7000 feet. Of special interest is the enormous faulted mass of rock on the eastern edge. To the southeast of Plato, well within the borders of the Mare Imbrium, lie the

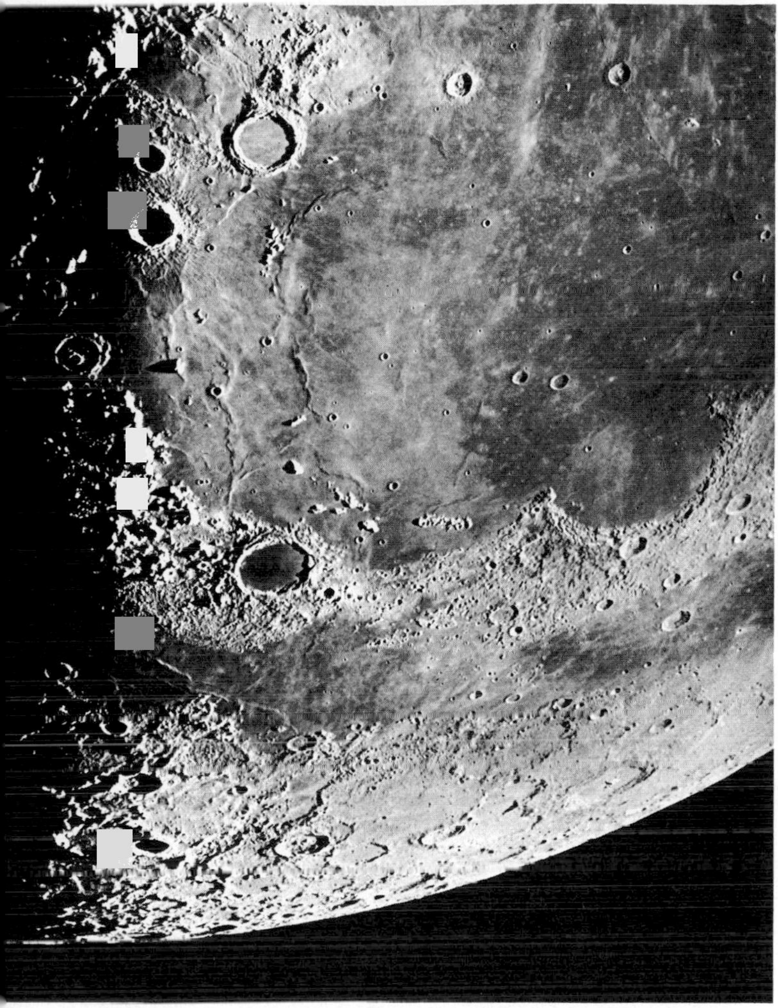

Teneriffe Mountains, the Straight Range, and to the south the isolated mountain peak Pico, which rises some 8000 feet above the plain. Piton, a similar isolated peak about 7000 feet high, lies to the southwest near the border of Mare Imbrium and near the complex crater Cassini, with its smooth inner crater that lunar observers sometimes call the "washbowl."

On the western border of Mare Imbrium we find a distinctive group of three craters. Archimedes is a walled plain, about 50

(*continued on p. 284*)

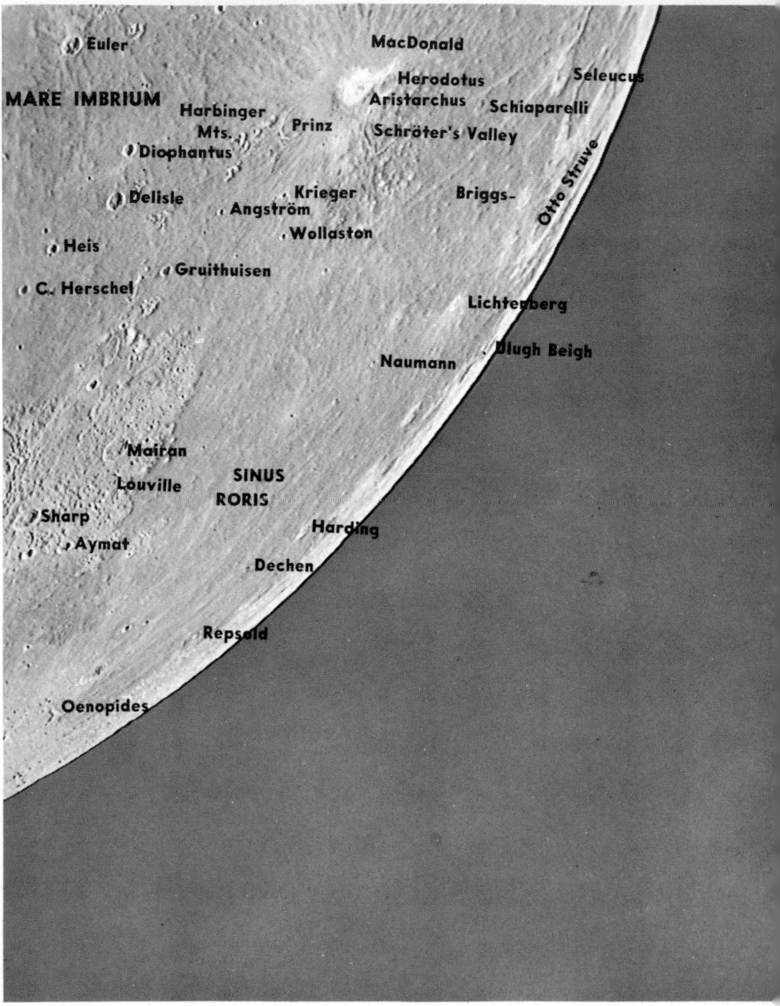

Moon Map 12 delineates the northeastern limb of the moon, the chief feature of which is an eastern extension of the Oceanus Procellarum. The crater Aristarchus, 30 miles in diameter, is by far the most brilliant formation in this region, if not on the entire moon. Note the complex ray system associated with the de-

pression. Schröter's Valley, on the shoulder of Aristarchus, is an excellent example of a sinuous rill. It is easily visible in a small telescope. Prinz is an incomplete semi-ghost ring in the Harbinger Mountains. Mairan, 25 miles across, has exceptionally precipitous walls that extend to 15,000 feet on the western edge.

Moon Map captions, continued

Moon Map 1 continued

Fracastorius, a partially ruined crater some 60 miles in diameter, might be described as a "bay" of Mare Nectaris, since the walls are missing on the northern boundary. The floor of this crater is a favorite with amateurs. The crater Orús, about 20 miles across, has a fine ray system.

Moon Map 2 continued

Werner, some 45 miles in diameter and surrounded by exceptionally lofty peaks ranging from 13,000 to 16,000 feet, possesses a fine central peak. Aliacensis, to the southwest, possesses similar characteristics. Apianus, 38 miles in diameter, has an extremely level floor. Piccolomini, 56 miles in diameter, surrounded by a terraced ring rising 15,000 feet, has a complex structure of internal peaks.

Azophi and Abenezra, about 25 miles across, form another interesting pair. The latter is the newer. Both possess very deep interiors for their size.

Catharina, 55 miles in diameter, is surrounded by irregular walls that occasionally attain heights of 16,000 feet. Its irregular rough floor is broken by a ghost crater about 25 miles in diameter. Abulfeda, 40 miles across, is exceptionally level, though it does possess a central peak. Note the chain of crater pits connecting it with Almanon on the southwest.

Cyrillus and Theophilus make another interesting crater pair, some 65 miles across. The former has irregular terraced walls that rise in a circular pattern. On the northwestern boundary of its companion, Theophilus has ruined the wall of its neighbor. This crater contains a complex central structure. It is one of the finest craters on the moon, with walls that reach as high as 18,000 feet on the west. These two craters and Mädler, also a fine crater with a central peak, lie on the eastern border of Mare Nectaris.

Moon Map 3 continued

rays may suggest meridians stretching out from the pole represented by Tycho. See Figure 25 (p. 253).

Walter, 100 miles in diameter, has a complex internal structure. Hell, which derives its name from the German signifying "bright," is a small perfect crater. Note the chain of crater pits curving from the northern boundary. Pitatus lies on the southern boundary of Mare Nubium. Its floor has apparently been inundated by the lava sea that produced the Mare, for it shows the same smooth dark characteristics. The Mare itself covers many partially ruined and ghost craters, such as Kies and Lubiniezky. Bullialdus, on the contrary, is a new and perfect crater, about 38 miles across. Near the southwestern boundary of Mare Nubium note the

Straight Wall, whose southern boundary points toward several ghost craters. This formation is about 60 miles long and consists of a steep cliff about 800 feet high, where the broken floor of the Mare has faulted. Arzachel, Alphonsus, and Ptolemaeus (see also Moon Map 7) make an interesting triple group with significant differences. The floor of Arzachel is rough; that of Ptolemaeus is relatively smooth. Alphonsus is one of the regions where gas has been observed (1958) to seep from the central peak.

Moon Map 4 continued

of Mare Humorum. Gassendi has the appearance of being very old. Though it is large, 55 miles in diameter, the walls are broken and seem to be partially inundated by the neighboring mare. This region is a favorite for amateur observation. It contains a fine central group of mountains and an interesting system of criss-crossed rills in the interior, best seen when the sun is rising or setting on this crater, but still visible at full phase.

Moon Map 5 continued

regions not screened by the walls of the eastward crater. Taruntius, 44 miles in diameter, has irregular terraced walls rising to 3500 feet. Note the tiny crater on the northern rim. Macrobius, 42 miles in diameter with relatively high walls extending to 13,000 feet and with a sharp central peak, is plainly visible at all illuminations because of its bright surface.

On July 20, 1969, as all the world listened and watched, two men landed their space module on the lunar surface, while a third man continued in orbit around the moon in a command ship. The two men, Neil Armstrong and Edwin Aldrin, later walked on the surface of the moon, picked up samples of rocks, took many photographs, and set out various pieces of scientific equipment. After a rest, they jettisoned unnecessary materials, fired their rocket engines, and rendezvoused with their companion, Michael Collins, who was still orbiting in the command ship. Then they returned to earth, landing safely in the Pacific Ocean.

The lunar landing point (Tranquillity Base from its location on Mare Tranquillitatis; see cross on Moon Map 6) still bears the footprints of the astronauts. Analysis proved that the rock samples returned to earth were older than any known terrestrial rocks.

Moon Map 6 continued

for Julius Caesar, lies appropriately just west of the larger crater named for the great Roman emperor. Plinius, 32 miles across, has a double central peak with craterlets. The western Apennines, which range 185 miles north to south and 165 miles east to west, form the southwestern border of Mare Imbrium. The terrain is extremely irregular, a mass of splattered lumps without valleys, emanating from a high plateau 6500 feet above the neighboring maria. Mount Hadley rises 15,000 feet.

Moon Map 7 continued

teresting object. Note the rugged or wrinkled ridges of Mare Imbrium in its vicinity.

Moon Map 10 continued

ary, Eudoxus, 35 miles across, and Aristoteles, some 50 miles in diameter. Both of these craters are rugged. They possess steep and terraced walls on the interior and radial ridges on the exterior.

The lunar Alps to the east lie on the northwestern border of Mare Imbrium. These mountains contain the famous Alpine Valley, a great gash 83 miles long and from 3 to 6 miles in width, cutting perpendicularly through the chain in almost unique fashion. This valley may have resulted from a glancing meteoric impact. Or a giant meteor may have crashed into Mare Imbrium, producing an intense splash whose fragments cut through the alpine chain and produced many splash craters in the neighborhood.

Moon Map 11 continued

miles in diameter, whose relatively smooth floor and terraced circumference extend above 7000 feet. The nearby Aristillus, a crater some 35 miles in diameter, shows a rugged and terraced interior that contrasts markedly with the smooth floors of Plato and Archimedes. The crater walls rise to 11,000 feet above and the rugged interior declines to 3000 feet below the surrounding Mare. The interior contains a mountain peak. The outer rim displays a marked system of radial ridges. Autolycus somewhat resembles Aristillus except that its diameter is less, about 24 miles. The lunar Apennines, whose mountains extend from 15,000 to 18,000 feet, mark the western border of the relatively smooth Mare Imbrium. Timocharis and Lambert, with respective diameters of 25 and 18 miles, are among the most perfect examples of lunar craters. Each contains a central peak with a smaller crater on the top.

The surface of Mare Imbrium along the line adjoining Plato and Aristillus displays what appear to be the edges of ancient lava flows — smooth curving ridges of a distinctive character. Conceivably they may mark the remnants of extensive mountain chains. Piton, Pico, and the Teneriffes protrude like islands above the surface of a wrinkled lava flow.

Note the curved boundary of Mare Imbrium, the Sinus Iridum, and especially the rugged slope of the surrounding mountains. The long, thin Mare Frigoris separates the Plato area from the mountainous regions near the moon's north pole. In this region, the many mountains and overlapping craters will repay careful examination, especially when the illumination is most favorable.

VIII
The Sun

For the astronomer the sun has a double significance. It is, of course, the primary body of our solar system, about which all the planets revolve. Also, the sun holds the unique distinction of being the nearest of all the stars, the only one we can study in great detail.

The sun is the source of our light and heat. Its very brightness forms an obstacle to observation, with or without telescopic aid. Never look directly at the sun unless you have taken special precautions to reduce its brilliance with the aid of smoked glass or special reflectors. On rare occasions, when the sun is low on the horizon and dimmed by smoke, fog, or haze, you can safely look at the sun directly and observe a few of its more obvious features.

Note first of all that the sun's disk is not uniformly bright. The edge, or *limb*, is distinctly less luminous than the center. When we look diagonally into the sun's atmosphere, near the limb, we see a cooler and therefore less luminous region than when we look straight down at the center of the disk. Occasionally we may even glimpse some of the dark areas called *sunspots*. The Chinese annals contain many records of spots observed on the solar surface long before the invention of the telescope.

Observing the sun through a telescope, even one of comparatively low power, can be tremendously exciting. The simplest — and safest — method employs the telescope to project the bright solar image on a white card or other convenient white surface. Try changing the distance of the card from the end of the telescope, refocusing the eyepiece until a clear, sharp image, 7 inches or more in diameter appears (Fig. 27). If possible, attach the card firmly to a support behind the eyepiece. Also, to keep the glare of sunlight off the card mount a second card pierced by a hole large enough to admit the telescope objective, so that it will cast a shadow on the other card.

Fig. 27. Observation of sun with telescope

This simple procedure should disclose the most conspicuous solar features: (1) the dark centers, or *umbrae*, of the spots; (2) the delicate fibrous structure of the *penumbrae* immediately surrounding the spot (Fig. 28); (3) the bright mottling, or *faculae*, that usually accompany spots and appear most clearly toward the sun's limb; (4) the granular structure of the solar surface most clearly visible near the center of the disk; and finally (5) the above-mentioned darkening to the limb (Fig. 29). If you wish, draw on the card itself to get a permanent record of sunspots and other visible solar features.

To view the sun directly through the telescope, one may employ a special dark glass somewhere in the optical beam. For best results set this dark filter close to the focus of the primary objective or immediately behind the eyepiece. I regard as dangerous the conventional substitutes of glass smoked in a candle flame or of heavily fogged photographic film. They are safe enough for very brief glimpses of the sun with the unaided eye. But do not use them with a telescope: the concentrated solar heat may crack the glass or ignite the film so that an intense flash would strike the eye. Ordinary dark glasses do not provide sufficient protection, although certain types of commercial dark glass are perfectly safe for this purpose. Your optician will be able to cut out some small circles and perhaps even repolish them, if you are unable to buy them ready for use. American Optical Company

Fig. 28. Large sunspot group of May 17, 1951

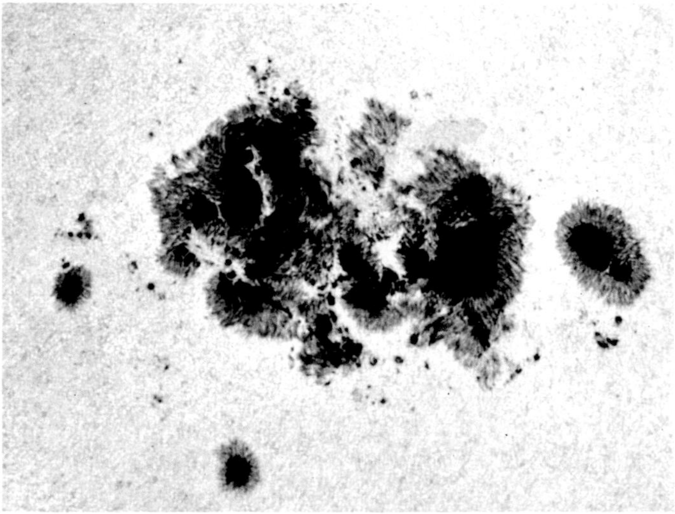

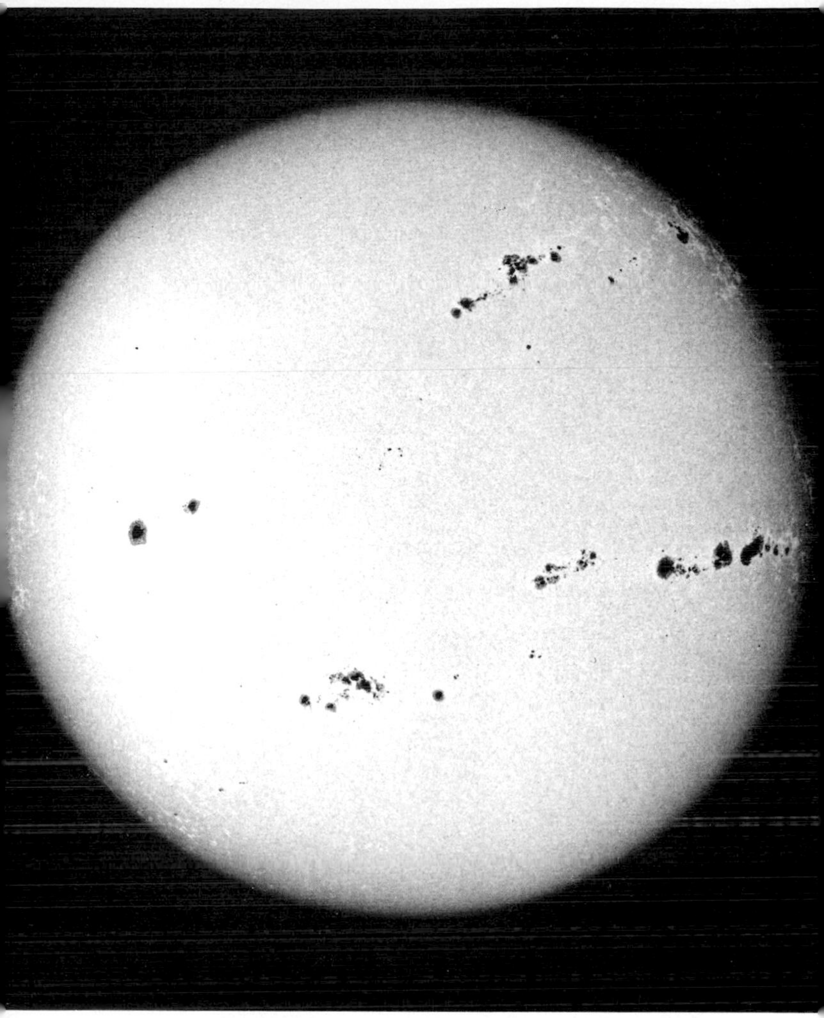

Fig. 29. The sun at sunspot maximum, Dec. 21, 1957

or Metal and Thermit Welding Glass No. 12 is recommended.

Many professional observers use a solar diagonal, which is simply an unsilvered mirror. The back surface of this mirror, however, preferably should be inclined with respect to front surface in order to eliminate the possibility of multiple reflections between them. Perhaps the most effective and inexpensive device is the miniature unsilvered *pentaprism* (Fig. 30a). The double reflection from glass surfaces effectively reduces the intensity of sunlight to a reasonable value, with marked losses at each re-

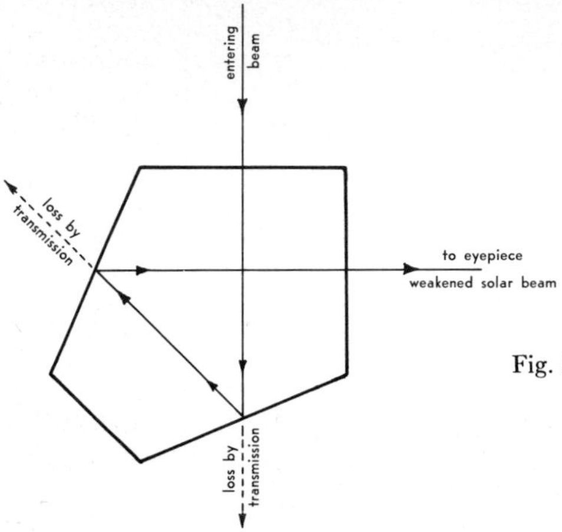

Fig. 30a. Pentaprism

flection (Fig. 30b). The observer will particularly appreciate the possibility of seeing the sun without the color distortion that absorbing glasses introduce. In addition to the pentaprism, some weak, neutral glass will usually suffice.

Sunspots stand out as the most conspicuous markings on the solar disk. Astronomers have generally supposed that these areas are storms, perhaps analogous to terrestrial tornadoes. Recent studies have shown that the intense magnetic fields existing in these spots inhibit convection in the region, so that they tend to be cooler than their surroundings. The number of spots visible

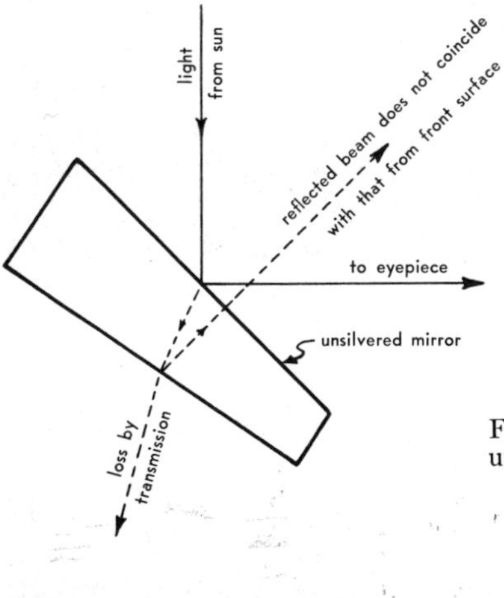

Fig. 30b. Wedge-shaped unsilvered mirror

at any one time varies over a long-range cycle, with times of maximum spottedness occurring at intervals of about 11 years and a well-pronounced minimum during the in-between periods. The American Association of Variable Star Observers (AAVSO, see page 120) maintains a solar section to instruct and coordinate the work of those who are willing to maintain a regular patrol to record sunspot numbers.

Bursts of activity in the solar atmosphere produce solar *prominences*, which appear as jets, loops, ribbons, or arches of luminous gas (Figs. 31, 32). Ordinary telescopes will not reveal

SOLAR PROMINENCE: Fig. 31. Hedgerow type
Fig. 32. Loop or sunspot type

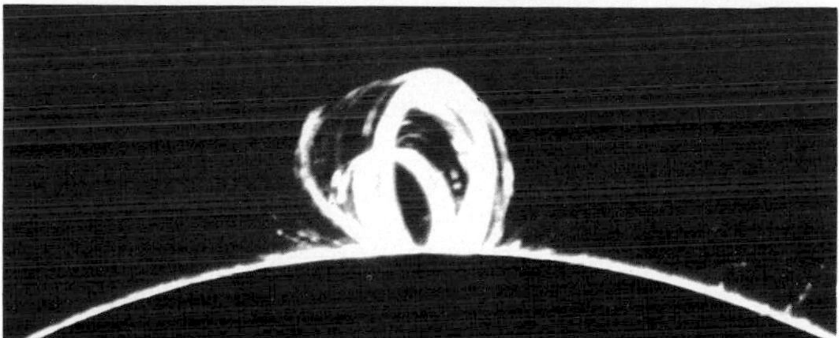

Fig. 33. Corona at eclipse

these interesting formations. Observing them requires complicated (and expensive) accessory equipment, of which the simplest is a special filter transmitting a very narrow band of color, identical with that radiated by hydrogen atoms.

Still fainter and more difficult to observe is the solar *corona*—white, fan-shaped streamers stretching radially outward from the sun. These are best seen at total solar eclipse, when the moon happens to move to a position directly between the earth and sun and momentarily hides the sun. Under ordinary conditions, both prominences and corona are completely concealed by the glare of bright sky surrounding the sun. The form of the corona changes with the degree of solar activity. Near sunspot minimum the corona possesses long equatorial streamers. Near maximum (Fig. 33), its outline is irregular.

Eclipses are among the most spectacular astronomical events. The sudden disappearance of the sun by day or the moon by night terrified the ancients, who did not understand what was happen-

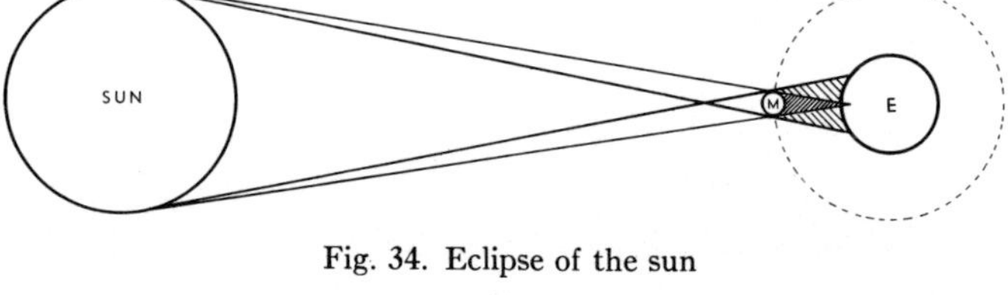

Fig. 34. Eclipse of the sun

ing. The explanation is simple. Figure 34 schematically illustrates the circumstances of a *solar eclipse*. The moon (M) moves between the sun and the earth (E) so that its shadow falls on the earth. Anyone located in the cone-shaped shadow will see the sun completely obscured, a total solar eclipse. If, as frequently happens, the cone does not reach the earth, someone directly under the tip of the shadow cone will see a ring of sunlight surrounding the darkened moon, an *annular eclipse*. A person in the lightly shadowed area will see the moon taking a bite out of the sun, a *partial eclipse*.

Since the diameter of the shadow cone where it strikes the earth is never more than 200 miles across, the fraction of the earth's surface swept out by the moving shadow is very small. From a given spot, a total eclipse is a rare event, happening only once every 300 years or so on the average, even though one occurs for some place on the earth about every year and a half. At the time of totality, the beautiful corona, ordinarily obscured by bright sunlight, reveals its fan-shaped streamers (Fig. 33).

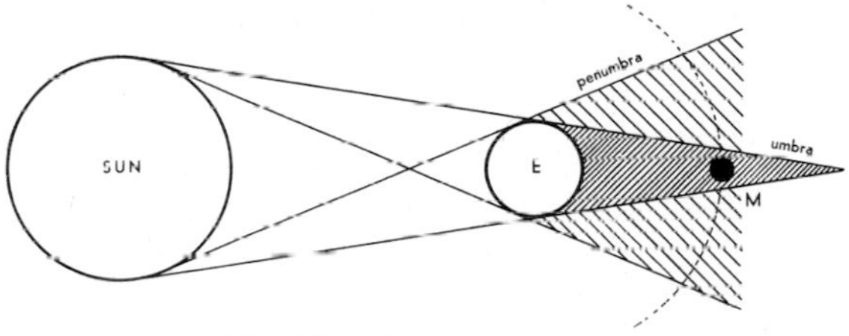

Fig. 35. Eclipse of the moon

Figure 35 shows a total *lunar eclipse*, which occurs when the moon moves into the shadow of the earth. The earth's shadow, however, is not completely black. Some of the red rays of sunset filter through, imparting a coppery color to the moon. Though lunar eclipses are rarer than solar eclipses for the earth as a whole, they are much more frequent from a given point because they are visible from a full hemisphere.

IX

The Planets and Their Positions

THE ANCIENTS distinguished between the *fixed* stars and the *wandering* stars. Among the wanderers they included the sun, the moon, Mercury, Venus, Mars, Jupiter, and Saturn — seven bodies that give their names, in English by way of Norse mythologies, to the seven days of the week. To the five naked-eye planets known from antiquity, we must add the three discovered in modern times: Uranus, Neptune, and Pluto. Table 18 summarizes the most important information about these planets, their orbits, and their satellites.

Figure 36 schematically shows the orbits of the earth, of an *inferior* planet (one inside the earth's orbit), and of a *superior* planet (one outside the earth's orbit). Each type of planet shows four definite *configurations*. An inferior planet may appear in the direction of the sun, (1) at *inferior conjunction* (IC) between the sun and the earth, or (2) at *superior conjunction* (SC) on the far side of the sun. Or it may appear (3) at *eastern elongation* (EE) or (4) *western elongation* (WE).

A superior planet may also be (1) in line with or in *conjunction* (C) with the sun. It may be 180° from the sun (2) at *opposition* (O). The planet may also lie at right angles to the sun, (3) at *eastern quadrature* (EQ) or (4) *western quadrature* (WQ). A superior planet is most conspicuous at opposition, when it rises just at sunset. It is invisible at C, but stands out clearly from EQ to WQ.

An inferior planet shows phases like the moon, being crescent from EE to WE and gibbous the rest of the time. It is invisible at IC or SC. A superior planet is never crescent. Only Mars is near enough to the earth to display a distinctly gibbous phase. The *synodic period* is the interval (usually expressed in days) between successive recurrences of the same phase for a given planet.

Except when very close to the horizon, planets rarely twinkle perceptibly. Unlike stars, which appear as minute points of light, planets are large enough so that the different areas of the surface twinkle independently. As a result, the light from planets seems to be relatively steady.

Mercury. The planet nearest the sun, Mercury, is difficult to see. In the Northern Hemisphere we can view it best at eastern elongation in spring just after sunset, or at western elongation in fall just

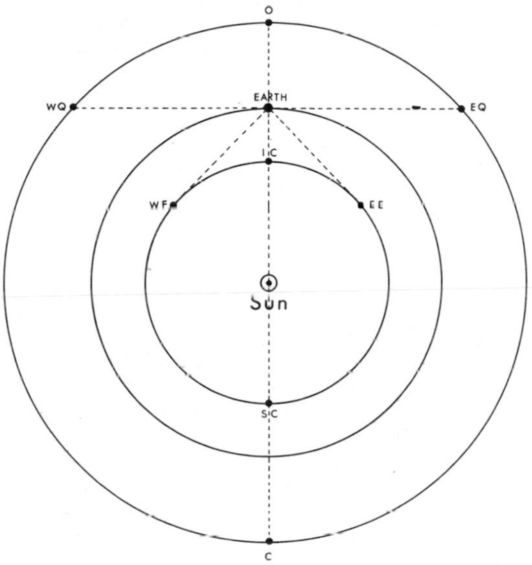

Fig. 36. Planetary configurations
O, opposition; C, conjunction; IC, inferior conjunction; SC, superior conjunction; EE, eastern elongation; WE, western elongation; EQ, eastern quadrature; WQ, western quadrature

before sunrise. Mercury moves so rapidly, circling the sun in a mere 88 days, that an observer catches only fleeting glimpses of it. One can see Mercury during the daytime, but a 3-inch telescope is necessary. The markings are somewhat indistinct. The planet rotates directly, with a period of 59 days, exactly two-thirds of its period of revolution. Mercury has no appreciable atmosphere.

Venus. Often Venus appears to be the most brilliant "star" in the sky, especially when near eastern or western elongation. The planet possesses an extensive atmosphere, in which carbon dioxide is one of the major constituents. Clouds perpetually obscure the surface. Dust, rather than water droplets, appears to be the major constituent. The atmosphere consists mainly of carbon dioxide. The surface pressure is about 90 times that of the earth and the surface temperature is about 600° F., hotter than the hottest bake oven. As far as size is concerned, Venus is essentially the earth's twin. Radar studies have shown that Venus rotates slowly backward, in 243 earth days, with respect to the sun. Venus has no satellites.

Mars. This nearest neighbor among the planets, and perhaps the most publicized, displays a remarkable reddish color. Every 26 months it returns to opposition, and dominates the evening sky. Since the orbit is much more eccentric than that of the earth or Venus, the distance of Mars from the earth at opposition can vary from 35,000,000 miles (56,000,000 km) to as much as 60,000,000 miles (97,000,000 km), and corresponding changes occur in apparent size and brightness. Mars has a diameter about half that of the earth. The planet has two small satellites, which are probably captured asteroids (lesser planets). They are visible only in large telescopes.

Even a 4-inch telescope will reveal prominent surface features. The conspicuous red color apparently comes from regions not too different from various deserts of the earth, such as the Painted Desert of Arizona. White "buttons" at the two poles consist of a thin layer of ice and perhaps frozen carbon dioxide. These polar caps vary in size with the Martian seasons. Photographs of Mars taken by the NASA Mariner spacecraft show no "canals." Instead, they reveal a heavily cratered, moonlike surface. Fig. 37a shows a sculptured canyon five times wider and almost twice as deep as the Grand Canyon of Arizona. Mars possesses a thin atmosphere. Fig. 37b is a map of Mars by Gérard de Vaucouleurs (south at the top).

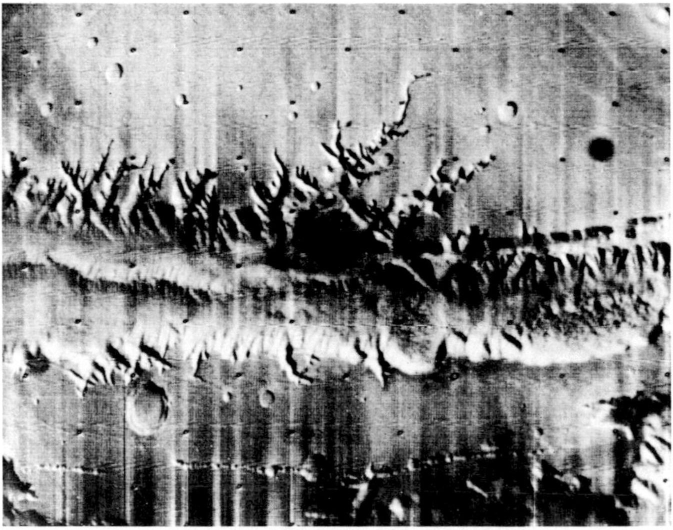

Fig. 37a. Mars photographed by Mariner 9
Jet Propulsion Laboratory, NASA

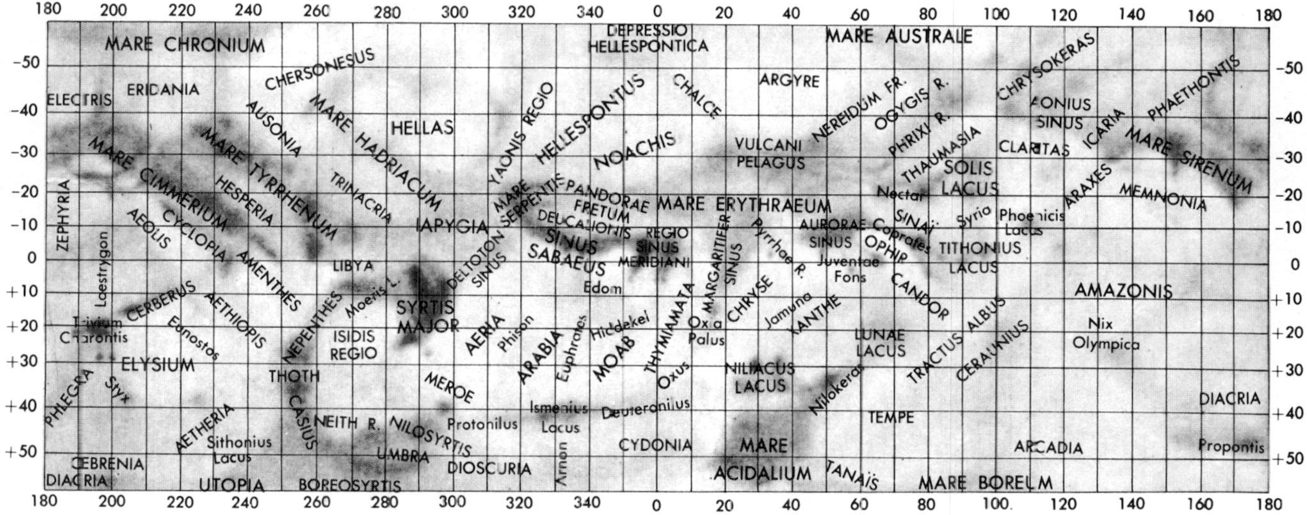

Fig. 37b. Map of Mars

Jupiter. The giant planet has a diameter about 11 times that of the earth. This planet bulges at the equator because of its rapid rotation. Jupiter, whose general color is yellowish, is crossed by many parallel dark bands colored from brown to violet. The famous "red spot" is one of the conspicuous features (see Fig. 38). The surface itself remains concealed under the atmospheric layers visible to us. The dark belts represent some sort of cloud formation in an atmosphere whose temperature approaches that of liquid air.

Jupiter has 12 satellites, but only 4 are conspicuous — those that Galileo first discovered. These moons, readily visible in a 2-inch telescope, change their positions rapidly. Sometimes they move into Jupiter's shadow and undergo eclipse. The disk of the planet may occult them. Occasionally they appear in transit across the face of Jupiter, accompanied by a conspicuous black spot, the shadow of the satellite.

Saturn. This planet with the rings (Fig. 39) is one of the most beautiful objects in the sky. To see the rings clearly one needs at least a 3-inch telescope. The planet resembles Jupiter except for the fact that the belts are less pronounced. The atmospheres of the giant planets contain methane and ammonia.

Fig. 38. Jupiter

Fig. 39. The rings of Saturn

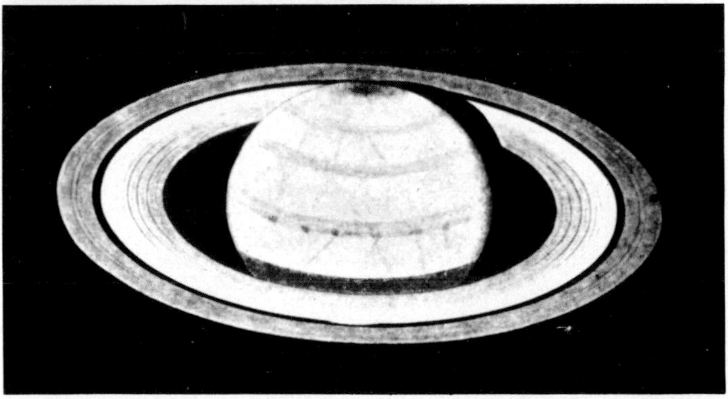

Saturn's rings are not solid like a phonograph record. They consist of billions of small particles, probably rocks coated with ice or frost. Each particle moves in an independent orbit. The fainter outer ring is separated from the brighter inner ring by a black dividing line, where particles are few. Toward the inner edge the grains again thin out, producing a partially transparent ring, the *gauze*, or *crepe*, ring. Saturn has 10 satellites. Titan, the largest moon in the solar system, has a methane atmosphere.

Uranus, Neptune, and **Pluto.** Uranus at opposition attains magnitude 5.7 and thus is just barely visible to the eye; the telescope shows it as a small greenish disk. Neptune, of magnitude 7.7 at opposition, has a delicate blue color. Pluto, of the 15th magnitude, can be seen only with the largest telescopes.

Because the planets move, calculation of their positions presents something of a problem. Almanacs, which list such information, are not always available. The planets more or less follow the *ecliptic*, the apparent path of the sun across the sky, so astronomers describe planetary positions in terms of *celestial longitude*, measured along the ecliptic, and *celestial latitude*, measured perpendicular to the ecliptic. Since the ecliptic is inclined to the celestial equator, celestial longitude and latitude are not the same as right ascension and declination (see p. 117), which are also used to describe the positions of celestial objects. The Ecliptic Star Maps (following p. 301) show the region of the ecliptic, labeled with longitude and latitude.

Figures 40, 41, and 42 enable one to calculate quickly and accurately the positions of the five principal planets. The zero of these figures specifies the position of each planet on December 31, 1957. For Mercury, Venus and Mars, the other numbers indicate the planet's position on successive days during its next revolution around the sun. Since February 9, 1958, is the 40th day of the year, for example, the planet lies at the point marked 40, and so on. Mercury completes one revolution in 88 days and we have to start counting over again. Venus takes 225 days and Mars 687. Table 18 gives exact figures and other planetary data.

Let us take a specific example. Suppose that we wish to calculate the position of Mercury for May 16, 1965. Between this date and the zero date of December 31, 1957, occur 5 full ordinary years, 2 leap years, and 136 days, or 2693 days in all. Now where will Mercury be on the date we have chosen? Divide 2693 by the period of Mercury:

$$\begin{array}{r} 30. \\ 87.969\overline{)2693.000} \\ \underline{2639.070} \\ 53.930 \end{array}$$

The result is 30, with a remainder of 53.93 days. This means that, between December 31, 1957, and May 16, 1965, Mercury will have circled the sun 30 times and will have completed almost 54 days of its 31st circuit. The number 30 is of no basic interest for our purpose. The remainder, 54, indicates the position of the planet on Figure 40, at the 54th point on the orbit. You can also find the position of the earth at May 16 and draw the line connecting the two bodies. Mercury lies near western elongation, visible in the morning sky. The outer circle, with the degree marks and constellation names, is supposed to represent the sky. However, the point where the line from the earth to Mercury inter-

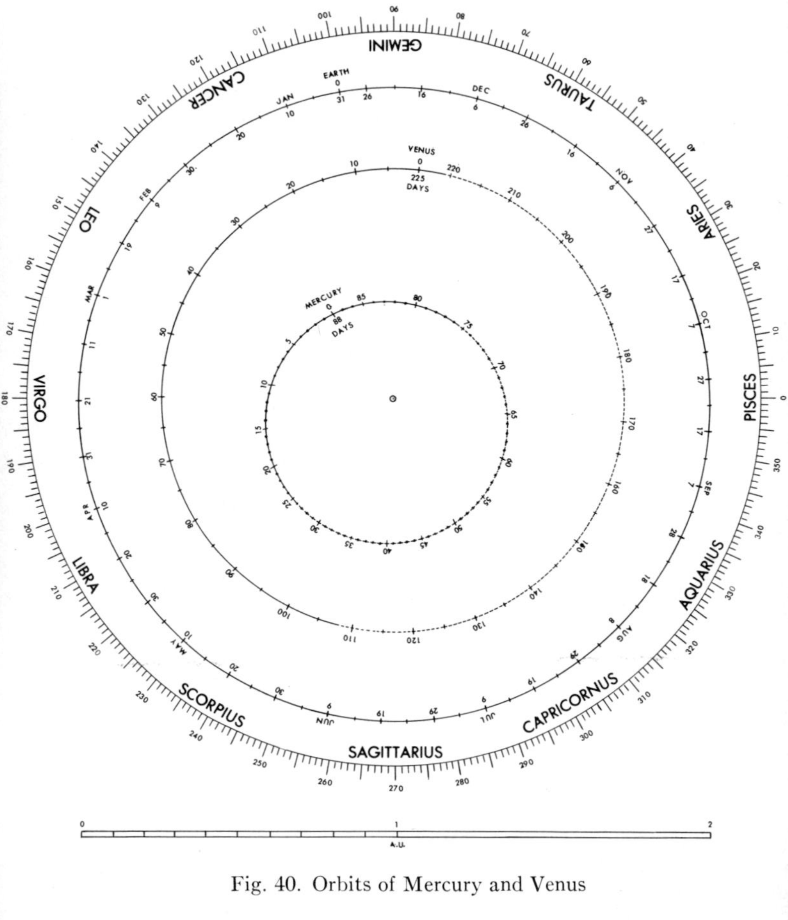

Fig. 40. Orbits of Mercury and Venus

sects this outer circle does not indicate the correct position of Mercury, because the circle does not have an infinite radius. Since the sky circle is centered about the sun, draw a second line through the sun, parallel to the first connecting the earth and Mercury. It intersects the outer circle in the constellation of Aries, at 30°. This number represents the celestial longitude of the planet along the ecliptic.

On the Ecliptic Star Maps (pp. 302–3) you will find a longitude scale. Indicate the position of Mercury by plotting a point at 30°

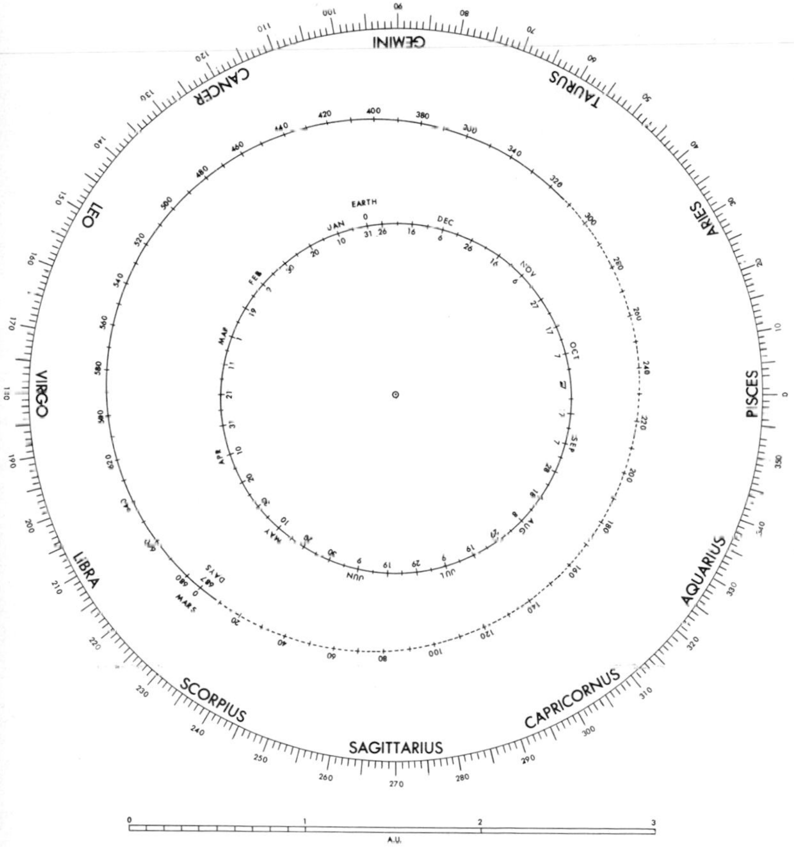

Fig. 41. Orbit of Mars

longitude on one of the Ecliptic Star Maps. This is still not the true position of the planet because it may not lie exactly in the ecliptic. In other words, it may not have zero latitude.

Planetary orbits are slightly tilted with respect to the earth's orbit. Note that in Figures 40–42 each orbit consists of two parts. The portion with the connected line lies north of the earth's orbit; the portion to the south is indicated by a broken line.

Since the point 54 in Figure 40 lies in the broken-line portion of Mercury's orbit, we infer that the planet lies south of the ecliptic.

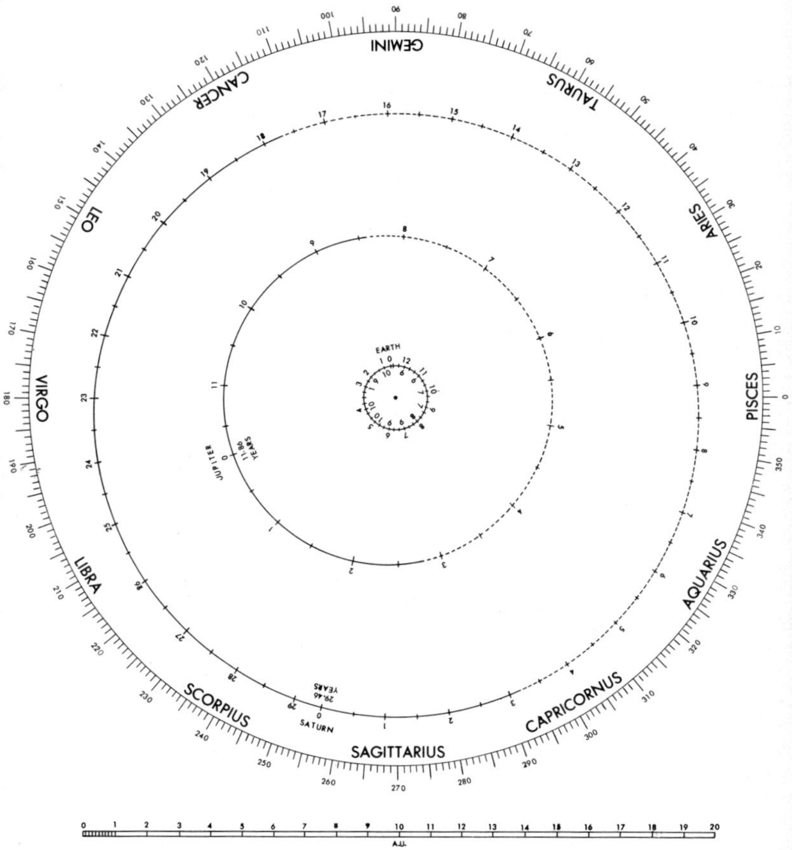

Fig. 42. Orbits of Jupiter and Saturn

To determine the approximate amount, first use the scale at the bottom of Figure 40 to measure the distance from the earth to Mercury. This turns out to be 1.01. Table 19 gives the latitudes (l_0) that the respective planets would have if they lay at the distance of 1 *astronomical unit*, the astronomer's name for the distance of the earth from the sun. From the table, for day 54 we find $l_0 = -2^\circ.9$. The measured distance happens to be so near 1 that the latitude is $-2^\circ.9$ also. If the distance (d) had been 1.3 astronomical units, we should have divided the number by *d* to get the latitude: $l = l_0 \div d$, or $-2^\circ.2$.

The planets Venus and Mars are treated in exactly the same manner. Jupiter and Saturn move more slowly, however, and we find it convenient to use years instead of days. Figure 42 shows the positions of these planets at yearly intervals. The small inner circle represents the earth's orbit. Here the outer figures denote the months, the inner the days of the month. Each division corresponds to 10 days.

Now let us determine the position of Jupiter for May 16, 1965. The interval corresponds to 7.37 years. Since it is less than Jupiter's orbital period of 11.86, found from Table 18, the planet has not yet made a complete revolution. Hence Jupiter lies at 7.37 in Figure 42. Its longitude is about 63°, in the constellation of Taurus, near conjunction with the sun.

Table 20 gives the latitudes of Jupiter and Saturn as seen from the sun. To reduce the given figure to the latitude as seen from the earth, multiply it by the distance of the planet from the sun and divide by the distance of the planet from the earth. The uncertainty is not great, however, even if one uses Table 20 without correction. Hence, for Jupiter at 7.37, the nearest tabular value is $-0^\circ.6$. To get a more accurate figure, estimate between 7.0 and 7.5 the appropriate value for 7.37. This turns out to be $-0^\circ.7$.

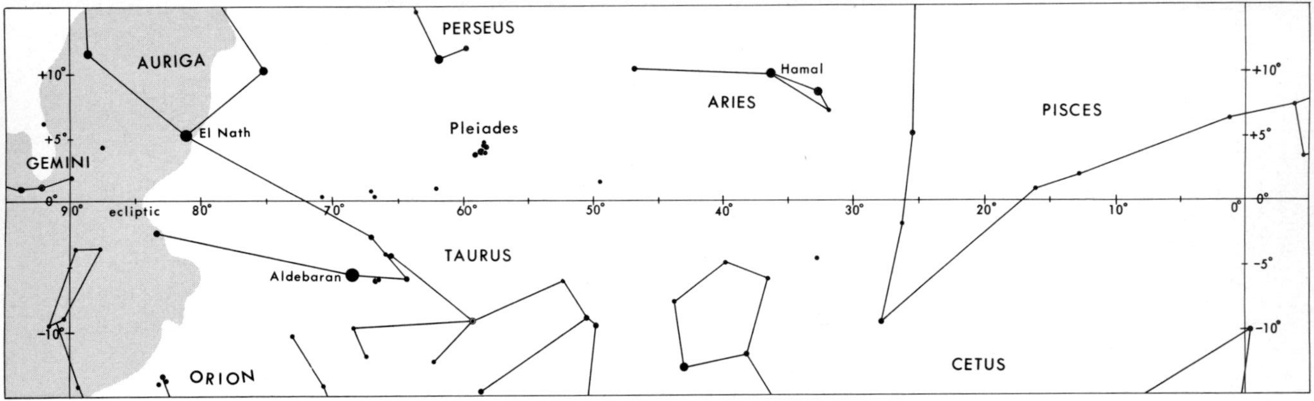
PERSEUS
AURIGA
Hamal
ARIES
PISCES
El Nath
Pleiades
GEMINI
ecliptic
TAURUS
Aldebaran
ORION
CETUS
+10°
+5°
0°
-5°
-10°
90°
80°
70°
60°
50°
40°
30°
20°
10°

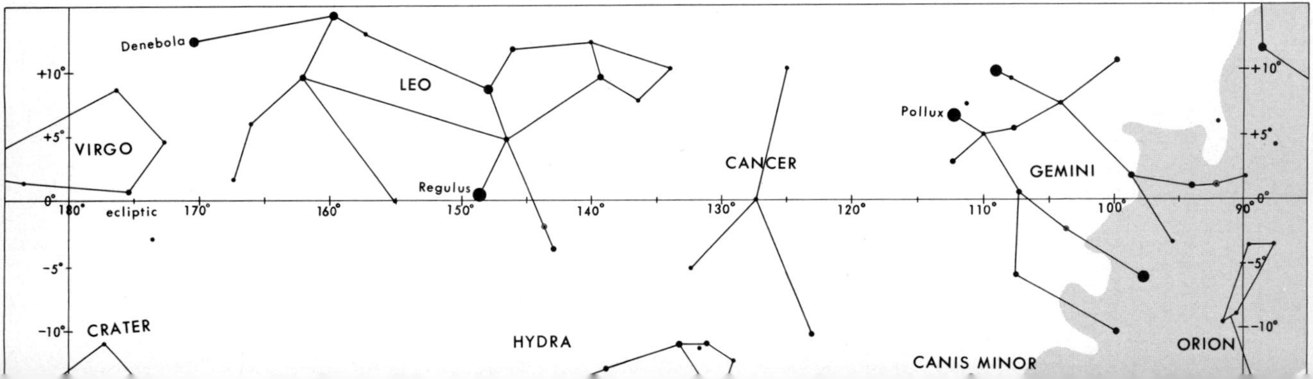
Denebola
LEO
Pollux
VIRGO
CANCER
GEMINI
Regulus
ecliptic
CRATER
HYDRA
ORION
CANIS MINOR
+10°
+5°
0°
-5°
-10°
180°
170°
160°
150°
140°
130°
120°
110°
100°
90°

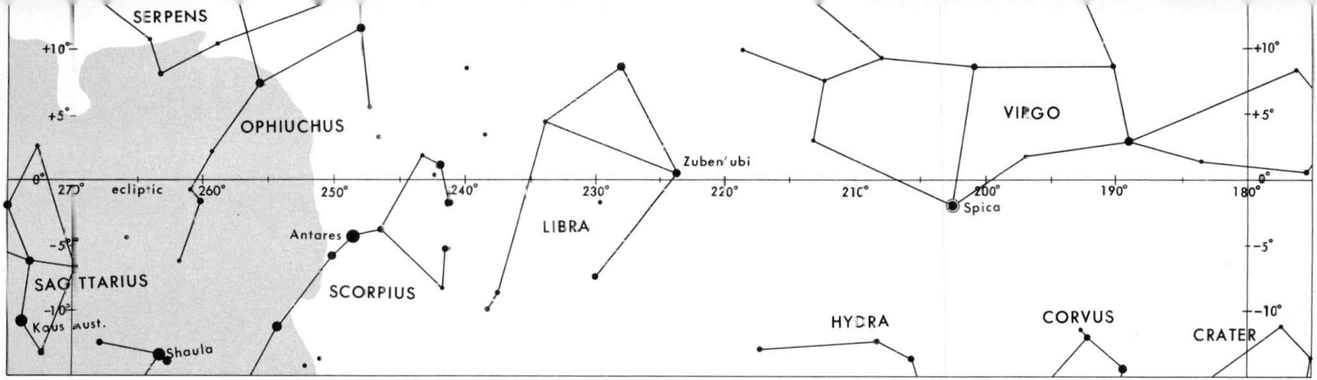
SERPENS
OPHIUCHUS
SAGITTARIUS
Kaus Aust.
Shaula
Antares
SCORPIUS
LIBRA
Zuben'ubi
HYDRA
VIRGO
Spica
CORVUS
CRATER
ecliptic
270°
260°
250°
240°
230°
220°
210°
200°
190°
180°
+10°
+5°
0°
-5°
-10°

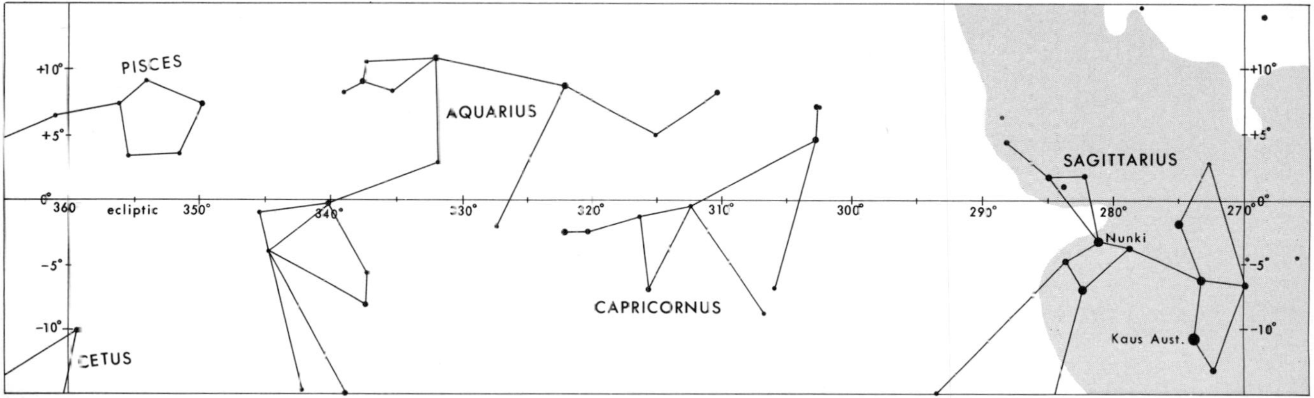
PISCES
CETUS
AQUARIUS
CAPRICORNUS
SAGITTARIUS
Nunki
Kaus Aust.
ecliptic
360
350°
340°
330°
320°
310°
300°
290°
280°
270°
+10°
+5°
0°
-5°
-10°

X

Other Bodies of the Solar System

Asteroids. The solar system contains, in addition to the 9 major planets and their satellites, a large number of lesser planets, or asteroids. These range in size from the 4 largest — Ceres (diameter 480 mi or 770 km), Pallas (300 mi or 490 km), Juno (110 mi or 180 km) and Vesta (240 mi or 380 km) — down to irregular chunks of rock a mile or two across, perhaps the wreckage of a planet that once plied its way between Mars and Jupiter. Astronomers have detected many thousands of these objects and there are doubtless many thousand others that have not come close enough to earth for us to discover them. Ceres and Vesta are easy telescopic objects; Vesta, the brightest, just reaches naked-eye visibility at opposition. The observer who manages to glimpse an asteroid through his telescope, usually with the aid of an almanac that lists its exact position (see *The American Ephemeris and Nautical Almanac*), can readily check his identification in a few hours, while the object slowly drifts with respect to the starry background. Photographs of the sky, especially of regions near the ecliptic, often show short trails — tracks of these drifting asteroids. They are hard to find and therefore are not particularly interesting to the amateur.

Comets and Meteors. The solar system has many comets, which probably are huge chunks of loosely packed ices — frozen gases such as carbon dioxide (dry ice), methane, cyanogen, and ammonia, in addition to ordinary water. Comets usually move in highly elliptical orbits, spending most of their time in the frigid regions far beyond the orbits of the giant planets — even beyond Pluto. Once every 10,000 years or so they come close to the sun, rapidly traverse the inner portion of their orbits, and then speed back out again to the depths of space. During this fleeting visit to the solar neighborhood, the comet encounters sunlight, which melts and evaporates some of the ices. The "wind" of atoms flowing out from the sun — the extremities of the solar corona — catches comet material and blows it out into a long, luminous tail, sometimes stretching millions of miles in a direction away from the sun (Fig. 43).

Bright and spectacular comets are fairly rare, one appearing on the average of every ten years or so. No record exists of former visits of most of these comets, so the time of their return cannot be

Fig. 43.
Halley's Comet

predicted. An exception is the famous Halley's Comet, next due to return about 1986. It now completes an orbital circuit once every 76 years. Presumably it once had a much longer period, but during one of its infrequent visitations to the sun the comet came close to a major planet whose gravitational pull deflected it into its present orbit.

A careful watch of the sky, mainly by amateur astronomers, results in the discovery of comets when they first appear as very faint objects. With this advance notice, we can forecast the appearance of most of the really bright ones.

Comets are named for their discoverers, who thus receive the reward of persistent vigilance. Most searchers use small telescopes called *comet seekers*, with objectives ranging in size from 3 to 5 inches. Use low power; 40 to 60 times is ample. Even good binoculars (7 × 50, for example) are useful. Clock-driven cameras are also effective; see page 320.

Since comets brighten as they near the sun, the best hunting ground for observers is in the west after sunset or in the east before sunrise. Sweep the sky slowly and systematically, looking for patches of haze. At great distances from the sun the comet has scarcely sprouted a perceptible tail. Check these patches with the Photographic Atlas Charts (following p. 139) to make sure that it is not just a nebula or star cluster. Also move the telescope to eliminate the possibility that the object you see is merely a chance reflection from a bright star nearby. If you are sure you have found a comet, notify (preferably by telegram) the nearest large observatory, giving your estimate of the position in right ascension and declination, together with the direction and rate of motion, if determined. As noted on page 320, photographic patrol is also an effective method of detecting comets.

A number of periodic comets are known, with times of revolution from about 3 to more than 10 years; but these comets, in returning to the sun so frequently, have lost most of their ices. A small blob of haze surrounds them, and their tails, if any, are short and stubby.

When the ices have completely evaporated, all that remains is a loose cloud of fine dust and rocks. These aggregates, not massive enough to hold together by mutual gravitational attraction, gradually disperse along the orbits of the burned-out comets. Interplanetary space is full of such debris, some in compact swarms from defunct comets and the remainder widely dispersed. Invisible in space, this cosmic matter becomes evident only when the earth in orbital motion chances to hit it. And then the fragment, heated by friction with the atmosphere, glows and vaporizes. A *shooting star* or *meteor* results. On a clear moonless night, one frequently sees such luminous objects. Most of them are starlike in character, though occasional fireballs, or *bolides*, are more brilliant than Venus. A few attain the brilliance of the full moon. Such spectacular meteors may leave luminous trains, persisting for many minutes. The colors of meteors range from reddish yellow to brilliant green. The latter shade is due to the presence of magnesium (an abundant constituent of many meteors), which glows green when heated to incandescence.

Meteors from a disintegrated comet tend to move parallel to one another in the path and position of the original comet orbit. As the earth in its circuit of the sun intersects the comet orbit, the high concentration of meteors leads to a great increase in the number of shooting stars. And the high incidence of meteors on this occasion is called a *meteor shower*. Moreover, since the meteor paths are essentially parallel, perspective makes us see the luminous paths as intersecting at a point. This point, termed the *radiant*, indicates the direction of the meteor orbit as seen from the earth.

Table 21, adapted from a table by P. M. Millman and D. W. R. McKinley, lists the radiants of various known meteor showers, named after the constellations in which they occur, and the dates of the showers. We now know that many showers also occur during the daylight hours, but we detect them only from the radio-radar-type echoes returned from the hot gases of the meteor track.

Artificial Satellites of the Earth. The advent of man-made satellites opens a new field for the amateur astronomer. Someone must observe these artificial moons if we are to determine their orbits. And here the amateurs, through "Moonwatch," sponsored by the Smithsonian Astrophysical Observatory in Cambridge, Massachusetts, have carried a major responsibility.

The satellites shine, of course, only by reflected sunlight. Since most of them revolve in orbits a mere few hundred miles from the earth's surface, they become invisible as soon as they enter the nighttime shadow. Alas, they are too small and faint to be visible during the day. The best time for observation is that brief time after sunset when the stars begin to appear in the darkened sky and the sun illuminates objects at high levels (Fig. 44).

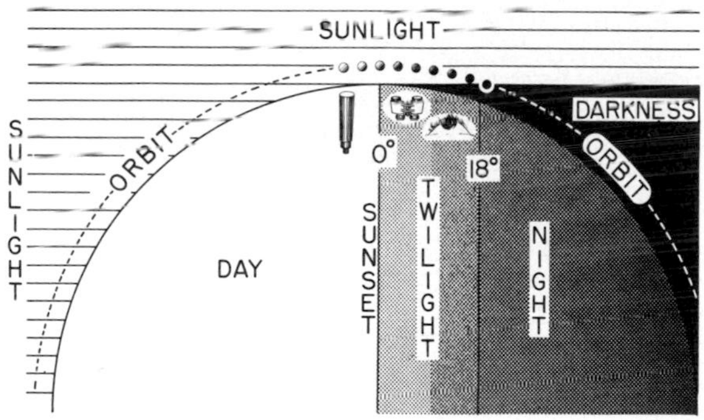

Fig. 44. Satellite observation

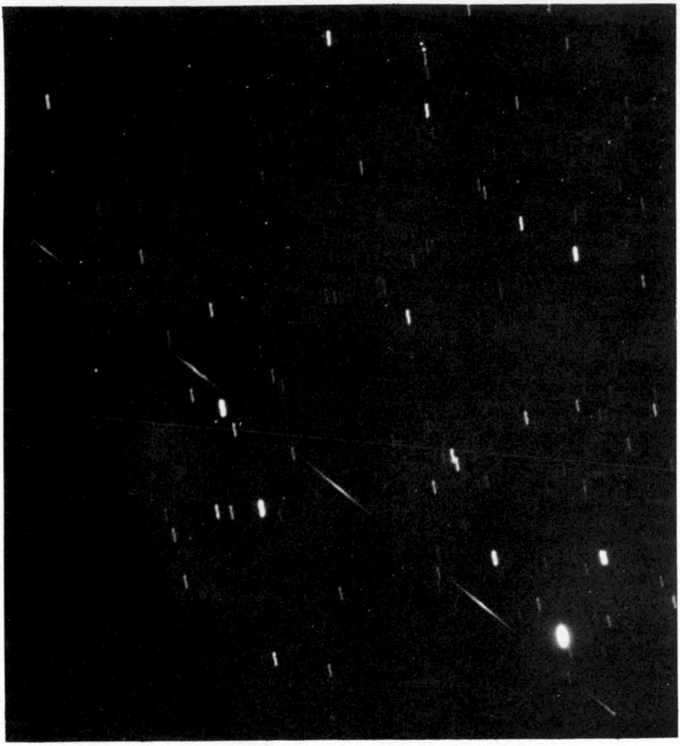

Fig. 45. Satellite photograph: track in Lyra

The brighter satellites to date have occasionally attained 1st magnitude. Although they are clearly visible to the eye under most favorable circumstances, crossing the sky in from 4 to 6 minutes, a binocular greatly assists in seeing them. I particularly recommend those labeled 7 × 50, which means a magnification of 7 and an objective diameter of 50 millimeters, or 2 inches. Low-power telescopes, such as the Balscope, are also useful before the sky has faded into dusk. Use the lowest available power.

Since most of the larger satellites are irregular in shape (cones, rockets, and so on), the amount of sunlight they reflect in the observer's direction varies as the object tumbles end over end. During the time the satellite takes to cross the sky, its brightness may fluctuate up and down by several magnitudes (see Fig. 45).

XI

The Telescope and How to Use It

EVERY PERSON sufficiently interested in the stars to read this book will gain deep satisfaction from owning and using some form of telescope. Don't scorn even the slightest optical aid. That antique opera glass that belonged to great-grandmother is far better than the unaided eye. A field glass or a binocular will reveal still fainter objects. Larger telescopes will disclose even more of the various wonders of the universe, outstanding examples of which appear in Tables 9 through 16.

The simplest telescope consists of two convex lenses, one of long and one of short focal length (Fig. 46). The first of these, the *objective*, forms an image of the objects in front of it. The second lens, or *eyepiece*, acts as a simple magnifier for viewing the image. The fact that the image actually is upside down does not concern the astronomer. To be consistent, he sometimes prints his maps upside down, like the maps of the moon and Mars in this *Field Guide*, where the South Pole is at the top.

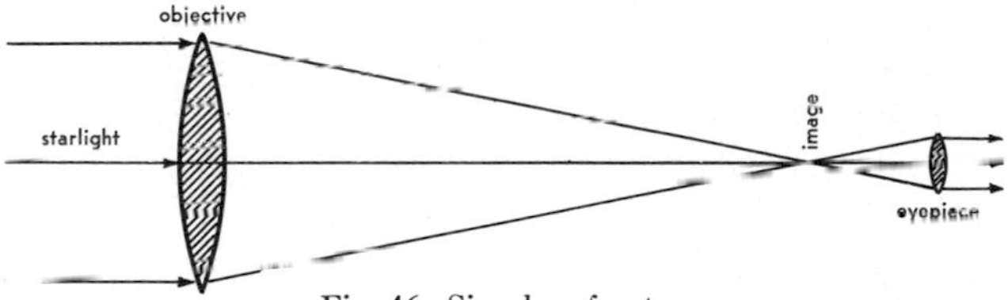

Fig. 46. Simple refractor

One obtains a much clearer view if the eyepiece itself consists of two lenses, the extra lens of focal length equal to or, preferably, slightly shorter than that of the one nearest the eye. This *field lens* lies close to the image formed by the objective (Fig. 47). To change magnification, one merely substitutes an eyepiece of different focal length.

An alternative arrangement employs a concave eyepiece set in front of the image plane. This combination, which gives an upright image, is commonly used in simple field or opera glasses (Fig. 48).

You can make an inexpensive telescope yourself with lenses

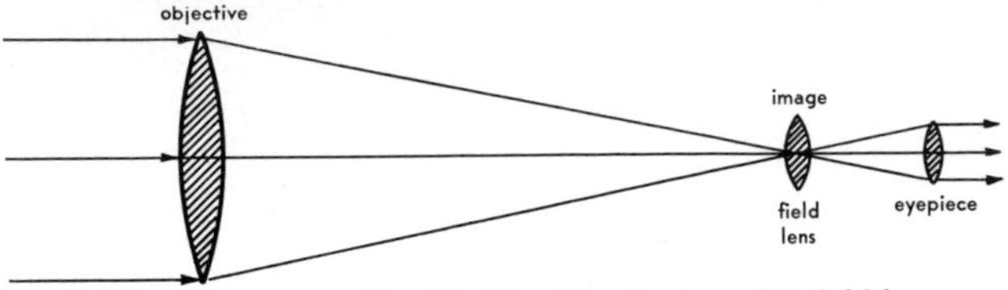

Fig. 47. Simple refractor with field lens

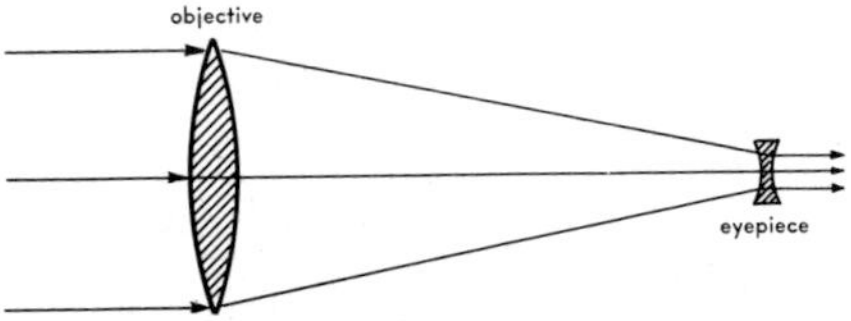

Fig. 48. Simple refractor with concave eyepiece

obtained from your local optician. For an objective, get a spherical positive eye-glass lens of 1-diopter strength (about 40-inch focal length). These blanks are standard, factory-ground, and cheap. Ask for one with its full original circular diameter. For an eyepiece, a twin-lens magnifier is excellent, though more expensive than a pair of simple lenses ½ inch or so in diameter and of 1-inch focal length. The foregoing figures are suggestions. Other focal lengths, especially for the objective, will be acceptable.

Find a strong cardboard mailing tube 38 inches long and slightly larger than the objective. For objectives of other power choose a tube about 2 inches shorter than the focal length. Paint the interior a dull black, say, with india ink; use glue, with small wood blocks or cardboard strips, to mount the objective at one end. If the two sides have unequal curvature, put the more convex one inward. Find or make a smaller cardboard or wooden tube about 8 inches long which will fit snugly into the larger tube, and mount the pair of eyepiece lenses as shown in Figure 47. A piece of velvet glued to the outside of the inner tube gives a firm joint, yet one that easily slides to allow for adjustment of the focus.

All telescopes constructed on the above plan possess a serious defect. A single lens focuses the various colors of light at different points. Hence a star image, instead of being sharp, has a fuzzy colored fringe around it. However, by using two (or more) lenses of different kinds of glass one can partially compensate for this effect and produce an image without color. We call such a lens *achromatic* (Fig. 49). The objective and eyepiece should be of this variety, but the field lens may be simple.

Telescopes constructed on any of the plans just described are known as *refractors*, because they employ bending (or refraction)

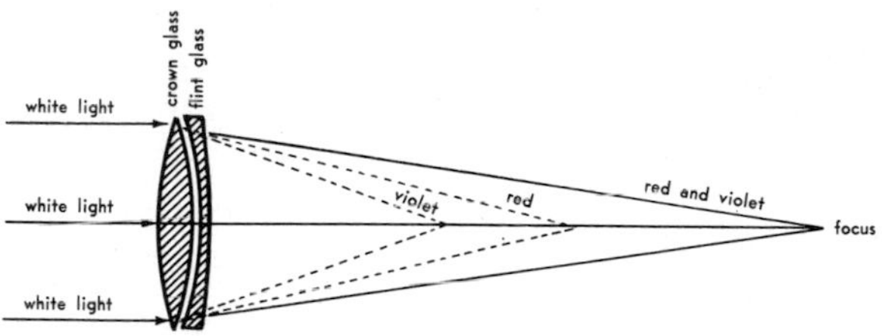

Fig. 49. Achromatic lens

of light to bring the rays to a focus. Instead of refraction, one may use *reflection* to form the primary image. Light falling on the silvered front surface of a concave mirror converges to a focus, where the conventional eyepiece and field lens magnify the image. To get light out of the telescope, one may use the Newtonian form, invented by Sir Isaac Newton himself, which has a diagonal mirror (as shown in Fig. 50).

In addition to refracting and reflecting telescopes there exists a third type, the so-called *catadioptric* variety, a combination of lenses and mirrors. These telescopes are particularly notable because they can produce remarkably sharp photographs, especially over large areas of the sky. The most familiar of these devices is the Schmidt camera (Fig. 51), but there are several other equally satisfactory combinations. Visual catadioptric instruments are generally much shorter than conventional telescopes of equal aperture. They are "folded" telescopes.

How should the amateur go about buying a telescope? He faces a confusingly large number of types, varieties, sizes, and costs. Furthermore, the claims of advertisements may be extravagant or misleading.

The prospective buyer should investigate three basic characteristics of the telescope he considers. They are, in order of importance: (1) perfection of the optical workmanship, (2) the light-gathering power, and (3) the degree of magnification. Of these,

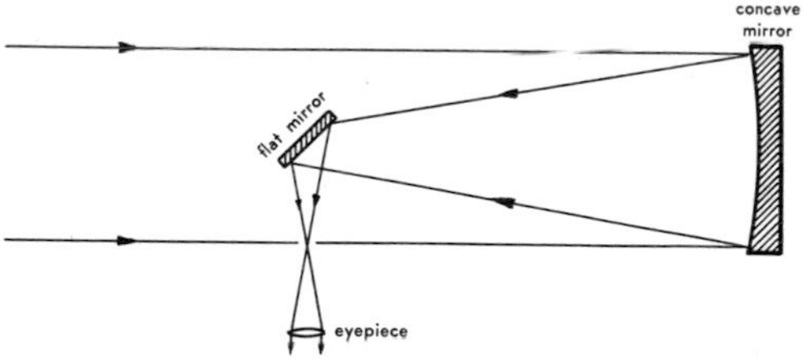

Fig. 50. Newtonian reflector

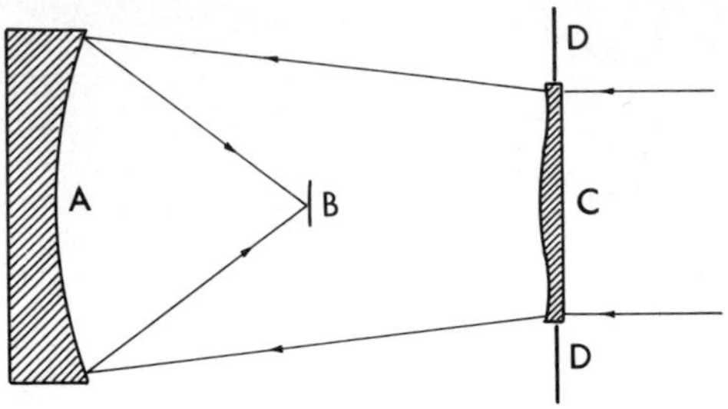

Fig. 51. Basic Schmidt telescope
A, spherical mirror; B, focal surface; C, correcting plate; D, diaphragm

the last is the easiest to determine. To find the magnifying power, divide the focal length of the objective (lens or mirror) by the focal length of the eyepiece. For example, a main lens of 40-inch focus with an eyepiece of 1-inch focus will give a magnification of 40. An eyepiece of ¼-inch focus gives a magnification of 160, and so on.

The maximum allowable magnification is about 100 times for each inch of objective diameter. Hence, with a 3-inch telescope do not employ a power greater than 300. And, under ordinary circumstances, a magnification of half this amount is optimum (see p. 139). The chief limitation is *atmospheric tremor*, an unsteadiness of the image caused by ascending and falling air currents. In short, "twinkling" blurs the stars and planets, limiting the performance of any telescope. Rarely would atmospheric conditions permit the use of a magnifying power greater than 1000, no matter how large or perfect the telescope.

The objective diameter is particularly important because it measures the light-gathering power of a telescope. A 4-inch telescope collects 4×4, or 16 times as much light as a 1-inch instrument, since it has 16 times the area. It will disclose stars 16 times, or 3 magnitudes, fainter than the 1-inch glass does. Large diameters have another advantage too — the ability to separate close double stars or reveal fine detail of the moon and planets. We term this property *resolving power*. To predict the closest double star that a given objective can possibly resolve, divide 4.5 by the lens diameter in inches. The result is the *separation*, in seconds of arc, that a perfect objective should resolve under best atmospheric conditions. For example, a 3-inch telescope should be able just to split the two equally bright components of a double 1″.5 apart.

I recommend that the beginner purchase, if he can afford it, a telescope of from 3 to 4 inches in aperture. Some excellent second-hand ones are available. And the more ambitious amateur can easily grind and polish his own reflector.

These two properties of a telescope — magnifying power and lens diameter — are basic and easy to evaluate. Naturally, a large telescope has advantages over a small one, and several alternative magnifications are highly desirable. The real problem is how to judge the quality of workmanship. One has two choices, reliance on the manufacturer's name or subjecting the instrument to rigorous tests. As an aid to a potential buyer I list in Table 22 several makes of telescopes that I have tested and can recommend. There are, of course, many other reliable products that I have not had the opportunity of testing.

To be safe, purchase your telescope on trial if you can, and then check the quality of its images. Stars are better test objects than the moon or planets. Turn the instrument toward a bright star, preferably one well above the horizon, and note the sharpness of the image with various eyepieces. You should see a sharp central core, perhaps encircled by one or more faint rings of light.

Elliptical or elongated images suggest that the lens may have astigmatism. By all means beware of instruments whose images are irregular, pointed, or flared. As a further test, try rotating the eyepiece alone. If the distorted pattern also turns, then the fault lies in the eyepiece. If the pattern remains stationary, rotate the telescope as a whole. Should the pattern now turn, the objective is definitely either astigmatic or improperly mounted; but if it remains fixed while you rotate the instrument, you had better see an oculist. In all probability the astigmatism or other defect is in your own eye.

Look at a bright star with your naked eye. Is the image sharp or distorted? Do not be concerned if the star appears to possess a few minute points. Most people have some minor eye imperfections that add points to the stars. In using a telescope, keep your eye as close as possible to the eyepiece. Nearsighted or farsighted persons will probably prefer to remove their glasses and compensate by refocusing the eyepiece. Those who have appreciable astigmatism, however, should keep their glasses on, because the telescope cannot compensate for this eye defect.

If the star image never appears sharp and displays a colored fringe that changes from red to blue as you rack the eyepiece in or out, the telescope is not properly corrected for *chromatic aberration*, as we term this defect characterizing simple lenses. As a substitute for a star you may use a pinhole, illuminated from behind with a candle or flashlight and set 200 to 300 feet away. Even then you will have to adjust the eyepiece to find the focus.

The so-called Foucault test is a sensitive and critical way of checking a telescope objective. Point the instrument at the distant illuminated pinhole, clamp it in position, and then remove the eyepiece. Now place your eye close to the position where the eyepiece was; you should see the objective illuminated as a bright circle of light.

One final step will complete your test of the telescope lens. To make a simple device for this purpose, tape an old razor blade to a book end, and place it at the height of the eyepiece. Bring your eye close to the position where the eyepiece was so that you see a brightly illuminated objective (lens or mirror). Now slide the sharp edge of the blade into the beam until it cuts out the light from the objective. Note whether the shadow cuts from the left or from the right. If the blade is inside the focus the shadow will move across in the same direction you are moving the blade; see *A* in Figure 52. If the blade is outside the focus, the shadow will cross in the opposite direction, as at *B* in Figure 52. Try to find some position where the illumination cuts

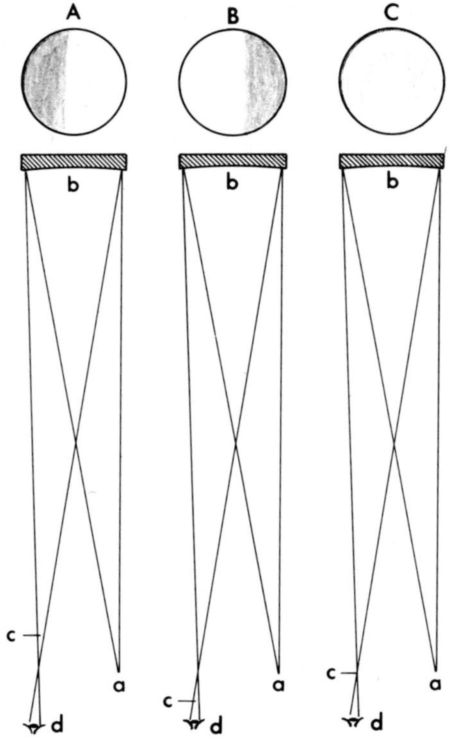

Fig. 52. Foucault test. A, B, C, images; a, point source of light; b, parabolic mirror; c, razor blade edge; d, eye

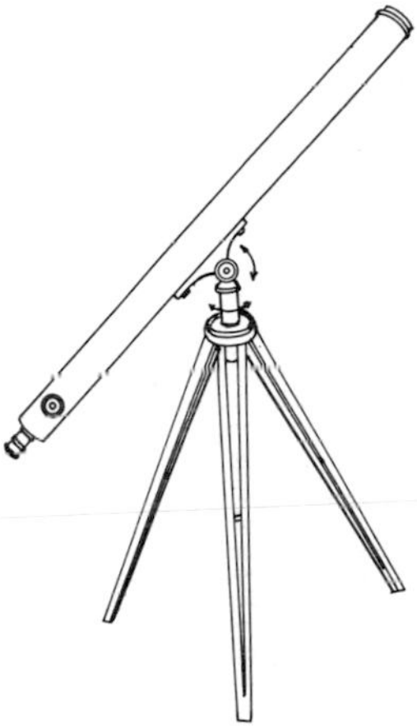

Fig. 53.
Altazimuth mounting

off simultaneously over the entire objective; see *C* in Figure 52. When you have done this, the razor blade is set at the focus. If you can find no place where the illumination disappears simultaneously and completely, the lens you are testing may be defective. This procedure is sometimes called the "knife edge" test. Some zones of the lens may remain bright while the others darken. Such an objective is nonuniform and should be rejected. For best results, test the lens in a cool, dark basement with a tube shielding the entire optical path. This procedure will reduce convection of the atmosphere and tend to eliminate the shimmering of the objective when the razor blade edge is at the focus, an effect due to air currents rather than to a faulty lens or mirror.

If the image shows colored fringes, the lens, as noted earlier, probably has chromatic aberration. A mirror possesses this one advantage over a lens in that it has no chromatic aberration. A mirror reflects all colors of light at the same angle. Firmly reject any telescope whose objective lens or mirror does not measure up to this Foucault test.

A steady mount is vital for the astronomer. A flimsy tripod that vibrates to the touch or sways in a light breeze will cause the image of star or planet to oscillate and blur. The higher the magnification of the telescope the greater will be the blurring from this source. Some amateurs provide a permanent base, an upright concrete pillar or a heavy wooden post, with only the telescope and its bearings as the portable components.

Mountings are of two basic types, *altazimuth* and *equatorial*. The altazimuth mounting, which is the simpler, serves adequately for telescopes with objectives up to 3 inches in diameter. It consists

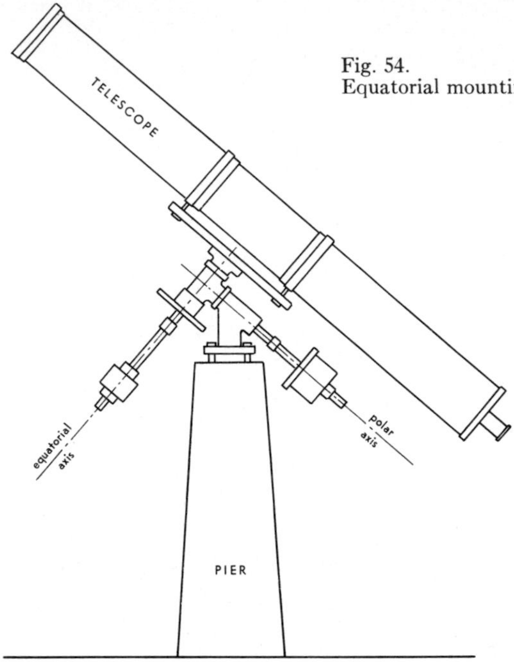

Fig. 54.
Equatorial mounting

of an upright post and vertical axis so that the telescope can be pointed in any desired compass direction (azimuth). The instrument is held in some sort of yoke that permits the observer to tilt the telescope at different angles above the horizon. Since the stars are in apparent motion, because of the earth's rotation, the observer must be able to change the position of the telescope in both altitude and azimuth.

However, if the originally vertical axis is tilted until it points toward the celestial pole, about which all stars appear to circle, then merely turning the telescope about this polar axis keeps a given star in the field of view (Fig. 54). With some ingenuity one may attach a clockwork drive to the polar axis, so that it revolves once in 24 hours sidereal time or about 23 hours and 56 minutes solar time. Most large telescopes use some form of this equatorial mounting, and long-exposure photographs require its use.

XII

Photography in Astronomy

In this day of Sputniks and Outer Space, man is just beginning to become acquainted with his universe. The starry skies have suddenly captured his imagination. In those skies man finds beauty and wonder, some part of which he can record with his camera.

Most people do not know that the astronomer's photographic telescope is in every sense a camera — often of gigantic size, but still a camera. Also, few persons realize that astronomers may employ small cameras for special purposes. Anyone may take interesting pictures of the sky and various celestial phenomena with his own camera.

The sky contains approximately 40,000 square degrees as we measure it. A 35-millimeter camera with standard 50-millimeter lens covers an area some 38° by 27°, roughly 1000 square degrees. Theoretically, one can cover the sky with about 40 photographs. To provide for overlap, the minimum practical number is 54, with centers chosen as in the Photographic Atlas Charts (see Table 17).

As a first experiment, load your camera with the fastest black-and-white film available, focus for infinity, set the shutter for a time exposure, and open the lens to its maximum. Choose a clear, moonless night. Mount your camera on a steady tripod and tilt it, or lay it against a slanted brick or book so that it points toward the Pole Star. Then open the shutter and expose for 6 or 8 hours — all night if possible. Be sure, however, to close the shutter before the first glow of dawn appears.

The stars, slowly circling the pole as the earth turns on its axis, trace long arcs which record on the photograph. The short bright arc near the center is the Pole Star (see Fig. 55). Many of the tracks representing stars below the limit of naked-eye visibility are revealed by the sensitivity of the photograph.

On a second night, point the camera at right angles to its original position and repeat the process, with an exposure of only 1 hour. In this region of the sky the trails appear as essentially straight lines.

A simple experiment will give you the basic information you will need for other types of pictures. Set up the camera as described above, point it toward some interesting region of the sky — one with many bright stars — and cover the lens with a piece of

Fig. 55. Circumpolar star trails

cardboard, preferably black. Open the shutter and keep it open for the duration of the test. You will use the card, instead of the camera's shutter, to control exposures. Now, taking care not to shake the camera, uncover the lens for about 1 second. Wait 2 minutes, then uncover it for 2 seconds; wait another 2 minutes and uncover the lens for 4 seconds. Continue the process, doubling the time of exposure with each photograph until you have made 7 pictures; then close the shutter. You now have a variety of exposures (1, 2, 4, 8, 16, 32, and 64 seconds) made at 2-minute intervals. When this film is developed and printed, the stars will appear as dots on the first short exposures and as streaks on the later ones. From your own records you can decide what exposure would give the most effective pictorial rendition of the sky. The decision is a compromise. The stars will appear rounder and therefore more natural in the shorter exposures. But the longer exposures will reveal fainter stars. Even though the star images may not be quite round, this simple procedure enables one to get fairly good pictures of the constellations. Fast black-and-white film like Eastman Tri-X or Royal Pan will give the best rendition of the constellations. But various color films, especially the high-speed Ektachrome, Kodachrome, Agfachrome, and Polacolor, will give unusual effects. The brighter red stars give colored images. Lights on a distant horizon can also produce beautiful effects.

I have found the Polaroid Land camera with the 3000 film very useful. For astronomical work I prefer Model 180, with the lens wide open at *f*/4.5. A 4-second exposure records stars down to or even fainter than the naked-eye limit (see Fig. 56). Be sure to go where the sky is not brightened by city lights. Cameras of the 200 or 300 series have *f*/8.8 lenses and hence are about 4 times slower than the Model 180. Try a 16-second exposure and use a cable release so as not to shake the camera, which should be set on a very firm base. Place a card in front of the window that controls the exposure, to keep the shutter open. Or shine a small flashlight

Fig. 56. Photograph of Big Dipper taken with Polaroid Land camera

into the window to close the shutter. A number of camera manufacturers supply special cassettes to hold Polaroid film packs.

If you happen to know that one of the brighter man-made satellites is due to make a transit, set your camera on a sturdy base, turned toward the predicted region of the sky, and open the shutter a minute or so before the scheduled time and close it after the satellite has passed by. The stars will make only short trails, whereas the satellite will appear as a streak across the picture, perhaps variable in intensity as the object tumbles in its orbit.

For most effective photography, one may wish to expose the film an hour or more. For long exposures there is no substitute for a polar axis and clock drive (see Fig. 54, p. 316), to make the camera follow the moving stars. Although you can now buy at reasonable prices small telescopes fitted with such mountings, with a little ingenuity you can "do-it-yourself," and devise a satisfactory camera mount and drive.

As previously noted, small clock-driven cameras are useful as comet seekers. Simply direct the camera toward the western horizon after sunset or toward the eastern horizon before sunrise and take a series of pictures. Comets will show up as fuzzy patches not identifiable with one of the nebulae or clusters shown on the Photographic Atlas Charts. Since occasional photographic defects occur, be sure to take overlapping photographs, to check the reality of any object detected.

A clock-driven telescope may also be adapted to other photographic uses. If possible, detach the standard camera lens and use the telescope objective as a telephoto lens. Attach the camera so that the film lies in the focal plane. To determine the position of accurate focus, provide some means of moving the camera in or out, and take a series of pictures of a bright star. You can readily determine which position gives the sharpest image.

With a reflex camera one can gauge the focus from the projected image on the ground glass. Leitz makes a device termed a Micro-Ibso that can be used to attach a Leica to either a microscope or a telescope and provide ground-glass focusing.

For certain types of astronomical photographs, especially of the moon, sun, or brighter planets, one may use the eyepiece of the telescope to project an enlarged image on the film, as described for observing the sun, on page 285. This procedure increases the effective *f* number of the telescope. To calculate this *f* number approximately, measure the diameter of the solar or lunar image, multiply by 120, and divide by the diameter of the objective. Be sure to use the same measuring unit (inches or centimeters) for the two diameters. For example, if a 3-inch telescope gives a 5-inch solar image, the effective *f* number is $5 \times 120/3 = 200$, or $f/200$. The larger this figure, of course, the longer the required exposure will be.

Photography of the stars does provide a severe test for the lens of your camera or of your telescope. The smaller the *f* number the more severe the test. Do not be disturbed, therefore, if the images toward the edge of the field are slightly distorted, as long as the central ones are in sharp focus. You may improve the definition by using a smaller aperture. Try several settings.

To photograph the full moon with film speed ASA 200 or 24 DIN at *f*/120 (image and lens diameters equal), the exposure should be about ¼ second. Try several different exposure times to ensure a picture of the right density. A crescent moon will require from 2 to 3 times longer exposure than the full moon. For other films, adjust the exposures to conform to the relative speed.

Photography of the sun is difficult only because the sun is so bright. It is necessary to dim the image by the use of a very dark filter, of density 3 or more. This figure signifies that only 1/1000 (1 part in 10^3 of the incident light) goes through. One should be able to view the sun comfortably through such a filter. The dark glass described on page 286 for direct solar observation is satisfactory, if set near the primary focus of the objective. I prefer a neutral Wratten gelatin filter of density 4 (an Eastman Kodak product) set loosely just in front of the film (or plate). Then use the shortest exposure time available on the camera. As for the moon, you must experiment to find the best settings. Do not put the filter in front of the objective unless it is of high optical quality.

One more word of caution. Never let your camera point directly toward the sun. The heat rays, concentrated by the camera lens, can melt the rubber blades of the shutter or burn a hole in a focal-plane shutter. With a telescopic lens system operating at *f*/120, however, you can safely allow the direct rays to enter the camera. Even then, however, use a lens cap to keep the excess heat out of the camera, except when you are actually focusing or snapping the picture.

Fast color film used with a 35-millimeter camera and fast lens can give remarkably beautiful pictures. The colors of the bright stars will generally register in time exposures up to a minute or so. The illuminated horizon of a distant city may provide a very interesting background. Trees silhouetted against the faint glow of a fading sunset can also be spectacular. The stars in a dark sky over a brightly lighted courtyard give a remarkable effect. If possible, try different lenses, from telephoto to wide-angle.

In Table 23 I list selected cameras, all of which I have tested and found satisfactory for photographing the sky. Most cameras, however, will yield interesting photographs.

XIII
Time

INHABITANTS of the earth possess two natural units of time: the day and the year. One complete rotation of the earth on its axis measures the day. One full revolution of the earth about the sun, in 365.242195 days, defines the year.

The Year and the Calendar. The fact that the year does not contain an exact number of days posed a problem for the makers of calendars. How were they to keep the seasons from slowly drifting forward or backward with passing time? They resolved the difficulty by saving the extra decimal fraction of a day, 0.2422, for 4 years, and every 4th year they added 1 day to the length of the year to make leap year, the extra day being given to February. The available excess of $4 \times 0.2422 = 0.9688$ day failed by 0.0312 — or about 45 minutes — of being a full day.

Three-quarters of an hour in four years may seem not worth while bothering about. And so concluded the astronomer Sosigenes, who invented the leap year at Julius Caesar's behest in 46 B.C. By 1582, however, the lost minutes had accumulated so that the calendar year had slipped backward a full 12 days from the year measured by the sun, at the rate of 3.12 days per 400 years. Pope Gregory, acting on scientific advice, decreed that 10 of these days should be restored to the calendar. Then, to minimize a similar drift in the future, he further ruled that the "century years," such as 1900 or 2000, would not contain their allotted extra day unless divisible by 400. Hence, 1700, 1800, and 1900 turned back three leap days, reducing the discrepancy to a mere 0.12 day per 400 years. Our distant descendants could reduce the discrepancy still further by deciding that, contrary to the present rules, 4000 and its multiples would not be leap years.

Solar Time. The year and the calendar are relatively simple compared with time itself. Man originally defined noon as the moment when the sun reached the meridian and attained its highest altitude for the day. He further divided the interval between that moment and the next meridian passage into 24 hours.

This definition introduced certain difficulties, because the days are not of equal length. The earth's orbit is elliptical, not circular, and the earth moves more rapidly when near the sun and more slowly when farther away. Also, the earth's axis of rotation is tipped relative to the orbit. These two phenomena cooperate to make the *apparent* sun first run ahead of and then lag behind its average position. To avoid the inconvenience of changing the rate of our clocks from day to day, we employ *mean* instead of *apparent solar time* for practical use.

Standard Time. The earth rotates from west to east. Hence, if the sun is exactly on your meridian, it will not yet have reached the meridian of a person a few miles to the west. His *local time* will be slower than yours by an amount that depends on the difference in longitude. To keep the correct time as measured by the sun, a man traveling westward would continuously have to move back the hands of his watch.

How confusing it would be if each community operated on its own local time! To simplify matters, nations have generally agreed to employ, on an earth-wide basis, only 24 varieties of time, differing by whole hours from one another. We subdivide the 360° of longitude around the equator into 24 intervals of 15°, one for each hour. Thus the *zero meridian*, the meridian of Greenwich, is centered on one of these 15-degree strips. Within each segment of the earth, time remains constant; across the boundary of the neighboring 15° segment, time shifts by an hour, and so on, in orderly progression. A glance at a map of *standard time* zones, however, shows that the boundaries actually adopted are often quite irregular, to suit the convenience of individual communities. Eastern Standard Time is 5 hours earlier than Greenwich Standard Time, Central Standard is 6 hours earlier, Mountain Standard 7, Pacific Standard 8, and so on. These figures, which represent the difference between Greenwich Standard Time and your own standard time, we shall refer to as the Standard Longitude Difference, abbreviated as SLD.

Rules for Computing Time. Greenwich Mean Time (GMT), which is also called Greenwich Standard Time, Greenwich Civil Time, and Universal Time (UT), is the basic reference time used for most astronomical work. To avoid the cumbersome A.M. and P.M., astronomers number the hours from 0 to 24, starting with midnight.

To derive GMT, add to your own Standard Mean Time (SMT) the number of whole hours corresponding to the Standard Longitude Difference (SLD): GMT = SMT + SLD. For example, if you live anywhere in the zone of Central Standard Time, SLD = 6 hours. If your SMT is $21^h\,32^m$, GMT = $21^h\,32^m + 6^h\,00^m = 27^h\,32^m$. Subtract 24 hours if necessary, in order to give a result in the normal range from 0 to 24: $27^h\,32^m = 3^h\,32^m$.

If you live *east* instead of west of Greenwich, the SLD is negative. To add a negative number to the SMT, simply subtract the number itself. If this (negative) SLD has a larger value than the SMT, SMT + SLD will then be negative. In such cases, add 24 hours to the SMT before subtracting, in order to keep the GMT within the normal range.

To calculate your Local Mean (solar) Time (LMT), first find your longitude from a map, the encyclopedia, or your post office. Divide the longitude by 15, and obtain the result in hours and minutes (also seconds, if desired). Call this figure the Local Longitude Difference (LLD). Then LMT = GMT − LLD. If you live east of Greenwich, the LLD is negative and you would add rather than subtract.

The following alternative formula is often useful: LMT = SMT + SLD − LLD. Here is an example. Suppose that you live in Cleveland, Ohio, longitude 81° 45′ W. Divide by 15 to obtain the Local Longitude Difference of $5^h\,27^m$. Since Cleveland operates on Eastern Standard Time, SLD = 5^h. So if the standard clock reads 3:49 P.M. ($15^h\,49^m$), the Local Mean Time is: LMT = $(15^h\,49^m) + (5^h) - (5^h\,27^m) = 15^h\,22^m$. If necessary, add or subtract 24 hours, as explained above.

Daylight Time. Many areas of the country adopt Daylight Time (DT, originally Daylight Saving Time) during the summer months to provide an extra hour of daylight for recreation. To obtain Daylight Time add 1 hour to Standard Mean Time, or DT = SMT + 1^h.

Apparent Solar Time. The difference between Local Apparent Time (LAT), or "sundial" time, and Local Mean Time we call the Equation of Time (ET): LAT = LMT + ET. Table 24, column 2, indicates that this difference may amount to some 16 minutes. Although the tabulated values are precise only for 1959, they apply approximately to within a minute or two for any other year. If greater accuracy is desired, refer to one of the many available almanacs for the specific year.

Sidereal or Star Time. The earth's revolution in its orbit causes the sun to drift eastward approximately 1° per day with respect to the stars. Thus, at the end of a year, when the earth has completed 365.2422 rotations with reference to the sun, it has made exactly 1 additional rotation with reference to the stars; and so our year contains 366.2422 sidereal, or star, days. A *sidereal day*, in consequence, is about 4 minutes shorter than a solar day (24 hours divided by 366.2422).

On or about March 21, when the sidereal and solar years start, sidereal "noon" (or zero hours sidereal time) occurs at solar noon (or at 12 hours on the solar clock). Hence the sidereal and solar clocks, whose time scales run consecutively from 0 hours to 23 hours and 59 minutes, do not really coincide until 6 months later, at the autumnal equinox, on or about September 21. By this time the sidereal clock has gained 12 hours on the solar clock and both read midnight, or $0^h\,00^m$. The sidereal clock continues to gain on a mean-time solar clock by $3^m\,56^s.555$ of sidereal time for each solar day.

To calculate the sidereal time approximately, figure out the number of months and days that have elapsed since the autumnal equinox. Allow 2 hours for each month and 4 minutes for each day in excess of the months. Add this figure to the Greenwich Mean Time to get a rough estimate of the Greenwich Sidereal Time. For example, on May 29 at $17^h\ 30^m$ GMT, 8 months and 8 days after the autumnal equinox, figure as follows:

8 months at 2^h per month	=	16^h
8 days at 4^m per day	=	$0^h\ 32^m$
GMT	=	$17^h\ 30^m$
Total	=	$33^h\ 62^m$

Subtract 24 hours to obtain the approximate Greenwich Sidereal Time: GST = $10^h\ 02^m$.

Sidereal time is the elapsed sidereal interval since the *vernal equinox* was last on the meridian. This equinox is the point on the celestial sphere where the sun crosses the equator in the spring. Sometimes called the first point of Aries, it is often designated by the zodiacal symbol ♈.

To obtain a more accurate value of the sidereal time at a given instant we have recourse to Tables 24 and 25. As noted on page 329, precession causes the equinox to drift slowly with reference to the stars. This book employs *uniform* sidereal time, corrected for precession but not for the irregularities of nutation (see p. 331). The error so introduced cannot exceed 3/5 of a second of time.

The extra fraction of a sidereal day poses difficulties similar to those encountered in connection with the calendar. Let us take an actual example. Refer to Table 24, which was calculated for an arbitrarily chosen year, 1959. At 0^h GMT (Greenwich mean midnight) for January 0, 1959, the tabulated sidereal time is $6^h\ 35^m\ 39^s\!.05$. January 0 is shorthand for December 31 of the previous year and 0^h refers to the midnight that *started* the day of December 31. Hence the indicated time is 1 day before actual New Year's. Exactly 365 days later, again at Greenwich midnight prior to December 31, 1960, the solar year has 0.242195 days still to run (as was pointed out on p. 322). One full day yields a sidereal discrepancy of $3^m\ 56^s\!.555$ or $236^s\!.555$ (the gain per day of sidereal time over mean solar time). Multiply this figure by the fractional daily excess of the solar year, or $236.555 \times 0.242195 = 57.29$ seconds. Thus, the sidereal time of Greenwich mean midnight for January 0, 1960, occurs at $6^h\ 34^m\ 41^s\!.76$ (the tabulated value for 1959 less $57^s\!.29$) in agreement with the second entry of Table 25.

This discrepancy would accumulate indefinitely, except for the provision that 1960 is a leap year. Suddenly, on February 29, we insert an extra day of $3^m\ 56^s\!.55$. The discrepancy now becomes an excess for the rest of the year, $3^m\ 56^s\!.55 - 0^m\ 57^s\!.29 = 2^m\ 59^s\!.26$, as shown in the second column of Table 25. For 1961, this difference is further reduced by $0^m\ 57^s\!.29$ to $2^m\ 1^s\!.97$, as tabulated in the second column of Table 25. And so on — indefinitely. The tabulated figures could be similarly extended for many years into the future. The other index corrections given in Table 25 can be applied to the values of Table 24 to reduce them to the year specified.

Let us determine the local sidereal time (LST) at Cleveland, Ohio, longitude $5^h\ 27^m$ W (see p. 324) at $12^h\ 30^m\ 21^s$ Eastern Standard Time, May 29, 1974. Add 5 hours (the Standard Longitude Difference) to obtain the GMT, or Universal Time (UT), of $17^h\ 30^m\ 21^s$. From Table 24, we note the sidereal time of 0^h GMT for May 25, the previous tabulated day. The interval between the tabulated day and the given sidereal time is $4^d\ 17^h\ 30^m\ 21^s$. During this interval the sidereal time has been advancing at the rate of $3^m\ 56^s.555$ per day, $9^s.8565$ per hour, $0^s.1643$ per minute, or $0^s.0027$ per second.

To find the Local Sidereal Time, make the following calculation, step by step:

Tabulated value for May 25, 1959 (from Table 24)	$16^h\ 07^m\ 19^s.58$
Code for 1974 (Table 25)	1 26.84
4^d at $3^m\ 56^s.555$ per day	15 46.22
17^h at $9^s.8565$ per hour	2 47.56
30^m at $0^s.1643$ per minute	4.93
21^s at $0^s.0027$ per second	0.06
GMT or UT	17 30 21.00
Total	$33^h\ 57^m\ 46^s.19$
Less	24
Greenwich Sidereal Time	$9^h\ 57^m\ 46^s.19$
Local Longitude Difference	$5^h\ 27^m$
Local Sidereal Time	$4^h\ 30^m\ 46^s.19$

It is also interesting to compare the exact value GST of $9^h\ 57^m\ 46^s$ with the approximate figure of $10^h\ 02^m$ previously calculated.

How to Tell Time by the Stars. He who knows the constellations can estimate the time of night by a casual glance at the sky. The stars provide an accurate clock, which an astronomer can use to gauge the time to within a thousandth of a second or better — if he has the technical skill and telescopic equipment available. Anyone, however, can tell time to within a few minutes by applying some simple rules.

The sky rotates once every 24 hours — or, rather, 23 hours and 56 minutes. The 4 minutes a day (actually 3 minutes and 56 seconds) amount in the course of 365 days to enough to account for the fact that the earth, in revolving around the sun, has one more axial rotation with respect to the stars than to the sun.

For observers in the Northern Hemisphere, the Pointers of the Big Dipper provide an excellent star clock (Fig. 57). Imagine an hour hand running from the North Star through the Pointers. Around the North Pole further imagine the face of a clock, with 24 hours indicated on it rather than the conventional 12; and this clock runs backward, you might say, since the hour hand turns in a counterclockwise fashion.

Whenever these Pointers are straight up — i.e., on the meridian — the *star time*, or *sidereal time*, is 11 hours (actually 10 hours 56 minutes). During an ordinary year of 365 days the hour hand of the star clock makes 366 complete turns. This means that our star clock appears to run fast, by the 4 minutes per day described

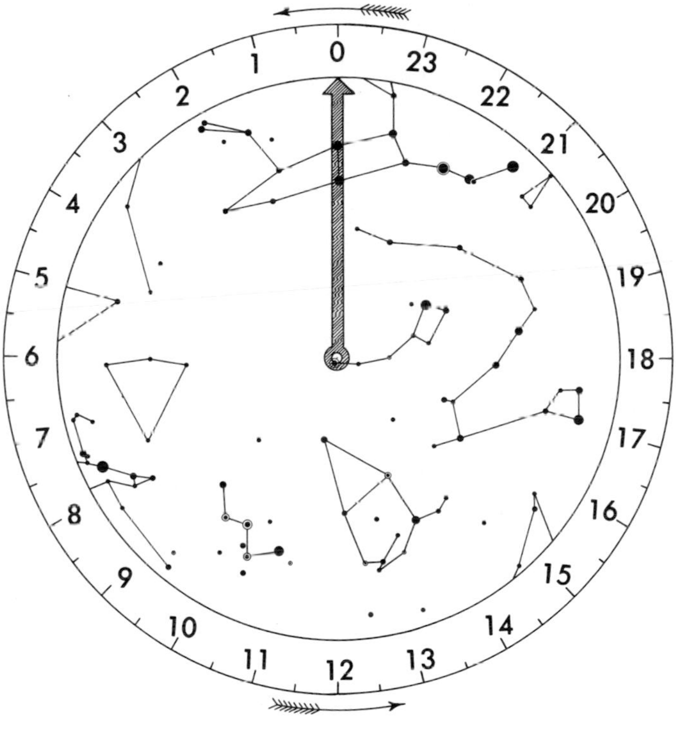

Fig. 57. Northern star clock

above. As explained on page 324, star time agrees with ordinary solar time on only one day each year, about September 21.

At midnight of March 4 of any year, our sky clock appears to read 0 hours (or midnight) and therefore corresponds with ordinary time (see Fig. 57). From then on, however, the clock gains a steady 4 minutes a day (or 1 hour every 2 weeks, or 2 hours every month) in terms of solar, or ordinary, time. Hence, on March 21, when the star clocks reads 23^h, the solar time is 1 hour earlier, or 22^h (10 P.M.), and so on. If the Pointers indicate 10 hours, 15 minutes on November 2, we subtract 14 hours for the 7-month March 21–October 21 interval and 48 minutes for the additional 12 days. The sun time becomes 19 hours, 27 minutes, or 7:27 P.M.

The Southern Hemisphere provides no conspicuous hour hand for our clock. The one here drawn, Figure 58, indicates the zero

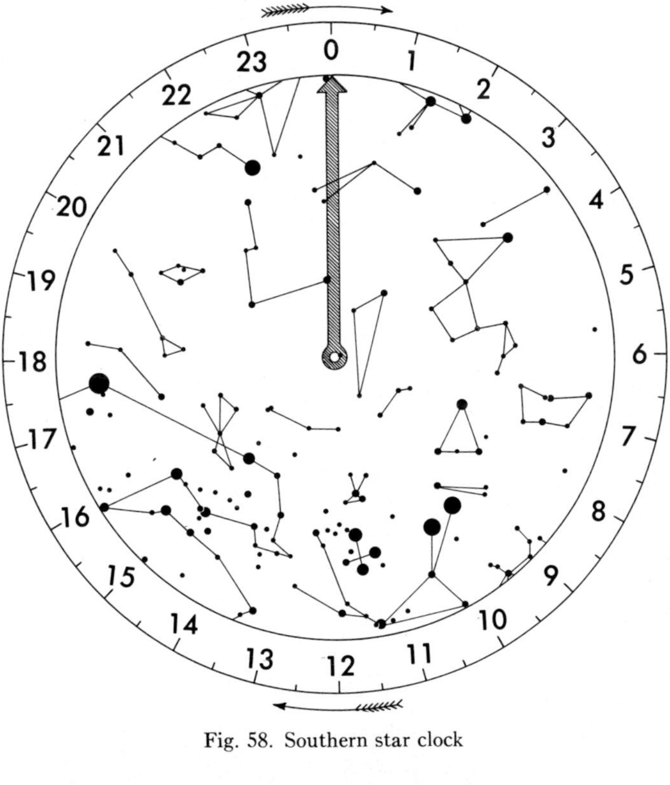

Fig. 58. Southern star clock

of sidereal time, which agrees with solar time on September 21. The hour hand points toward the star α Andromedae, which is common to both Andromeda and the Square of Pegasus. In the Northern Hemisphere, if we regard the star clock as registering solar time, the sun clock loses 4 minutes per day, from September 21 onward. The southern sky clock, one should note, revolves clockwise and in like manner gains 4 minutes a day in relation to ordinary time.

Latitude and Longitude. Geographers designate positions on the surface of the earth in terms of latitude and longitude. The earth's axis of rotation and the equator, which lies midway between the North and South Poles, define the basic reference system. The latitude of a place is the angular distance, measured north or south of the equator, along a circle drawn through the poles and the point in question, the meridian. We measure longitude east or west along the equator to the foot of the meridian of the given place, zero of course being Greenwich meridian.

Declination and Right Ascension. These have been defined on page 117 but are here taken up as integral to our present concern with time in this chapter. Astronomers describe the positions of stars in the sky by a system closely resembling that of latitude and longitude on the earth. The apparent rotation of the heavens allows us to locate two celestial poles, with the celestial equator midway between them. Declination, like terrestrial latitude, is the angular distance, measured north or south of the celestial equator along a meridian or "hour circle" drawn through the poles and the given star. To measure the equivalent of longitude, we arbitrarily select as our fundamental zero the point where the apparent path of the sun (called the *ecliptic*) crosses the celestial equator in the spring (the so called vernal equinox) and measure right ascension eastward from the vernal equinox to the foot of the star's hour circle. The ecliptic coincides with the plane of the earth's orbit.

Precession. Determining the positions of stars presents one complication we do not encounter in the finding of positions on the earth's surface. The earth's pole of rotation is relatively fixed *in the earth*, except for a very minor variation of about 50 feet. Hence the latitudes and longitudes of points on the earth's surface remain essentially constant. The celestial poles, however, are not fixed. They slowly move, relative to the stars. Since these poles lie directly above the earth's pole of rotation, we conclude that the earth's axis of rotation is tilting with reference to the stars. The gravitational pull of the sun and moon on the earth's equatorial bulge produces the motion, which is similar to that of the axis of a tilted spinning top (Fig. 59).

This motion of the earth's axis causes each celestial pole to follow a circular path, completing a full cycle every 26,000 years (see Fig. 60). It also causes the celestial equator to shift slowly

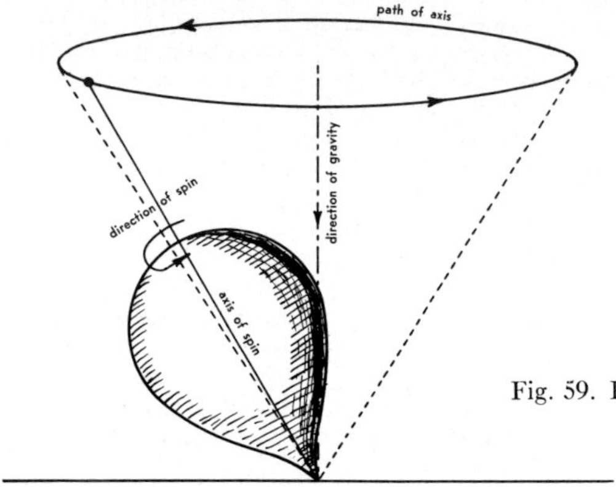

Fig. 59. Precession

among the stars with the pole of the ecliptic as a center. The point of intersection between the celestial equator and the ecliptic moves slowly westward among the stars at the rate of 50″.27 per year. As a result, the right ascensions of the stars gradually in-

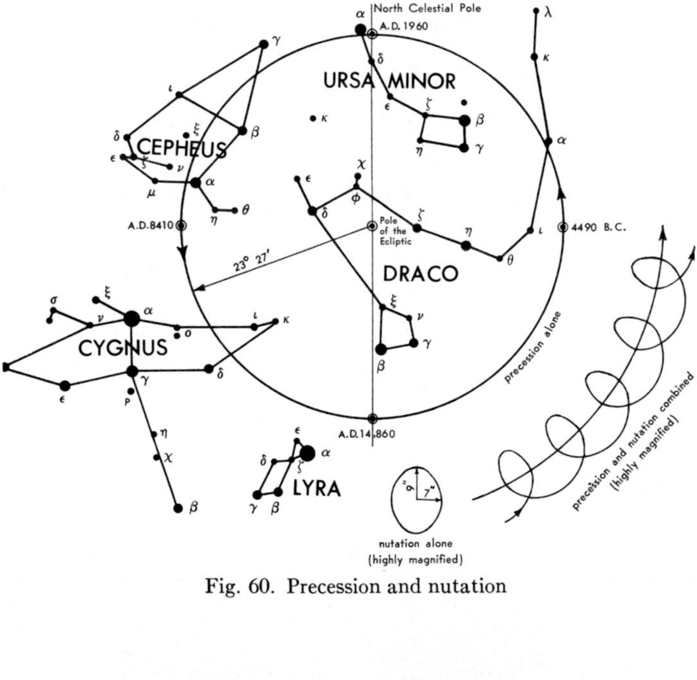

Fig. 60. Precession and nutation

crease with time, because the reference zero point is not stationary. The declinations also change. The annual variations in the two coordinates depend on the positions of the stars. We refer to this motion and its effect upon the positions of stars as *precession of the equinoxes* or, more simply, as *precession.*

The eminence we give Polaris as the Pole Star is therefore only a temporary distinction. In 3000 B.C. α Draconis was the Pole Star. About A.D. 14,000 α Lyrae (Vega) will take over that role, although this star will be much further from the pole than Polaris is at the present time.

Nutation. The axis of the earth's rotation possesses, in addition to the slow regular precession, a small "nodding" movement called *nutation*, caused chiefly by changes in the position of the moon's orbit. This extra motion, which is small, superposes small "rosettes" on the pattern of precession. Nutation, in effect, is a small ellipse with axes 18″.5 and 13″.7. The celestial pole completes a single nodding movement once every 19 years or so. Hence there are more than 1300 loops in a complete precession cycle.

Aberration of Light. A person walking rapidly through a heavy rainstorm with the drops falling straight downward, will have to tilt his umbrella slightly forward to compensate for his own motion (Fig. 61). In the same way and for the same reason, an astronomer on the rapidly moving earth must tilt his telescope slightly forward in the direction of the earth's motion in order to have the starlight fall exactly down the center of his tube. As a result of this motion, the apparent position of a star does not ordinarily coincide with its true position. We term this phenomenon the *aberration* or "wandering" of light. The maximum shift from true to apparent position is 20″.47.

Fig. 61. Aberration of light

Star Catalogs. Various catalogs exist which describe the positions of the stars. Lists of right ascensions and declinations of the heavenly bodies usually carry the date for which these positions are correct. Unless otherwise specified, the positions are *average* or *mean* values, corrected for precession but not for nutation or aberration. These two additional corrections, if successively applied, are said to reduce the mean position first to its true and then to its apparent position.

Since nutation and aberration are such small corrections, however, 9″.2 and 20″.5 respectively at maximum, neglecting them can produce at most an error of about ½ minute of arc, which is negligible for most purposes. The maximum error that would result in navigation, if one used mean instead of apparent positions in his calculations, would be about half a mile, which is usually less than the accuracy of an ordinary sextant. For greater accuracy one should refer to *The American Ephemeris and Nautical Almanac*, which lists apparent star positions to 0′.1 for each day of a given year.

This *Field Guide* gives the mean positions of astronomical objects as referred to the equinox of 1950. The changes in right ascension and declination are so slow, however, that for most astronomical purposes, the tabulated values will be amply accurate for identification purposes at least until 1980.

Hour Angle of Heavenly Bodies. Most large telescopes possess graduated circles, which enable the astronomer to turn his instrument toward some object he wishes to observe. From the printed tables he notes the right ascension and declination of that object. Then, employing the methods described on page 326, he determines his Local Sidereal Time.

We may aptly compare the stars, which move uniformly across the sky from east to west, to railroad trains running on circular tracks. In this analogy, we may think of the tabulated right ascensions of stars as representing the times that the "train" is due at the "station"—that is, the meridian running from north to south through the zenith. Thus, if a star has a right ascension of $7^h\,43^m$ and if the sidereal time is $9^h\,27^m$, we know that the "train" has left the "station" $1^h\,44^m$ beforehand. We say that its hour angle is $1^h\,44^m$ *west* of the meridian. But if the time is $3^h\,41^m$, the "train" has not yet reached the "station," and the hour angle is $4^h\,2^m$ *east.*

Set the graduated circles of an equatorially mounted and properly aligned telescope to the calculated hour angle and the declination, and you should find the sought-for object within or near the field of view.

Appendixes
Index

APPENDIX I

Glossary

Aberration, chromatic. Imperfect correction of a compound lens, producing a colored fringe around the image of a star.

Aberration, spherical. A lens imperfection in which the focal plane for rays passing near center of lens differs from that for rays close to the edge.

Aberration of light. Shift in position of a star as the result of the compounding of motion of earth with the finite velocity of light.

Almagest. One of the early star catalogs, by the Alexandrian astronomer Claudius Ptolemy. Its name, which signifies *The Syntax* or *The Ordering of Nature*, comes to us from the Arabs, who preserved it for posterity.

Altitude. Angle of elevation of a star above horizon.

Aperture. Diameter of a telescope lens or mirror.

Aphelion. Point on an orbit farthest from the sun.

Apogee. Point in orbit of the moon or artificial satellite farthest from the earth.

Apparent Solar Time (AST). Sundial time. To derive the Local Solar Time (LST), add the Equation of Time (ET).

Asterism. Distinctive group of stars not one of the 88 recognized constellations.

Asteroid. Minor planet.

Astrology. A pseudo-science, based on ignorance and superstition, wherein the stars are supposed to influence events on the earth.

Astronomical unit. Mean distance of the earth from the sun — one half of the major axis of the earth's orbit.

Astronomy. Science of the heavenly bodies and the universe.

Azimuth. Angle measured from the south point of the horizon toward the west to the foot of the star's vertical circle. *See also* Bearing.

Bearing. Angle from the north point of the horizon eastward to foot of the star's vertical circle. Bearing = Azimuth $\pm$ 180°.

Binary star. A double star, with the components in orbital revolution around one another.

Bolide. A bright meteor; a fireball.

Celestial latitude. Angle north or south of the ecliptic to the object.

Celestial longitude. Angle measured eastward along ecliptic from the vernal equinox to foot of the circle perpendicular to ecliptic and passing through the object.

Comet. A body of the solar system, composed of ices and rocks. The frozen material evaporates as comet approaches the sun and, driven away by the solar wind, forms comet's tail.

Conjunction. Having the same right ascension or, more precisely, the

Position angle (PA). For a binary star, position angle is the angle (measured counterclockwise) at the brighter star between a line drawn due north to a line drawn to the fainter star.

Precession of the equinoxes. The axis of the earth's rotation slowly changes its direction, maintaining a constant tilt with respect to the ecliptic and making a complete rotation once every 26,000 years. In consequence, the equinoxes on the intersections of the celestial equator and the ecliptic slowly drift westward.

Prime vertical. A great circle on the celestial sphere from east to west and passing through the zenith.

Prominence. Cloud of gas elevated above the solar surface.

Proper motion. Apparent drift of a star across the sky, compounded from the motion of the star and the sun.

Quadrature. A planet is in quadrature when its longitude differs from that of the sun by 90°

Radiant. Point on the celestial sphere from which meteors of a given shower appear to radiate. The effect is due to perspective.

Retrograde motion. Westward or backward drift of a planet resulting from the forward motion of the earth as it passes the planet.

Revolution. Orbital motion of a body around its primary.

Rift. An apparent division of the Milky Way caused by dark clouds of superposed dust.

Right ascension (RA). Angle, measured eastward, from the vernal equinox to the foot of a star's hour circle.

Rotation. Turning of an object on its axis.

Satellite. Object in orbit around a planet.

Shooting star. Popular term for meteor.

Sidereal time. Star time; the hour angle of the vernal equinox.

Solar wind. Gases driven from the sun as the result of boiling activity.

Solstice. Position of the sun when farthest north (summer solstice) or farthest south (winter solstice).

Spectral type. The classification of stellar spectra. *See* Spectrum.

Spectrum. Starlight, dispersed according to color or wave length by means of a prism or other device, is known as a spectrum.

Star. A hot ball of glowing gas. *See also* Sun.

Stereographic projection. Map of a sphere as projected upon a tangent plane from a point diametrically opposite to the point of tangency.

Sun. Star of type G0, center of solar system. Sometimes used generally as synonym for star.

Sunspot. Cooler dark area of visible solar surface.

Telescope, catadioptric. A telescope using a lens-mirror combination for compound objective.

Telescope, reflecting. A telescope with mirror objective.

Telescope, refracting. A telescope using a lens (usually achromatic) as an objective.

Telescope mounting, altazimuth. A mounting with two axes (one horizontal and one vertical, employing azimuth and altitude) to facilitate scanning in the horizon system.

Telescope mounting, equatorial. One rotatable axis, set parallel to the earth's axis of rotation, compensates for diurnal motion of the heavenly bodies in right ascension. The other axis permits adjustment in declination.

Terminator. Dividing line between the illuminated and shadowed portions of the lunar or planetary disk.

Time, Local Mean (LMT). Solar time for a given location reduced from apparent to mean by application of the equation of time. Also called Local Solar Time (LST).

Time, Standard Mean (SMT). Mean solar time for some standard adopted longitude (for example, Eastern, Central, Mountain, or Pacific Standard Times).

Transit. When a small celestial body moves in front of a much larger one (as when Mercury or Venus appears in silhouette against the solar disk or when a satellite passes in front of Jupiter or Saturn), the event is termed transit rather than eclipse. The shadow of a satellite may also transit the disk of its primary.

Variable star. A star whose brightness varies.

Vertical circle. An arc of a great circle drawn from the zenith through a star and perpendicular to the horizon.

Wane. The waning moon refers to that portion of the lunar revolution between full and new. Opposite of wax.

Wax. The waxing moon refers to that portion of the lunar revolution between new and full, when the illuminated area is increasing.

Zenith. The point directly overhead, as determined by the indefinite upward extension of a plumb line.

Zodiac. Twelve constellations along the ecliptic through which the sun, moon, and planets move.

APPENDIX II

Bibliography

1. HISTORICAL

Allen, Richard H.	Star-Names and Their Meanings	G. E. Stechert, 1899
Berry, Arthur	A Short History of Astronomy	Dover, 1961
Crew, H., and A. de Salvio	Two New Sciences by Galileo	Dover, 1914
Dreyer, John L. E.	A History of Astronomy (2nd ed.)	Dover, 1953
Graves, Robert	The Greek Myths (2 vols.)	Penguin Books, 1955
Lodge, Sir Oliver	The Pioneers of Science	Dover, 1960
Pannekoek, Anton	A History of Astronomy	Barnes and Noble, 1961
Reichen, Charles-Albert	A History of Astronomy	Hawthorn, 1963
Rosen, Edward	Three Copernican Treatises	Dover, 1959
Shapley, Harlow	Source Book in Astronomy, 1900–1950	Harvard, 1961
——, and H. E. Howarth	A Source Book in Astronomy	McGraw-Hill, 1929
White, Andrew D.	A History of the Warfare of Science with Theology (2 vols.)	Dover, 1960
Whittaker, Sir Edmund	From Euclid to Eddington	Dover, 1961

2. TEXTBOOKS

Gamow, George	Matter, Earth, and Sky	Prentice-Hall, 1958
Menzel, Donald H., Fred L. Whipple, and Gérard de Vaucouleurs	Survey of the Universe	Prentice-Hall, 1971
Payne-Gaposchkin, Cecilia	Introduction to Astronomy	Prentice-Hall, 1954

Struve, Otto, Beverly Lynds, and Helen Pillans	Elementary Astronomy	Oxford (New York), 1959

3. GENERAL

Bates, David R., ed.	Space Research and Exploration	Sloane, 1958
Beet, E. A.	The Sky and Its Mysteries	Dover, 1961
Burgess, Eric	Satellites and Spaceflight	Macmillan, 1957
Gamow, George	The Creation of the Universe	Viking, 1952
Hawkins, Gerald S.	Splendor in the Sky	Harper, 1961
Hubble, Edwin	The Realm of the Nebulae	Dover, 1958
Jeans, Sir James	The Mysterious Universe	Macmillan, 1930
——	The Stars in Their Courses	Macmillan, 1931
——	The Universe Around Us	Macmillan, 1929
——	Through Space & Time	Macmillan, 1934
Johnson, Martin	Astronomy of Stellar Energy and Decay	Dover, 1959
Lovell, A. C. B.	The Individual and the Universe	Harper, 1959
Martin, Martha Evans, and Donald H. Menzel	The Friendly Stars	Dover, 1964
Menzel, Donald H.	Astronomy	Random House, 1971
Moore, Patrick	The Atlas of the Universe	Philip (London), 1971
Oppolzer, T. von	Canon of Eclipses	Dover, 1962
Pfeiffer, John	The Changing Universe	Random House, 1956
Ryabov, Y.	An Elementary Theory of Celestial Mechanics	Dover, 1961
Sciama, D. W.	The Unity of the Universe	Doubleday, 1960
Scientific American	The New Astronomy	Simon and Schuster, 1955
Smart, W. M.	Some Famous Stars	Longmans, Green, 1950
Vaucouleurs, Gérard de	Discovery of the Universe	Macmillan, 1957
Whitney, Charles A.	The Discovery of Our Galaxy	Knopf, 1971

4. SOLAR SYSTEM

Abetti, Giorgio	The Sun	Macmillan, 1957
Baldwin, Ralph B.	The Face of the Moon	Chicago, 1949

Kopal, Zdeněk	The Moon	Chapman and Hall, 1960
Kuiper, Gerard P., ed.	The Earth as a Planet	Chicago, 1954
——, ed.	The Sun, Vol. I in The Solar System	Chicago, 1953
——, and B. M. Middlehurst	Planets and Satellites	Chicago, 1960
Moore, Patrick	The Planet Venus	Macmillan, 1957
——	Moon Flight Atlas (new ed.)	Philip (London), 1970
Nininger, Harvey H.	Out of the Sky	Dover, 1959
Peek, Bertrand M.	The Planet Jupiter	Macmillan, 1958
Vaucouleurs, Gérard de	The Planet Mars	Faber and Faber (London), 1950
Wilkins, H. P., and Patrick Moore	The Moon	Faber and Faber (London), 1958

5. HANDBOOKS

Bernhard, H. J., D. A. Bennett, and H. S. Rice	New Handbook of the Heavens	New American Library, 1950
Callataÿ, Vincent de	Atlas of the Sky	St. Martin's Press, 1958
McKready, Kelvin	A Beginner's Star-Book (3rd ed.)	Putnam's, 1929
Olcott, William T.	Field Book of the Skies (4th ed., rev. and ed. by R. N. and M. L. Mayall)	Putnam's, 1954
Rey, H. A.	Find the Constellations	Houghton Mifflin, 1954
——	The Stars (enl. world-wide ed.)	Houghton Mifflin, 1962
Sidgwick, John B.	Observational Astronomy for Amateurs	Faber and Faber (London), 1955

6. TELESCOPES AND CAMERAS

Howard, Neale E.	Standard Handbook for Telescope Making	Crowell, 1959
King, Henry C.	The History of the Telescope	Sky Publishing Corp. (Cambridge, Mass.), 1955

Mayall, R. Newton, and M. L. Mayall	Skyshooting	Ronald Press, 1949
Neal, Harry Edward	The Telescope	Messner, 1958
Texereau, J.	How to Make a Telescope	Dover, 1961
Selwyn, E. W. H.	Photography in Astronomy	Eastman Kodak, 1950

7 *The Harvard Books on Astronomy* SERIES
(all now available under Harvard University Press imprint)

Bok, Bart J., and Priscilla F. Bok	The Milky Way (3rd. ed.)	Harvard, 1957
Campbell, Leon, and Luigi Jacchia	The Story of Variable Stars	Blakiston (Philadelphia), 1941
Dimitroff, George Z., and James G. Baker	Telescopes and Accessories	Blakiston (Philadelphia), 1945
Goldberg, Leo, and Lawrence H. Aller	Atoms, Stars, and Nebulae	Blakiston (Philadelphia), 1943
Menzel, Donald H.	Our Sun (rev. ed.)	Harvard, 1959
Miczaika, Gerhard R., and William M. Sinton	Tools of the Astronomer	Harvard, 1961
Payne-Gaposchkin, Cecilia	Stars in the Making	Harvard, 1952
Shapley, Harlow	Galaxies (rev. ed.)	Harvard, 1961
Watson, Fletcher G.	Between the Planets (rev. ed.)	Harvard, 1956
Whipple, Fred L.	Earth, Moon, and Planets (3rd ed.)	Harvard, 1968

8. SPECIAL REFERENCES

Aitken, Robert G.	New General Catalogue of Double Stars, Vols. I and II	Carnegie Inst. of Wash. Pub. No. 417, 1932
U. S. Nautical Almanac Office	The American Ephemeris and Nautical Almanac	Gov't Printing Office
——	American Nautical Almanac	Gov't Printing Office
Blagg, Mary A., and K. Müller	. . . Named Lunar Formations (for the Internat'l Astronomical Union, 2 vols.)	Lund, Humphries (London), 1935

Cederblad, S.	Studies of Bright Diffuse Galactic Nebulae (in Lund Meddelanden, Vol. 12, No. 119)	Lunds Astronomiska Observatorium (Sweden)
Dreyer, John L. E.	New General Catalogue of Nebulae and Clusters of Stars (in Memoirs of Royal Astronomical Society)	Royal Astronomical Society, 1888

Also published by the Royal Astronomical Society:

	Index Catalogue (1895), Second Index Catalogue (1908), which was reprinted and bound in one vol. in 1953	
Innes, R. T. A., comp., with B. H. Dawson and W. H. van den Bos	Southern Double Star Catalog (4 vols.)	Union Observatory, Johannesburg, 1927
Kukarkin, B. V., P. P. Parenago, and Others	Catalog of Variable Stars (2nd ed.), Vols. I and II	U.S.S.R. Nat'l Acad. Sci., 1958
Lohmann, W.	"Die Entfernungen der Kugelförmigen Sternhaufen" (in Zeitschrift für Astrophysik, Vol. 30, pp. 234–47)	1952
Melotte, P. J.	A Catalogue of Star Clusters Shown on Franklin-Adams Chart Plates (in Memoirs of Royal Astronomical Society, Vol. 60, Pt. 5)	Royal Astronomical Society, 1915
Schlesinger, Frank, and Louise F. Jenkins	Yale Catalogue of Bright Stars	New Haven Printing Co., 1940
Shapley, Harlow	Star Clusters	McGraw-Hill, 1930
——, and Adelaide Ames	"A Survey of the External Galaxies Brighter than the Thirteenth Magnitude" (in Annals of Harvard College Observatory, Vol. 88, No. 2)	Harvard, 1932
Vorontsov-Velyaminov, Boris A.	Gaseous Nebulae and Novae (in Russian)	U.S.S.R. Nat'l Acad. Sci., 1948
	German translation by Otto Singer: Gasnebel und Neue Sterne	Kultur und Fortschritt (Berlin), 1953

9. POPULAR MAGAZINES

The Review of Popular Astronomy (bimonthly)	Sky Map Publishers, Inc. (St. Louis)
Scientific American (monthly)	Scientific American, Inc.
Sky and Telescope (monthly)	Sky Publishing Corp. (Cambridge, Mass.)

APPENDIX III
Tables

TABLE 1
Curves Defining Mask for Different Latitudes

(see page 4)

	For Northern Observers		For Southern Observers	
	Facing North	Facing South	Facing North	Facing South
Aa	55°	15°	− 5°	−45°
Bb	50°	20°	−10°	−40°
Cc	45°	25°	−15°	−35°
Dd	40°	30°	−20°	−30°
Ee	35°	35°	−25°	−25°
Ff	30°	40°	−30°	−20°
Gg	25°	45°	−35°	−15°
Hh	20°	50°	−40°	−10°
Ii	15°	55°	−45°	− 5°

TABLE 2
The Greek Alphabet

(see page 7)

α	alpha	η	eta	ν	nu	τ	tau
β	beta	θ	theta	ξ	xi	υ	upsilon
γ	gamma	ι	iota	o	omicron	ϕ	phi
δ	delta	κ	kappa	π	pi	χ	chi
ϵ	epsilon	λ	lambda	ρ	rho	ψ	psi
ζ	zeta	μ	mu	σ	sigma	ω	omega

TABLE 3

Stars Fainter than Magnitude 4.55 Included in the Sky Maps

(*see page 6*)

	m		m		m
π Aqr	4.64	λ Gru	4.60	χ_2 Ori	4.71
λ Cet	4.69	κ Leo	4.61	τ Psc	4.70
ζ CrA	4.85	σ Lup	4.60	ϕ Psc	4.64
δ CrA	4.66	κ Boo	4.60	υ Psc	4.67
θ Crt	4.81	σ Oct	5.48	κ Psc	4.94
ζ Crt	4.90	c Ori	4.65	λ Psc	4.61
ϵ Crt	5.07	θ_1 Ori	4.76	η UMi	5.04
η Crt	5.16	χ_1 Ori	4.62	ζ Aur	5.0–5.7

TABLE 4

Sky Map Numbers, Standard Times

(*see page 6*)

Time	PM				AM			
	5:30	7:30	9:30	11:30	1:30	3:30	5:30	7:30
Jan 1	19, 20	21, 22	23, 24	1, 2	3, 4	5, 6	7, 8	9, 10
Feb 1	21, 22	23, 24	1, 2	3, 4	5, 6	7, 8	9, 10	11, 12
Mar 1	23, 24	1, 2	3, 4	5, 6	7, 8	9, 10	11, 12	13, 14
Apr 1	1, 2	3, 4	5, 6	7, 8	9, 10	11, 12	13, 14	15, 16
May 1	3, 4	5, 6	7, 8	9, 10	11, 12	13, 14	15, 16	17, 18
June 1	5, 6	7, 8	9, 10	11, 12	13, 14	15, 16	17, 18	19, 20
July 1	7, 8	9, 10	11, 12	13, 14	15, 16	17, 18	19, 20	21, 22
Aug 1	9, 10	11, 12	13, 14	15, 16	17, 18	19, 20	21, 22	23, 24
Sept 1	11, 12	13, 14	15, 16	17, 18	19, 20	21, 22	23, 24	1, 2
Oct 1	13, 14	15, 16	17, 18	19, 20	21, 22	23, 24	1, 2	3, 4
Nov 1	15, 16	17, 18	19, 20	21, 22	23, 24	1, 2	3, 4	5, 6
Dec 1	17, 18	19, 20	21, 22	23, 24	1, 2	3, 4	5, 6	7, 8

For the Southern Hemisphere, add 24 to the numbers tabulated above.

TABLE 5
The Constellations
(*see page 7*)

Abbre-viation	Latin Name	Possessive	English Name	Page
And	Andromeda	Andromedae	Andromeda	110
Ant	Antlia	Antliae	Air Pump	114
Aps	Apus	Apodis	Bird of Paradise	113
Aqr	Aquarius	Aquarii	Water Carrier	109
Aql	Aquila	Aquilae	Eagle	111
Ara	Ara	Arae	Altar	112
Ari	Aries	Arietis	Ram	109
Aur	Auriga	Aurigae	Charioteer	110
Boo	Boötes	Boötis	Bear Driver	108
Cae	Caelum	Caeli	Graving Tool	114
Cam	Camelopardalis	Camelopardalis	Giraffe	108
Cnc	Cancer	Cancri	Crab	110
CVn	Canes Venatici	Canum Venaticorum	Hunting Dogs	108
CMa	Canis Major	Canis Majoris	Larger Dog	112
CMi	Canis Minor	Canis Minoris	Smaller Dog	112
Cap	Capricornus	Capricorni	Sea Goat	109
Car	Carina	Carinae	Keel	113
Cas	Cassiopeia	Cassiopeiae	Cassiopeia	110
Cen	Centaurus	Centauri	Centaur	112
Cep	Cepheus	Cephei	Cepheus	110
Cet	Cetus	Ceti	Whale	110
Cha	Chamaeleon	Chamaeleontis	Chameleon	113
Cir	Circinus	Circini	Compasses	114
Col	Columba	Columbae	Dove	113
Com	Coma Berenices	Comae Berenices	Berenice's Hair	108
CrA	Corona Australis	Coronae Australis	Southern Crown	112
CrB	Corona Borealis	Coronae Borealis	Northern Crown	108
Crv	Corvus	Corvi	Crow	111
Crt	Crater	Crateris	Cup	111
Cru	Crux	Crucis	Cross	112
Cyg	Cygnus	Cygni	Swan	111

TABLE 5 (CONT'D)

Abbre-viation	Latin Name	Possessive	English Name	Page
Del	Delphinus	Delphini	Dolphin	112
Dor	Dorado	Doradus	Goldfish	113
Dra	Draco	Draconis	Dragon	108
Equ	Equuleus	Equulei	Little Horse	112
Eri	Eridanus	Eridani	River	113
For	Fornax	Fornacis	Furnace	114
Gem	Gemini	Geminorum	Twins	110
Gru	Grus	Gruis	Crane	113
Her	Hercules	Herculis	Hercules	111
Hor	Horologium	Horologii	Clock	114
Hya	Hydra	Hydrae	Sea Serpent	111
Hyi	Hydrus	Hydri	Water Snake	113
Ind	Indus	Indi	Indian	113
Lac	Lacerta	Lacertae	Lizard	110
Leo	Leo	Leonis	Lion	109
LMi	Leo Minor	Leonis Minoris	Smaller Lion	109
Lep	Lepus	Leporis	Hare	112
Lib	Libra	Librae	Scales	109
Lup	Lupus	Lupi	Wolf	112
Lyn	Lynx	Lyncis	Lynx	108
Lyr	Lyra	Lyrae	Lyre	111
Men	Mensa	Mensae	Table Mountain	114
Mic	Microscopium	Microscopii	Microscope	114
Mon	Monoceros	Monocerotis	Unicorn	112
Mus	Musca	Muscae	Fly	114
Nor	Norma	Normae	Level	114
Oct	Octans	Octantis	Octant	114
Oph	Ophiuchus	Ophiuchi	Serpent Holder	111
Ori	Orion	Orionis	Orion	112
Pav	Pavo	Pavonis	Peacock	113

TABLE 5 (CONT'D)

Abbre-viation	Latin Name	Possessive	English Name	Page
Peg	Pegasus	Pegasi	Pegasus	110
Per	Perseus	Persei	Perseus	110
Phe	Phoenix	Phoenicis	Phoenix	113
Pic	Pictor	Pictoris	Easel	114
Psc	Pisces	Piscium	Fish	109
PsA	Piscis Austrinus	Piscis Austrini	Southern Fish	113
Pup	Puppis	Puppis	Stern	113
Pyx	Pyxis	Pyxidis	Mariner's Compass	113
Ret	Reticulum	Reticuli	Net	114
Sge	Sagitta	Sagittae	Arrow	111
Sgr	Sagittarius	Sagittarii	Archer	109
Sco	Scorpius	Scorpii	Scorpion	109
Scl	Sculptor	Sculptoris	Sculptor's Apparatus	114
Sct	Scutum	Scuti	Shield	111
Ser	Serpens	Serpentis	Serpent	111
Sex	Sextans	Sextantis	Sextant	111
Tau	Taurus	Tauri	Bull	109
Tel	Telescopium	Telescopii	Telescope	114
Tri	Triangulum	Trianguli	Triangle	111
TrA	Triangulum Australe	Trianguli Australis	Southern Triangle	112
Tuc	Tucana	Tucanae	Toucan	113
UMa	Ursa Major	Ursae Majoris	Great Bear	107
UMi	Ursa Minor	Ursae Minoris	Little Bear	108
Vel	Vela	Velorum	Sails	113
Vir	Virgo	Virginis	Virgin	109
Vol	Volans	Volantis	Flying Fish	113
Vul	Vulpecula	Vulpeculae	Fox	111

TABLE 6
Asterisms
(*see page 7*)

Beehive, in Cancer. Also called Praesepe or M 44, open star cluster faintly visible to the naked eye. With γ and δ Cancri it forms another asterism: the Asses and the Manger.

Belt of Orion, δ, ϵ, ζ Orionis; the Three Marys, in Latin America.

Bier, α, β, γ, δ Ursae Majoris.

Big Dipper, α, β, γ, δ, ϵ, ζ, η Ursae Majoris, also known as the Wain (Wagon) or Charles's Wain, with the Dipper handle representing the wagon tongue.

Bull of Poniatowski, 66, 67, 68, and 70 Ophiuchi, a T-shaped asterism just east of γ Ophiuchi.

Circlet, γ, b, θ, ι, 19, λ, and κ Piscium, the western fish.

Coalsack, not a true asterism, but a dark patch on the Milky Way, in Crux. The African Bushmen call it the "Old Bag."

Frederik's Glory, ι, κ, λ, and ψ Andromedae.

Guardians of the Pole, β and γ Ursae Minoris.

Head of Cetus, α, γ, ξ_2, μ, and λ Ceti.

Heavenly G, a G-shaped group of 9 bright stars, 7 of them 1st-magnitude. In order they are: Aldebaran, Capella, Castor, Pollux, Procyon, Sirius, Rigel, Bellatrix, and Betelgeuse.

Hyades, V-shaped group superposed on α, γ, δ, and ϵ Tauri (open cluster).

Hydra Head, δ, ϵ, ζ, η, ρ, and σ Hydrae.

Job's Coffin, α, β, γ, and δ Delphini.

Keystone, ϵ, ζ, η, and π Herculis.

Kids, ϵ, ζ, and η Aurigae.

Lozenge, β, γ, ξ, and ν Draconis.

Milk Dipper, ζ, τ, σ, ϕ, and λ Sagittarii, an inverted dipper in the Milky Way.

Northern Cross, α, β, γ, δ, and ϵ Cygni.

Northern Fly, small triangle over the rear of Aries.

Pleiades, dipper-shaped group in Taurus (open cluster). Also called the Seven Sisters or, in Latin America, the Seven Little Goats.

Segment of Perseus, an arc comprising η, γ, α, δ, ϵ, and ζ Persei.

TABLE 6 (CONT'D)

Sickle, α, η, γ, ζ, μ, and ϵ Leonis.

Square of Pegasus, α, β, and γ Pegasi and α Andromedae.

Sword of Orion, θ and ι Orionis.

Venus' Mirror, Orion's belt and sword, plus η Orionis; the sword forms the handle of a diamond-shaped mirror.

Y of Aquarius, γ, η, π, and ζ Aquarius (also called the Water Jar).

TABLE 7

The Brightest Stars

(*see pages 116–18*)

Star	Name	RA h	RA m	Dec °	Dec ′	Mag	Spec†	Dist L-Y
α CMa	Sirius	6	42.9	−16	39	−1.42	A0*	8.7
α Car	Canopus	6	22.8	−52	40	−0.72	F0	230
α Cen	Rigil Kent	14	36.2	−60	38	−0.27	G0*	4.3
α Boo	Arcturus	14	13.4	+19	27	−0.06	K0	38
α Lyr	Vega	18	35.2	+38	44	0.04	A0	27
α Aur	Capella	5	13.0	+45	57	0.05	G0	46
β Ori	Rigel	5	12.1	− 8	15	0.14	B8p	500
α CMi	Procyon	7	36.7	+ 5	21	0.38	F5	11
α Eri	Achernar	1	35.9	−57	29	0.51	B5	73
β Cen	Hadar	14	0.3	−60	8	0.63	B1	190
α Aql	Altair	19	48.3	+ 8	44	0.77	A5	16
α Ori	Betelgeuse	5	52.5	+ 7	24	Var.	Ma	300
α Tau	Aldebaran	4	33.0	+16	25	0.86	K5	64
α Cru	Acrux	12	23.8	−62	49	0.9	B1*	220
α Vir	Spica	13	22.6	−10	54	0.91	B2	190
α Sco	Antares	16	26.3	−26	19	0.92	Ma*	230
β Gem	Pollux	7	42.3	+28	9	1.16	K0	33
α PsA	Fomalhaut	22	54.9	−29	53	1.19	A3	23
α Cyg	Deneb	20	39.7	+45	6	1.26	A2p	650
β Cru	Becrux	12	44.8	−59	25	1.28	B1	500

† Asterisks indicate spectral type of brighter component where star is double.

7-2-5

TABLE 7 (CONT'D)

Star	Name	RA h	RA m	Dec °	Dec ′	Mag	Spec	Dist L-Y
α Leo	Regulus	10	5.7	+12	13	1.36	B8*	78
α Gem	Castor	7	31.4	+32	0	1.58	A0*	47
γ Cru	Gacrux	12	28.4	−56	50	1.61	Mb	
ε CMa	Adhara	6	56.7	−28	54	1.63	B1	330
ε UMa	Alioth	12	51.8	+56	14	1.68	A0p	49
γ Ori	Bellatrix	5	22.4	+ 6	18	1.70	B2	230
λ Sco	Shaula	17	30.2	−37	4	1.71	B2	200
ε Car	Avior	8	21.5	59	21	1.74	K0*	330
ε Ori	Alnilam	5	33.7	− 1	14	1.75	B0	
β Tau	El Nath	5	23.1	+28	34	1.78	B8	130
β Car	Miaplacidus	9	12.7	−69	31	1.80	A0	
α TrA	Atria	16	43.4	−68	56	1.88	K2	130
α Per	Mirfak	3	20.7	+49	41	1.90	F5	270
η UMa	Alkaid	13	45.6	+49	34	1.91	B3	190
γ Vel		8	8.0	−47	11	1.92	Oap	
γ Gem	Alhena	6	34.8	+16	27	1.92	A0	78
ε Sgr	Kaus Aust.	18	20.9	−34	25	1.95	A0	160
α UMa	Dubhe	11	0.7	+62	1	1.95	K0	105
δ CMa	Al Wazor	7	6.4	−26	19	1.98	F8p	650
β CMa	Murzim	6	20.5	−17	56	1.99	B1	300
δ Vel		8	43.3	−54	31	2.01	A0	70
θ Sco		17	33.7	−42	58	2.04	F0	140
ζ_1 Ori	Alnitak	5	38.2	− 1	58	2.05	B0*	400
β Aur	Menkalinan	5	55.9	+44	57	2.07	A0p	84
α Pav	Peacock	20	21.7	−56	54	2.12	B3	160
α UMi	Polaris	1	48.7	+89	2	2.12	F8	470
α Oph	Rasalhague	17	32.6	+12	36	2.14	A5	67
σ Sgr	Nunki	18	52.2	−26	22	2.14	B3	160
α And	Alpheratz	0	5.8	+28	49	2.15	A0p	120
ζ UMa	Mizar	13	21.9	+55	11	2.16	A2p*	190

TABLE 7 (CONT'D)

Star	Name	RA h	RA m	Dec °	Dec ′	Mag	Spec	Dist L-Y
α Hya	Alphard	9	25.1	− 8	26	2.16	K2	200
α Gru	Al Na'ir	22	5.1	−47	12	2.16	B5	91
κ Ori	Saiph	5	45.4	− 9	41	2.20	B0	550
λ Vel	Suhail	9	6.2	−43	14	2.22	K5	220
β Per	Algol	3	4.9	+40	46	Var	B8	100
β Leo	Denebola	11	46.5	+14	51	2.23	A2	42
α Ari	Hamal	2	4.3	+23	14	2.23	K2	74
β Cet	Diphda	0	41.1	−18	16	2.24	K0*	57
β Gru		22	39.7	−47	9	2.24	Mb	270
β UMi	Kochab	14	50.8	+74	22	2.24	K5	120
γ Cas		0	53.7	+60	27	Var	B0p	200
ι Car		9	15.8	−59	4	2.25	F0	
θ Cen	Menkent	14	3.7	−36	7	2.26	K0	56
ζ Pup		8	1.8	−39	52	2.27	Od	800
γ_1 And	Almach	2	0.8	+42	5	2.28	K0*	400
α CrB	Alphecca	15	32.6	+26	53	2.31	A0	67
γ Cyg	Sadr	20	20.4	+40	6	2.32	F8p	470
ε Sco		16	46.9	−34	12	2.36	K0	69
β And	Mirach	1	6.9	+35	21	2.37	Ma	76
γ Cen		12	38.7	−48	41	2.38	A0	130
γ Dra	Eltanin	17	55.4	+51	30	2.42	K5	150
β Cas	Caph	0	6.5	+58	52	2.42	F5	45
η CMa	Aludra	7	22.1	−29	12	2.43	Bsp	270
β UMa	Merak	10	58.8	+56	39	2.44	A0	76
α Phe	Ankaa	0	23.8	−42	35	2.44	K0	76
α Cas	Schedar	0	37.7	+56	16	Var	K0	230
δ Ori	Mintaka	5	29.5	− 0	20	2.48	B0*	600
κ Sco		17	39.0	−39	0	2.51	B2	360
ε Peg	Enif	21	41.7	+ 9	39	2.54	K0	250
γ UMa	Phecda	11	51.2	+53	58	2.54	A0	88

TABLE 7 (CONT'D)

Star	Name	RA h	RA m	Dec °	Dec ′	Mag	Spec	Dist L-Y
α Peg	Markab	23	2.3	+14	56	2.57	A0	100
η Oph	Sabik	17	7.5	−15	40	2.63	A2	76
γ Crv	Gienah	12	13.2	−17	16	2.78	B8	130
α Cet	Menkar	2	59.7	+ 3	54	2.82	Ma	250
α_2 Lib	Zuben'ubi	14	48.1	−15	50	2.90	A3*	62
θ_1 Eri	Acamar	2	56.4	−40	31	3.42	A2*	120

TABLE 8

Spectral Classes and Star Colors

(*see page 118*)

Type of Spectrum	Star Color	Symbol for Color	Surface Temperature
O	very Blue	v Bl	50,000°
B	Blue	Bl	25,000
A	Green	Gr	11,000
F	White	W	7,600
G0	Yellow	Yl	6,000
G5-K	Orange	Or	5,100
M, R, N, S	Red	Rd	3000–3600

TABLE 9

Variable Stars with Maxima Brighter than Magnitude 6.0

(*see pages 118–20*)

Name	RA h m	Dec ° ′	Range	Per days	Spec	Type
γ Peg	0 10.7	+14 54	2.8 – 2.82	0.152	B2	βC
AO Cas	0 15.0	+51 10	5.8 – 6.0	3.52	08+08	EB
T Cet	0 19.2	−20 21	5.5 – 6.9	162	M5	SR
R And	0 21.3	+38 18	5.0 –15.3	409	Se	LP
TV Pis	0 25.2	+17 37	4.6 – 5.2	49.1	M3	SR

TABLE 9 (CONT'D)

Name	RA h m	Dec ° ′	Range	Per days	Spec	Type
ζ And	0 44.6	+24 00	5.1 – 5.2	17.8	G8	Ell
γ Cas	0 53.6	+60 27	1.6 – 3.0		B0	NI
ζ Phe	1 06.3	−55 31	4.0 – 4.50	1.67	B8+B8	E
δ Cas	1 22.5	+59 59	2.8 – 2.88	759	A4n	EA?
R Scl	1 24.7	−32 48	5.8 – 8.7	376	N	LP
φ Per	1 40.6	+50 26	4.3 – 4.5	18.1?	B0 ne	
α UMi	1 42.7	+89 02	2.48– 2.62	3.97	cF7v	Cep
o Cet	2 16.8	− 3 12	2.0 –10.1	331	M6e	LP
ι Cas	2 24.9	+67 11	4.57– 4.60	1.74	A5p	αCV
R Tri	2 34.0	+34 03	5.4 –12.0	266	M4e	LP
R Hor	2 52.3	−50 06	6.0 –14.0	401	M7e	LP
ρ Per	3 02.0	+38 39	3.2 – 3.8	50	M4	SR
β Per	3 04.9	+40 46	2.2 – 3.47	2.87	B8	EA
SX Ari	3 09.3	+27 04	5.76– 5.82	0.73	AOp	αCV?
λ Tau	3 57.9	+12 21	3.5 – 4.00	3.95	B3	EA
b_1 Per	4 14.5	+50 10	4.6 – 4.66	1.53	A1	Ell
ν Eri	4 33.8	− 3 27	3.38– 3.52	0.174	B2S	βC
π_5 Ori	4 51.6	+ 2 22	3.6 – 3.65	3.70	B3	Ell
R Lep	4 57.3	−14 53	5.5 –10.7	436	Ne	LP
ε Aur	4 58.5	+43 45	3.73– 4.53	9883	cF2	EA
ζ Aur	4 59.0	+40 59	5.0 – 5.6	972	K5+B9	EA
AR Aur	5 15.0	+33 44	5.82– 6.49	4.13	A0+A0	EA
η Ori	5 22.0	− 2 26	3.2 – 3.35	7.99	B1	EB
δ Ori	5 29.4	− 0 20	2.40– 2.55	5.73	B0	EA
VV Ori	5 31.0	− 1 11	5.10– 5.46	1.49	B2n+B8?	EA
β Dor	5 33.0	−62 31	4.5 – 5.7	9.84	F6–G5	Cep
α Ori	5 52.5	+ 7 24	0.4 – 1.3	2070	cM2	SR
U Ori	5 52.9	+20 10	5.2 –12.9	373	M8e	LP
β Aur	5 55.9	+44 57	2.07– 2.16	3.96	A2+A2	EA
S Lep	6 03.6	−24 11	6.0 – 7.4	95	M6	SR

TABLE 9 (CONT'D)

Name	RA h m	Dec ° ′	Range	Per days	Spec	Type
η Gem	6 11.8	+22 32	3.1 – 3.9	234	F7–G3	SR
V Mon	6 20.1	− 2 07	6.0 –14.0	334	M5e	LP
T Mon	6 22.5	+ 7 07	5.8 – 6.8	27.02	F7–K1	Cep
RT Aur	6 25.3	+30 32	5.37– 6.55	3.73	F1–G5	Cep
WW Aur	6 25.9	+32 31	5.70– 6.36	2.525	A7+A7	EA
ξ_1 CMa	6 29.8	−23 23	4.3 – 4.36	0.21	B1s	βC
UU Aur	6 33.1	+38 29	5.1 – 6.8	3400	N	SR
ζ Gem	7 01.2	+20 39	3.7 – 4.1	10.15	F7–G3	Cep
R Gem	7 04.3	+22 47	5.9 –14.1	370	Se	LP
L_2 Pup	7 12.0	−44 33	3.4 – 6.2	140	M5e	LP
UW CMa	7 16.6	−24 28	4.47– 4.81	4.39	08+08	EB
U Gem	7 52.2	+22 08	8.8 –13.8	103?	Pec	UG
V Pup	7 56.7	−49 04	4.53– 5.14	1.45	B1+B3	EB
X Cnc	8 52.6	+17 25	5.9 – 7.3	165?	N	SR
R Car	9 30.9	−62 34	4.6 –10.1	309	M5e	LP
R LMi	9 42.6	+34 45	6.0 –13.3	372	M8e	LP
l Car	9 43.9	−62 17	4.97– 5.99	35.5	F8–K0	Cep
R Leo	9 44.9	+11 40	4.4 –11.6	313	M8e	LP
TX Leo	10 32.4	+ 8 55	5.70– 5.80	2.445	A0	EA
U Hya	10 35.1	−13 07	4.8 – 5.8		N	I?
VY UMa	10 41.6	+67 41	6.0 – 6.6		N	I
V Hya	10 49.1	−20 59	6.0 –12.5	532	N	LP
R Crv	12 17.0	−18 59	5.9 –14.4	317	M5e	LP
Y CVn	12 42.8	+45 43	5.2 – 6.6	158	N	SR
ϵ UMa	12 51.8	+56 14	1.7 1.73	5.09	AO	αCV
α_2 CVn	12 53.6	+38 35	3.0 – 3.1	5.47	A0p	αCV
α Vir	13 22.2	−10 54	1.2 – 1.30	4.014	B2+B5	EA
R Hya	13 26.9	−23 02	3.6 –10.9	387	M7e	LP
S Vir	13 30.4	− 6 57	6.0 –13.0	377	M7e	LP
T Cen	13 38.8	−33 21	5.2 –10.0	91	M0e	LP

TABLE 9 (CONT'D)

Name	RA h m	Dec ° ′	Range	Per days	Spec	Type
μ Cen	13 46.6	−42 13	3.0 – 3.2		B2n3	I
R CVn	13 46.8	+39 47	6.1 –12.8	326	M6e	LP
CS Vir	14 16.0	−18 29	5.7 – 5.75	9.30	A3s	αCV
γ Boo	14 30.0	+38 32	3.20– 3.25	0.29	A7	βC
R Boo	14 35.0	+26 57	5.9 –13.1	223	M4e	LP
δ Lib	14 58.3	− 8 19	4.76– 5.90	2.33	A1s	EA
S CrB	15 19.3	+31 33	5.8 –13.9	361	M7e	LP
α CrB	15 32.6	+26 53	2.24– 2.35	17.36	A0	EA
χ Ser	15 39.4	+13 00	5.25– 5.28	2.68	A0p	αCV
R CrB	15 46.5	+28 19	5.8 –14	Irr	cG0ep	R CrB
R Ser	15 48.3	+15 16	5.6 –14.0	357	M7e	LP
T CrB	15 57.4	+26 03	2.0 –10.6	Irr	Q+gM3	RN
S Her	16 49.7	+15 02	5.9 –13.6	307	M6e	LP
χ Oph	16 24.1	−18 21	4.3 – 5.1		B3pe	
α Sco	16 26.3	−26 20	0.9 – 1.8	1733	cM1	SR
g Her	16 27.0	+41 59	4.4 – 6.0	80	M6	SR
μ_1 Sco	16 48.5	−37 58	3.00– 3.31	1.446	B3p+B6	EB
ε UMi	16 51.0	+82 08	5.0 – 5.14	39.48	G1+A5±	E
RS Sco	16 51.9	−45 01	6.0 –12.7	320	M6e	LP
RR Sco	16 53.4	−30 30	5.0 –12.2	279	M6e	LP
α Her	17 12.4	+14 27	3.0 – 4.0	100?	M5	SR
U Oph	17 14.0	+ 1 16	5.80– 6.52	1.677	B5n+B5n	EA
u Her	17 15.5	+33 9	4.6 – 5.14	2.05	B3+B3	EB
VW Dra	17 15.9	+60 43	6.0 – 6.5		K0	I
X Sgr	17 44.5	−27 49	5.0 – 6.1	7.01	F5–G9	Cep
RS Oph	17 47.5	− 6 42	4.0 –12.0		Ocp	RN
W Sgr	18 02.0	−29 35	5.0 – 6.4	7.59	F5–G6	Cep
μ Sgr	18 10.8	−21 05	4.01–14.15	180.5	cB8e	EA
RS Sgr	18 14.6	−34 08	6.0 – 6.97	2.42	B5	EA
Y Sgr	18 18.5	−18 53	5.82– 6.92	5.77	F6–G5	Cep

TABLE 9 (CONT'D)

Name	RA h m	Dec ° ′	Range	Per days	Spec	Type
d Ser	18 24.7	+ 0 10	5.2 – 5.5		G0+A2	
X Oph	18 36.0	+ 8 47	5.9 – 9.3	335	M6e	LP
δ Sct	18 39.5	− 9 11	4.9 – 5.15	0.194	F4s	δ Sct
R Sct	18 45.8	− 5 46	5.0 – 8.4	144?	G0–M5	Flare
β Lyr	18 48.3	+33 18	3.4 – 4.3	12.91	B8p	EB
R Lyr	18 53.8	+43 53	4.0 – 5.0	50	M5	SR
FF Aql	18 56.0	+17 18	5.66– 6.17	4.47	F5–F8	Cep
R Aql	19 03.9	+ 8 09	5.1 –12.0	300	M7e	LP
Y Aql	19 04.9	+10 59	5.1 – 5.13	1.30	B8	Ell
RY Sgr	19 13.3	−33 37	6.0 –14.0		G0ep	I
υ Sgr	19 18.9	−16 03	4.26– 4.41	137.94	B8+F2p	EB
RR Lyr	19 23.9	+42 41	7.13– 8.03	0.567	A2–F0	Cl
R Cyg	19 35.4	+50 06	5.9 –14.6	425	Se	LP
σ Aql	19 36.7	+ 5 17	5.0 – 5.18	1.95	B3+B3	EB
QS Aql	19 38.8	+13 42	5.80– 5.91	2.51	B3	EA
χ Cyg	19 48.7	+32 47	2.3 –14.3	407	Mpe	LP
η Aql	19 49.9	+00 52	3.69 –4.40	7.18	F6–G4	Cep
RR Sgr	19 52.8	−29 19	5.5 –14.0	336	M5e	LP
S Sge	19 53.8	+16 30	5.73– 6.88	8.38	F6 G5	Cep
RR Tel	20 00.3	−55 52	6.0 –14.0		F5e	I
o_1 Cyg	20 12.1	+46 35	4.91– 5.29	3803	K0+B8	EA
o_2 Cyg	20 13.9	+47 33	5.2 – 5.4	1140.7	cK5+A	EA
P Cyg	20 14.9	+37 52	3 – 6		B1	NI
AF Dra	20 32.2	+74 47	5.12– 5.16	20.27	A2p	αCV
U Del	20 43.2	+17 54	5.6 – 7.5		M5	I
T Vul	20 49.5	+28 04	5.87– 6.75	4.44	F5–G1	Cep
BW Vul	20 52.2	+28 20	6.2 – 6.38	0.20	B1s	βC
DT Cyg	21 04.4	+30 59	5.90– 6.32	2.50	F5.5–F7	Cep
T Cep	21 08.9	+68 17	5.2 –11.2	388	M7e	LP
β Cep	21 28.0	+70 20	3.32– 3.37	0.19	B2	βC

TABLE 9 (CONT'D)

Name	RA h m	Dec ° ′	Range	Per days	Spec	Type
W Cyg	21 34.0	+45 09	5.0 – 7.6	131	M4e	SR
SS Cyg	21 40.8	+43 22	8.1 –12.0	50?	APec	UG
μ Cep	21 42.0	+58 33	3.6 – 5.1		cM2e	SR
π_1 Gru	22 19.7	−46 12	5.8 – 6.4		S	I
δ Cep	22 27.2	+58 10	3.78– 4.63	5.37	F5–G2	Cep
DD Lac	22 39.2	+39 58	4.6 – 4.80	0.193	B2	βC
EN Lac	22 54.1	+41 20	5.0 – 5.11	0.169	B2	βC
o And	22 59.6	+42 03	3.59– 3.72	1.58	B6+A1	
β Peg	23 1.4	+27 49	2.4 – 2.8	40±	M2	SR
AR Cas	23 27.7	+58 16	4.7 – 4.83	6.07	B3	EA
λ And	23 35.1	+46 11	5.0 – 5.28	54	G4–G9	SR
ρ Cas	23 51.9	+57 13	4.1 – 6.2		cF8–M2	R CrB?
R Cas	23 55.8	+51 07	4.8 –13.6	430	M7e	LP

TABLE 10

Double Stars (A) between Declinations +90° and −30°

(see page 125)

Aitken Number	RA h m	Dec ° ′	Mag	PA °	Sep ″	Dist L-Y
a 48	0 02.8	+45 32	9.4 & 9.4	160	5	
a 191	0 12.4	+ 8 33	5.9 & 8.1	150	11	250
a 434	0 29.0	+54 15	5.5 & 5.8	170	0.6	470
a 475	0 31.9	− 4 49	7.0 & 9.3	260	0.8	
			9.8	45	20	
a 671	0 46.1	+57 33	3.6 & 7.2	290	10	18
a 746	0 51.9	+18 55	6.1 & 7.2	240	0.5	330
a 755	0 52.3	+23 21	6.1 & 6.7	180	0.7	160
a 940	1 06.6	+46 59	4.5 & 6.0	150	0.4	470
a 1081	1 17.2	− 0 46	6.4 & 7.3	10	2	330
a 1087	1 17.5	−16 04	7.5 & 7.7	350	2	

TABLE 10 (CONT'D)

Aitken Number	RA h	m	Dec °	′	Mag	PA °	Sep ″	Dist L-Y
a 1254	1	33.4	+ 7	23	7.7 & 7.7	50	1.6	
a 1359	1	41.0	+57	17	6.4 & 7.8	25	1	360
a 1457	1	47.4	+22	02	6.2 & 7.4	160	3	300
a 1459	1	47.6	+64	36	7.2 & 9.2	35	35	
a 1477	1	48.8	+89	02	2.1 & 9.1	220	18.3	460
a 1507	1	50.8	+19	03	4.7 & 4.8	0	8	150
a 1538	1	53.3	+ 1	36	6.9 & 6.9	50	1.5	120
a 1598	1	57.8	+70	40	4.7 & 7.0	330	0.6	125
a 1615	1	59.4	+ 2	31	4.3 & 5.2	300	2	130
a 1630	2	00.8	+42	06	2.3 & 5.1	65	10	400
a 1697	2	09.5	+30	04	5.5 & 6.9	75	4	330
a 1709	2	10.8	+47	15	6.4 & 7.3	230	0.7	130
a 1953	2	31.6	− 5	51	8.0 & 8.2	340	3	
a 2004	2	35.9	+33	21	7.5 & 8.2	160	2	
a 2080	2	40.7	+ 3	02	3.7 & 6.2	290	3	80
a 2257	2	56.4	+21	8	5.2 & 5.5	205	1.5	400
a 2294	2	59.2	+79	13	5.7 & 8.9	235	4.5	470
a 2616	3	31.5	+24	18	6.6 & 6.7	20	0.5	360
a 2628	3	32.5	+31	31	7.6 & 7.6	40	0.9	
a 2850	3	51.8	− 3	06	5.0 & 6.3	345	7	300
a 2984	4	03.4	+62	12	7.0 & 7.1	305	18	
a 3169	4	19.9	+14	56	7.2 & 9.2	10	1	
a 3390	4	38.8	+37	25	8.5 & 8.5	40	1.5	
a 3711	5	5.2	+ 8	26	5.9 & 6.7	80	0.9	200
a 3823	5	12.1	− 8	15	0.3 & 7.3	200	9.5	550
a 4002	5	22.0	− 2	26	3.8 & 4.8	75	1	550
a 4186	5	32.8	− 5	25	6.8 & 7.9	32	8.7	550
					5.4	130	13	
					6.8	95	21.6	
a 4188	5	32.9	− 5	27	5.4 & 6.8	95	53	
a 4193	5	33.0	− 5	56	2.9 & 7.3	140	11.5	150
a 4208	5	34.0	+26	54	6.4 & 6.5	330	0.9	410

TABLE 10 (CONT'D)

Aitken Number	RA h	m	Dec °	′	Mag	PA °	Sep ″	Dist L-Y
a 5012	6	21.1	+ 4	37	4.5 & 6.5	30	14	130
a 5106	6	26.4	− 7	00	4.7 & 5.2	130	7.5	470
					5.6	105	3	
a 5166	6	29.4	+17	49	7.1 & 8.0	210	20	
a 5197	6	31.5	+14	48	8.2 & 8.3	320	2	
a 5400	6	41.8	+59	30	5.3 & 6.2	100	2	180
					8.5	310	9	
a 5586	6	53.0	+58	29	4.9 & 6.0	30	0.8	230
a 5871	7	09.7	+27	19	7.1 & 7.1	330	1	160
a 5958	7	14.9	+ 9	23	7.6 & 7.9	90	1.5	
a 6012	7	18.8	+55	23	5.6 & 6.5	315	15	270
a 6175	7	31.4	+32	00	2.0 & 2.8	180	2	47
a 6255	7	36.8	−26	41	4.5 & 4.6	315	10	360
a 6263	7	37.5	+ 5	21	6.1 & 6.4	160	1	330
a 6319	7	41.2	+65	17	7.7 & 7.7	5	15	
a 6348	7	43.2	−14	34	6.1 & 6.8	340	17	
a 6381	7	45.6	−12	4	5.5 & 8.2	10	3	120
a 6425	7	50.1	− 3	31	7.5 & 8.0	20	1	
a 6650	8	09.3	+17	48	5.6 & 6.0	20–0	1	78
					6.3			
a 6988	8	43.7	+28	57	4.2 & 6.6	310	31	170
a 7203	9	06.0	+67	20	4.9 & 8.1	20	2	65
a 7292	9	15.8	+37	01	3.9 & 5.9	230	3	110
a 7307	9	17.9	+38	24	7.6 & 7.8	220	1.3	
a 7390	9	25.8	+ 9	17	5.9 & 6.7	170	0.4	110
a 7545	9	48.7	+54	19	5.1 & 5.7	20–50	0.5–0.3	170
a 7704	10	13.6	+17	59	7.3 & 7.5	180	1	230
a 7724	10	17.2	+20	06	2.6 & 3.8	120	4	160
a 7979	10	52.9	+25	01	4.5 & 6.3	110	6	200
a 8119	11	15.6	+31	49	4.4 & 4.9	170	2	25
a 8202	11	29.8	−28	59	5.8 & 5.9	210	9	105
a 8539	12	21.9	+25	52	6.6 & 7.8	310	0.8	250
a 8575	12	28.0	+10	00	8.5 & 8.8	240	1	

TABLE 10 (CONT'D)

Aitken Number	RA		Dec		Mag	PA	Sep	Dist
	h	m	°	′		°	″	L-Y
a 8600	12	32.6	+18	39	5.2 & 6.7	270	20	270
a 8630	12	39.1	− 1	11	3.6 & 3.7	300	5	34
a 8695	12	50.8	+21	32	5.2 & 8.0	150	0.7	190
a 8706	12	53.7	+38	35	2.9 & 5.4	230	20	135
a 8739	12	58.6	+56	38	4.9 & 8.5	35	1	93
a 8814	13	09.7	+32	21	7.2 & 7.6	340	2	
a 8891	13	21.9	+55	11	2.4 & 4.0	150	14.5	78
a 8974	13	35.2	+36	33	5.1 & 7.0	110	1.6	125
a 9031	13	46.8	+27	14	8.0 & 8.3	140	3	
a 9338	14	38.4	+16	38	4.9 & 5.8	105	5.5	360
a 9343	14	38.8	+13	57	4.4 & 4.8	310	1	230
a 9372	14	42.8	+27	17	2.7 & 5.1	335	3	220
a 9413	14	49.1	+19	19	4.8 & 6.7	350	6	22
a 9494	15	02.2	+47	51	5.5 & 5.9	260	3	41
a 9578	15	16.2	+27	01	7.1 & 8.6	230	0.7	
a 9617	15	21.1	+30	28	5.6 & 6.1	80–90	0.65	50
a 9626	15	22.6	+37	31	4.5 & 6.7	170	10.8	100
a 9701	15	32.4	+10	42	4.2 & 5.2	180	4	170
a 9716	15	34.3	+39	58	7.4 & 7.7	180	1	
a 9737	15	37.5	+36	48	5.1 & 6.0	305	6.3	300
a 9757	15	40.6	+26	27	4.0 & 7.0	280	0.5	140
a 9880	15	58.6	+13	25	7.5 & 8.0	160	1	
a 9979	16	12.8	+33	59	5.8 & 6.7	225	5	71
a 9982	16	13.5	+ 7	30	9.1 & 9.6	30	2	
a 10075	16	26.7	+18	31	7.7 & 7.7	170	0.7	
a 10087	16	28.4	+ 2	06	4.0 & 6.1	330	0.8	190
a 10129	16	35.0	+53	01	5.6 & 6.6	110	3.5	300
a 10157	16	39.4	+31	41	3.0 & 6.5	90	2	30
a 10312	16	59.6	+ 8	31	6.5 & 7.7	180	1	330
a 10345	17	04.3	+54	32	5.8 & 5.8	90	2	71

.ABLE 10 (CONT'D)

Aitken Number	RA h	m	Dec °	′	Mag	PA °	Sep ″	Dist L-Y
a 10418	17	12.4	+14	27	3.5 & 5.4	110	4.5	550
a 10526	17	22.0	+37	11	4.5 & 5.5	315	4	270
a 10576	17	26.5	− 9	58	8.5 & 8.6	60	1.6	
a 10628	17	31.2	+55	13	4.9 & 5.0	310	62	120
a 10759	17	42.8	+72	11	4.9 & 6.1	15	31	69
a 10850	17	49.7	+15	20	7.1 & 7.4	350	0.8	650
a 10990	17	59.2	+ 1	18	4.4 & 9.2	70	0.7	155
a 10993	17	59.4	+21	36	5.1 & 5.2	260	6	460
a 11005	18	00.4	− 8	11	5.3 & 6.0	270	2	100
a 11046	18	02.9	+ 2	32	4.1 & 6.1	100	6	17
a 11089	18	05.8	+26	05	5.9 & 6.0	180	14	230
a 11111	18	07.1	+ 3	59	5.9 & 7.4	40	0.5	180
a 11353	18	24.6	+ 0	10	5.4 & 7.7	315	4	270
a 11483	18	33.6	+16	56	6.8 & 7.2	170	2	155
a 11635	18	42.7	+39	37	4.6 & 6.3	0	3	200
					4.9 & 5.2	110	3	
a 11639	18	43.0	+37	33	4.3 & 5.9	150	44	135
a 11640	18	43.0	+ 5	27	6.3 & 6.7	115	2.5	330
a 11853	18	53.7	+ 4	08	4.5 & 5.4	105	22	140
a 11871	18	55.2	+32	50	5.3 & 7.5	230	1	52
a 12447	19	24.5	+27	13	8.5 & 8.7	290	1.5	
a 12540	19	28.7	+27	51	3.2 & 5.4	55	35	400
a 12815	19	40.6	+50	24	6.3 & 6.4	135	39	78
a 12880	19	43.4	+45	00	3.0 & 7.9	240	2	150
a 12962	19	46.4	+11	41	6.1 & 6.9	115	1.5	135
a 13007	19	48.3	+70	08	4.0 & 7.6	15	3	250
a 13082	19	51.7	+15	10	7.3 & 8.7	300	2	
a 13277	19	59.9	+24	48	5.9 & 6.3	120	0.8	230
a 14126	20	37.7	+40	24	6.5 & 6.8	10	0.8	650
a 14279	20	44.4	+15	57	4.5 & 5.5	270	11	110
a 14360	20	48.8	− 5	49	6.3 & 7.6	0	1	140

TABLE 10 (CONT'D)

Aitken Number	RA h m		Dec ° ′		Mag	PA °	Sep ″	Dist L-Y
a 14573	21	00.5	+ 1	20	7.0 & 7.5	140	1.5	
a 14636	21	04.4	+38	28	5.6 & 6.3	145	25	10.9
a 14787	21	12.8	+37	50	3.8 & 8.0	260	0.7	68
a 15176	21	36.9	− 0	17	7.3 & 7.8	280	0.6	
a 15270	21	41.9	+28	31	4.7 & 6.1	280	1.5	65
a 15769	22	12.3	+29	20	7.6 & 8.1	90	1.5	
a 15902	22	21.5	− 5	06	6.6 & 6.6	320	0.6	230
a 15971	22	26.2	− 0	17	4.4 & 4.6	270	2	140
a 15988	22	27.4	+ 4	10	5.7 & 7.1	120	1	150
a 16095	22	33.6	+39	23	5.8 & 6.6	185	22	1100
a 16345	22	51.4	+44	29	5.7 & 7.7	170	0.5	160
a 16665	23	16.3	+ 5	08	9.0 & 10.0	280	1	
a 16836	23	31.5	+31	03	5.9 & 5.9	240	0.5	460
a 16877	23	35.1	+44	09	6.3 & 7.2	340	0.6	540
a 16979	23	43.4	−18	57	5.7 & 6.9	135	6	200
a 17140	23	56.4	+55	29	5.0 & 7.1	330	3	800

TABLE 11

Double Stars (B) between Declinations −30° and −90°

(see page 125)

No.†	RA h m		Dec ° ′		Mag	PA °	Sep ″	Dist L-Y
b 1	0	29.3	−63	14	4.5 & 4.5	170	27	150
b 2	1	03.9	−46	59	4.1 & 4.1	350	1.4	
b 3	1	14.0	−69	09	5.1 & 7.3	330	5	70
b 4	1	37.9	−56	27	6.0 & 6.0	210	9	21
b 5	2	56.4	−40	30	3.4 & 4.4	90	8	
b 6	3	46.7	−37	47	4.9 & 5.4	210	8	140
b 7	4	19.5	−25	51	6.5 & 6.9	300	0.8	155

† The right ascension and declination are omitted for stars having more than two components.

TABLE 11 (CONT'D)

No.	RA h	RA m	Dec °	Dec ′	Mag	PA °	Sep ″	Dist L-Y
b 8	4	49.8	−53	33	5.6 & 6.4	58	12	
b 9	6	28.6	−50	12	5.3 & 8.3	310	12	135
					6.0 & 6.1	120	1	
					9.1 & 9.2		0.5	
b 10	6	37.3	−48	10	5.1 & 7.4	325	13	360
b 11	7	09.2	−70	25	3.9 & 5.8	300	14	130
b 12	7	14.5	−23	13	4.8 & 6.8	60	27	300
b 13	7	32.2	−23	22	5.9 & 6.0	115	9	130
b 14	8	08.0	−47	11	2.2 & 4.8	220	41	
b 15	8	21.0	−48	20	5.2 & 6.6	140	1	800
b 16	8	54.8	−52	32	4.9 & 7.7	340	2	32
b 17	9	28.8	−40	15	3.9 & 5.1	170	0.8	50
b 18	9	31.9	−48	47	5.8 & 6.4	215	2	270
b 19	9	45.9	−64	50	3.2 & 6.0	130	5	550
b 20	10	29.8	−44	49	6.2 & 6.5	220	13.5	
b 21	10	35.2	−47	58	4.6 & 5.1		0.3	105
b 22	10	36.9	−58	55	4.7 & 7.7	20	15	3000
b 23	11	29.8	−28	59	5.8 & 5.9	210	9	105
b 24	11	50.4	−33	38	5.0 & 5.4	5	1	270
b 25	11	57.0	−77	57	5.4 & 6.2	180	1	220
b 26	12	11.4	−45	27	5.6 & 6.8	245	3	360
b 27	12	23.7	−62	49	1.6 & 2.1	120	5	220
b 28	12	38.8	−48	41	3.1 & 3.2	200	0.3	34
b 29	12	43.2	−67	49	3.9 & 4.2	20	1.3	270
b 30	12	51.7	−56	54	4.3 & 5.5	20	35	360
b 31	13	19.4	−60	44	4.6 & 6.5	345	60	360
b 32	13	48.9	−32	45	4.7 & 6.2	110	8	300
b 33	14	36.6	−60	38	0.3 & 1.7		4	4.3
b 34	15	01.7	−46	51	4.7 & 4.8	75	2	330
b 35	15	05.5	−45	05	5.0 & 5.4		1	360
b 36	15	08.5	−48	33	4.1 & 6.0	145	27	155

TABLE 11 (CONT'D)

No.	RA h m		Dec ° ′		Mag	PA °	Sep ″	Dist L-Y
b 37	15	15.0	−47	42	5.0 & 5.2	145	1.5	
b 38	15	19.3	−44	31	3.9 & 5.6	250	1	300
b 39	15	19.4	−59	09	5.2 & 5.4	45	1	270
b 40	15	31.8	−41	00	3.7 & 4.0		0.5	230
b 41	15	53.7	−33	49	5.4 & 5.7	50	11	160
b 42	15	59.5	−57	38	5.4 & 5.8		0.6	130
					4.9	245	11	
b 43	16	02.5	−19	40	2.9 & 5.1	25	14	410
b 44	16	09.1	−19	21	4.5 & 6.0	10	1	300
					6.5	335	41	
b 45	16	22.6	−23	20	5.1 & 5.9	345	3.5	220
b 46	16	26.5	−26	20	1.2 & 7.0	270	3	230
b 47	17	12.3	−26	30	5.3 & 5.3	165	4	18
b 48	17	15.0	−24	14	5.4 & 6.9	355	11	300
b 49	18	18.6	−61	31	4.3 & 8.1	150	3.3	220
b 50	18	22.4	−20	34	5.0 & 8.3	290	2	330
b 51	18	59.4	−29	57	3.4 & 3.6		0.3	100
b 52	19	03.1	−37	8	5.0 & 5.0	30	3	58
b 53	21	43.5	−82	57	5.5 & 7.7	55	3	300
b 54	23	04.1	−43	47	4.5 & 7.0	60	2	160

TABLE 12

Open Star Clusters (oc)

(*see page 125*)

NGC	RA h m		Dec ° ′		Diam ′	Stars	Mag 5★†
oc 129	0	27.1	+59	56	11	50	11.0
oc 133	0	28.4	+63	04	7	50	11.7
oc 146	0	30.3	+63	00	6	50	10.3
oc 457	1	15.9	+58	02	10	100	8.6
oc 559	1	26.2	+63	02	7	60	11.7

† ★ means "star"; here, magnitude of 5th brightest star in group.

TABLE 12 (CONT'D)

NGC	RA h	RA m	Dec °	Dec ′	Diam ′	Stars	Mag 5★
oc 6633	18	25.2	+ 6	32	20	65	8.0
oc I 4725	18	28.7	−19	16	40	50	9.3
oc I 4756	18	36.5	+ 5	25	70	80	8
oc 6705	18	48.4	− 6	20	10	200	12
oc 6755	19	05.3	+ 4	09	10	50	11.0
oc 6838	19	51.5	+18	39	4	100	11.0
oc 6866	20	02.2	+43	52	6	50	9.4
oc 7209	22	03.2	+46	15	20	50	10.1
oc 7380	22	45.0	+57	50	10	50	9.0
oc 7654	23	22.0	+61	19	12	50	11.0

TABLE 13

Globular Star Clusters (gc)

(*see page 125*)

NGC	RA h	RA m	Dec °	Dec ′	Diam ′	Mag	Dist L-Y
gc 104	0	21.8	−72	22	23	3	17,000
gc 288	0	50.2	−26	52	10	7.2	43,000
gc 362	1	00.6	−71	07	5.3	6.0	34,000
gc 1261	3	10.8	−55	25	2.0	8.5	78,000
gc 1851	5	12.4	−40	06	5.3	6.0	43,000
gc 1904	5	22.2	−24	34	3.2	8.1	54,000
gc 2298	6	47.2	−35	57	1.8	10.1	71,000
gc 2808	9	10.8	−64	39	6.3	5.7	22,500
gc 3201	10	15.5	−46	08	7.7	7.4	9,500
gc 4147	12	07.6	+18	50	1.7	10.3	71,000
gc 4372	12	22.9	−72	23	12	7.8	10,000
gc 4590	12	36.9	−26	28	2.9	7.6	36,000
gc 4833	12	55.9	−70	36	4.7	6.8	15,000
gc 5024	13	10.5	+18	27	3.3	6.9	62,000

TABLE 13 (CONT'D)

NGC	RA h	m	Dec °	′	Diam ′	Mag	Dist L-Y
gc 5053	13	14.0	+17	58	3.5	10.5	49,000
gc 5139	13	23.8	−47	02	23	3	15,000
gc 5272	13	39.9	+28	38	9.8	4.5	31,000
gc 5286	13	43.0	−51	07	1.6	8.5	28,000
gc 5466	14	03.2	+28	46	5.0	10.0	47,000
gc 5634	14	27.0	− 5	45	1.3	10.4	71,000
gc 5824	15	00.9	−32	52	1.0	9.3	91,000
gc 5897	15	14.6	−20	50	7.3	7.3	39,000
gc 5904	15	16.0	+ 2	16	13	3.6	30,000
gc 5927	15	24.4	−50	30	3.0	8.8	10,000
gc 5986	15	42.7	−37	37	3.7	7.0	47,000
gc 6093	16	14.1	−22	51	3.3	6.8	34,000
gc 6101	16	19.9	−72	05	3.8	9.5	30,000
gc 6121	16	20.5	−26	24	14	5.2	7,200
gc 6144	16	24.1	−25	56	3.3	10.3	33,000
gc 6139	16	24.4	−38	43	1.3	9.8	31,000
gc 6171	16	29.7	−12	56	2.2	8.9	7,200
gc 6205	16	39.9	+36	33	10	4.0	22,000
gc 6218	16	44.6	− 1	51	9.3	6.0	19,000
gc 6229	16	45.6	+47	37	1.2	9.7	82,000
gc 6254	16	54.6	− 4	02	8.2	5.4	18,000
gc 6266	16	58.1	−30	03	4.3	7.0	21,000
gc 6273	16	59.6	−26	12	4.3	6.8	19,000
gc 6284	17	01.4	−24	32	1.5	10.0	57,000
gc 6287	17	02.1	−22	39	1.7	10.4	28,000
gc 6293	17	07.2	−26	30	1.9	8.8	45,000
gc 6304	17	11.4	−29	24	1.6	9.2	16,000
gc 6316	17	13.5	−28	05	1.1	9.9	27,000
gc 6341	17	15.6	+43	12	8.3	5.1	31,000
gc 6333	17	16.2	−18	28	2.4	7.4	21,000

TABLE 13 (CONT'D)

NGC	RA h	RA m	Dec °	Dec ′	Diam ′	Mag	Dist L-Y
gc 6356	17	20.7	−17	46	1.7	8.6	33,000
gc 6352	17	21.3	−48	25	2.5	7.9	15,000
gc 6362	17	26.6	−67	01	6.7	7.1	19,000
gc 6388	17	32.6	−44	42	3.4	7.1	27,000
gc 6402	17	35.0	− 3	13	3.0	7.4	21,000
gc 6397	17	36.7	−53	39	19	4.7	7,800
gc 6441	17	46.8	−37	02	2.3	8.4	23,000
gc 6496	17	55.4	−44	15	2.2	9.7	20,000
gc 6541	18	04.4	−43	44	6.3	5.8	14,000
gc 6553	18	06.3	−25	55	1.7	10.0	4,200
gc 6569	18	10.4	−31	50	1.4	10.2	23,000
gc 6584	18	14.6	−52	14	2.5	8.3	41,000
gc 6624	18	20.5	−30	23	2.0	8.6	39,000
gc 6626	18	21.5	−24	54	4.7	6.8	13,000
gc 6638	18	27.9	−25	32	1.4	9.2	47,000
gc 6637	18	28.0	−32	23	2.8	7.5	20,000
gc 6652	18	32.4	−33	02	1.7	8.7	52,000
gc 6656	18	33.4	−23	58	17	3.6	9,800
gc 6681	18	39.9	−32	20	2.5	7.5	65,000
gc 6712	18	50.3	− 8	46	2.1	9.9	25,000
gc 6715	18	51.9	−30	32	2.1	7.1	34,000
gc 6723	18	56.1	−36	42	5.8	6.0	34,000
gc 6752	19	06.4	−60	03	13	4.6	16,000
gc 6779	19	14.7	+30	05	1.8	8.8	45,000
gc 6809	19	36.9	−31	03	10	4.4	23,000
gc 6864	20	03.1	−22	04	1.9	8.6	71,000
gc 6934	20	31.7	+ 7	14	1.5	9.4	54,000
gc 6981	20	50.7	−12	44	2.0	8.6	62,000
gc 7078	21	27.6	+11	57	7.4	5.2	39,000
gc 7089	21	30.9	− 1	03	8.2	5.0	47,000
gc 7099	21	37.5	−23	24	5.7	6.4	39,000

TABLE 14

Diffuse Galactic Nebulae (gn)

(see page 126)

NGC†	RA h	RA m	Dec °	Dec ′	Mag★‡	Dimen ′	Dist L-Y
gn 281	0	50.3	+56	19	7.9	27 × 33	5500
gn 59	0	54.4	+60	49	2.25	18 × 12	360
gn I 1805	2	28.2	+61	16	6.0	50 × 44	2500
gn I 1848	2	47.4	+60	13	7.11	60 × 30	2200
	3	41.1	+32	07	3.94	150 × 90	650
gn 1432	3	42.9	+24	13	4.02	30 × 30	390
	3	43	+23	54		400 × 300	430
gn 1435	3	43.4	+23	37	4.25	30 × 30	430
	3	44.5	+23	57	2.96	27 × 27	430
gn 1499	4	00.2	+38	16	4.05	157 × 54	1300
gn I 2118	5	04	− 7	18	0.34	149 × 63	460
gn I 405	5	13.0	+34	16	5.81	30 × 19	2700
	5	19.1	+ 8	23	5.71	39 × 30	1300
gn I 410	5	19.3	+33	28	7.3	23 × 20	2200
	5	20	− 5			180 × 60	
	5	22.5	+ 6	19	1.70	40 × 40	300
gn 1952	5	31.5	+21	59	15.2	6 × 4.5	3900
	5	32.4	+ 9	54	3.49	30 × 30	1600
gn 1976	5	32.9	− 5	25	4.8	66 × 60	1000?
gn 1977	5	32.9	− 4	52	4.65	42 × 26	
gn 1980	5	32.9	− 5	56	3.0	14 × 14	
gn 1982	5	33.1	− 5	18	9.1		
gn 1990	5	33.6	− 1	14	1.75	50 × 50	
gn I 430	5	36.2	− 7	06	4.88	11 × 11	
gn I 434	5	38.5	− 2	26	1.91	60 × 10	
gn 2070	5	39.1	−69	09			100,000
gn 2024	5	39.3	− 1	52	1.91	31 × 31	
	5	42	+ 2		1.91	300 × 30	

† Nebulae without NGC numbers were discovered after publication of original catalogs.
‡★ means magnitude of the brightest associated star.

TABLE 14 (CONT'D)

NGC	RA h	RA m	Dec °	Dec ′	Mag★	Dimen ′	Dist L-Y
gn 2068	5	44.1	+ 0	02	10.3	8 × 6	
gn 2174-5	6	06.7	+20	30	7.40	29 × 25	3300
gn I 443	6	14.9	+22	48		27.5 × 4.5	
gn 2244	6	30	+ 5		5.3	64 × 61	3600
gn 2264	6	38.2	+ 9	57	4.4	60 × 30	1200
gn I 2177	7	02	−10	38	7.1	60 × 15	1800
gn 3372	10	43.1	−59	24	5.0	120 × 120	3600
gn I 2944	11	33.5	−62	44	3.34	66 × 36	200
	13	21.3	−63	46	1.05	150 × 150	220
	15	51.0	−25		5.44	90 × 30	
	15	55.8	−25	58	3.00	40 × 10	
gn I 4592	16	09.1	−19	19	4.16	195 × 95	520
gn I 4591	16	09.2	−27	47	4.70	23 × 15	390
	16	18.1	−25	28	3.08	130 × 80	550
gn I 4604	16	22.6	−23	20	4.76	124 × 124	520
	16	26.2	−26	20	1.22	126 × 78	390
gn I 4605	16	26.9	−25	02	4.87		550
gn 6193	16	37.5	−48	40	5.35	19 × 12	2700
gn 6514	17	59.6	−23	02	6.91	29 × 27	2200
gn 6523	18	00.7	−24	23	5.4	90 × 40	2500
gn 6611	18	16.0	−13	48	7.3	35 × 28	4600
gn 6618	18	17.9	−16	12	7.7	46 × 37	3300
gn I 1287	18	28.7	−10	50	5.80	44 × 34	780
	19	58.1	+35	08	9.5	15.5 × 8.5	
gn 6888	20	10.6	+38	15	7.44	18 × 12	3000
gn I 1318a	20	14.2	+41	40	2.32	24.5 × 17	820
gn I 1318b	20	26.6	+39	47	2.32	10 × 10	820
gn 6960	20	43.6	+30	33		70 × 6	1300
gn I 5067	20	46.1	+44	11	1.33		240
gn I 5068	20	48.1	+42	21	1.33	43 × 20	
gn I 5070	20	49.2	+44	11	1.33	85 × 75	

TABLE 14 (CONT'D)

NGC	RA h m	Dec ° ′	Mag★	Dimen ′	Dist L-Y
gn 6992-5	20 54.4	+31		73 × 8	
gn 7000	20 57.0	+44 08	1.33	120 × 90	
gn 7023	21 01.2	+67 58	7.20	18 × 18	910
	21 10.6	+59 47	5.65	100 × 65	2000
	21 15.9	+58 22	6.41	15.5 × 13	1700
gn I 1396	21 38	+57 14	5.64	180 × 180	1600

TABLE 15

Planetary Nebulae (pn)

(*see page 130*)

NGC†	RA h m	Dec ° ′	Dimen ″	Mag n	Mag★‡	Dist L-Y
pn 40	0 10.3	+72 14	35 × 38	10.2	11.4	3300
pn 246	0 44.5	−12 09	210 × 240	8.5	11.3	1500
pn 650-1	1 39.1	+51 19	42 × 87	12.2	16.6	8200
pn 1514	4 06.2	+30 38	90 × 120	10.8	9.7	4300
pn 1535	4 11.9	−12 52	17 × 20	9.3	11.8	2200
pn I 2149	5 52.6	+46 06	6 × 12	9.9	14	2900
pn 2392	7 26.2	+21 01	15 × 19	8.3	10.5	1400
pn 2438	7 39.5	−14 36	68	11.3	16.8	5400
pn 2440	7 39.7	−18 05	20 × 54	11.7?	(16.5)	6600
pn 2867	9 19.9	−58 06	8	9.7		2600
pn 3132	10 04.9	−40 12	30	8.2	10.6	1300
pn 3242	10 22.3	−18 23	16 × 26	9.0	11.4	1900
pn 3587	11 11.9	+55 18	200	12.0	14.3	7500
pn 3918	11 47.8	−56 53	10	8.4		14000
pn 4361	12 21.9	−18 29	39 × 44	10.8	12.8	4300
pn 5189	13 29.9	−65 44	20	11		
	15 47.4	−51 21	72	8.4	13.6	14000

† Nebulae without NGC numbers were discovered after publication of original catalogs.
‡ ★ means magnitude of the central star. Parentheses around figures in this column mean they are uncertain.

TABLE 15 (CONT'D)

NGC	RA h m	Dec ° ′	Dimen ″	Mag n	Mag★	Dist L-Y
pn I 4593	16 09.3	+12 13	11 × 15	10.2	10.2	3300
pn 6210	16 42.4	+23 54	8	9.7	12.5	2600
pn II 4634	16 58.6	−21 45	7 × 10	10.8	(17.4)	8300
pn 6543	17 58.5	+66 38	16 × 22	8.8	11.1	1700
pn 6572	18 09.6	+ 6 51	13 × 16	9.6	12	2500
	18 52.0	−32 19	4	10.9		4400
pn 6720	18 51.7	+32 58	59 × 83	9.3	14.7	2200
pn 6803	19 28.9	+ 9 58	5.5	11.4	14.1?	5700
	19 32.8	+30 25	5	9.6	10.3	2500
pn 6818	19 41.1	−14 17	15 × 22	9.9	15.0	2800
pn 6826	19 43.5	+50 25	24 × 27	8.8	10.8	1700
pn 6853	19 57.5	+22 35	240 × 480	7.6	13.4	980
pn 6891	20 12.7	+12 35	7	11.4	11.6	5700
pn 7009	21 01.4	−11 34	12 × 25	8.4	11.7	1400
pn 7027	21 05.2	+42 02	11 × 18	10.4	(17.1)	3600
pn 7293	22 27.0	−21 06	720 × 900	6.5	13.3	590
pn I 1470	23 03.2	+59 59	45 × 70	8.1	11.9	6800
pn 7635	23 24.5	+60 54	180 × 205	8.5	8.5	1700
pn 7662	23 23.5	+42 15	14 × 17	8.9	12.5	1800

TABLE 16

Galaxies (eg)

(*see page 130*)

NGC	RA h m	Dec ° ′	Mag	Dimen ′	Type
eg 55	0 12.5	−39 30	7.8	25.0 × 3.0	S
eg 205	0 37.6	+41 25	10.8	8.0 × 3.0	Ep
eg 224	0 40.0	+41 00	5	160 × 40	Sb
eg 221	0 40.0	+40 36	9.5	2.6 × 2.1	E
eg 247	0 44.6	−21 01	10.7	18.0 × 5.0	S

TABLE 16 (CONT'D)

NGC	RA h	RA m	Dec °	Dec ′	Mag	Dimen ′	Type
eg 253	0	45.1	−25	34	7.0	22.0 × 6.0	Sc
eg SMC	0	50	−73		1.5	216 × 216	I
eg 598	1	31.1	+30	24	7.8	60 × 40	Sc
eg 1068	2	40.1	−00	14	10.0	2.5 × 1.7	Sb
eg 1097	2	44.3	−30	29	10.6	9.0 × 5.5	SB
eg 1291	3	15.5	−41	17	10.2	5.0 × 2.0	E
eg 1313	3	17.6	−66	40	10.8	4.5 × 3.0	SB
eg 1316	3	20.7	−37	25	10.1	3.5 × 2.5	S
eg 1399	3	36.6	−35	37	10.9	1.4 × 1.4	E
eg 1549	4	14.7	−55	42	11.0	3.0 × 2.7	E
eg 1553	4	15.2	−55	54	10.2	3.0 × 2.5	S
eg 1566	4	18.9	−55	04	10.5	8.0 × 6.0	S
eg 1792	5	03.5	−38	04	10.7	3.0 × 1.0	S
eg LMC	5	26	−69		0.5	432 × 432	I
eg 2403	7	32.0	+65	43	10.2	16 × 10	Sc
eg 2683	8	49.6	+33	38	10.8	10.0 × 1.0	Sc
eg 2841	9	18.6	+51	12	10.5	6.0 × 1.6	Sb
eg 2903	9	29.3	+21	44	10.3	11.0 × 5.0	Sc
eg 2997	9	43.5	−30	58	11.0	6.0 × 5.0	S
eg 3031	9	51.5	+69	18	8.9	16 × 10	Sb
eg 3034	9	51.9	+69	56	9.4	7.0 × 1.5	I
eg 3115	10	02.8	−07	28	9.8	4.0 × 1.0	E
eg 3310	10	35.7	+53	46	10.9	1.5 × 0.8	Sb
eg 3368	10	44.2	+12	05	10.4	7.0 × 4.0	Sa
eg 3379	10	45.2	+12	51	10.8	2.0 × 2.0	E
eg 3521	11	03.2	+00	14	10.3	4.5	Sc
eg 3556	11	08.7	+55	57	11.0	8.0 × 1.5	Sb
eg 3621	11	15.9	−32	32	10.6	5.0 × 2.0	S
eg 3623	11	16.3	+13	23	10.5	8.0 × 2.0	Sb
eg 3627	11	17.6	+13	17	9.9	8.0 × 2.5	Sb

TABLE 16 (CONT'D)

NGC	RA h	m	Dec °	′	Mag	Dimen ′	Type
eg 3893	11	46.1	+49	00	11.0	4.1	Sc
eg 4038	11	59.3	−18	35	11.0	2.5 × 2.5	S
eg 4214	12	13.1	+36	36	10.7	8.0 × 4.0	I
eg 4254	12	16.3	+14	42	10.5	4.5 × 4.5	Sc
eg 4258	12	16.5	+47	35	10.2	20.0 × 6.0	Sb
eg 4303	12	19.4	+04	45	10.4	6.0 × 6.0	SBc
eg 4321	12	20.4	+16	06	10.8	5.0 × 5.0	Sc
eg 4374	12	22.6	+13	10	10.9	2.9 × 2.6	E
eg 4382	12	22.8	+18	28	10.5	4.0 × 2.5	E
eg 4406	12	23.7	+13	13	10.9	3.8 × 2.9	E
eg 4449	12	25.8	+44	22	10.3	4.5 × 2.5	I
eg 4472	12	27.3	+08	16	10.1	4.5 × 4.0	E
eg 4490	12	28.3	+41	55	10.5	4.0 × 1.8	Sc
eg 4486	12	28.3	+12	40	10.7	3.3 × 3.3	E
eg 4494	12	28.9	+26	03	10.9	1.6 × 1.6	E
eg 4501	12	29.5	+14	42	10.9	6.0 × 3.0	Sc
eg 4526	12	31.6	+07	58	10.7	6.0 × 1.2	Sa
eg 4559	12	33.5	+28	14	10.7	8.0 × 2.0	Sc
eg 4565	12	33.9	+26	16	10.7	15.0 × 1.1	Sb
eg 4594	12	37.3	−11	21	8.1	7.0 × 1.5	Sa
eg 4605	12	37.8	+61	53	10.9	3.0 × 1.0	Sc
eg 4631	12	39.8	+32	49	9.6	12.0 × 1.2	Sc
eg 4636	12	40.3	+02	57	10.8	1.2 × 1.1	E
eg 4649	12	41.1	+11	49	10.6	3.9 × 3.1	E
eg 4697	12	46.0	−05	32	10.6	3.0 × 1.2	E
eg 4699	12	46.5	−08	24	10.5	3.7 × 2.0	SBb
eg 4725	12	48.1	+25	46	10.8	5.0 × 4.0	SBb
eg 4736	12	48.6	+41	23	9.0	5.0 × 3.5	Sb
eg 4753	12	49.8	−00	55	10.5	2.7	I
eg 4826	12	54.3	+21	47	8.0	8.0 × 4.0	Sb

TABLE 16 (CONT'D)

NGC	RA h m		Dec ° ′		Mag	Dimen ′	Type
eg 4945	13	02.4	−49	01	9.2	11.5 × 2.0	S
eg 5055	13	13.5	+42	17	10.5	8.0 × 3.0	Sb
eg 5102	13	19.1	−36	23	10.8	1.5 × 0.5	E
eg 5128	13	22.4	−42	45	7.2	10.0 × 8.0	I
eg 5194	13	27.8	+47	27	10.1	12.0 × 6.0	Sc
eg 5236	13	34.3	−29	37	8.0	10.0 × 8.0	Sc
eg 5248	13	35.1	+09	08	11.0	3.2 × 1.4	Sc
eg 5253	13	37.1	−31	24	10.8	4.0 × 1.5	E
eg 5457	14	01.4	+54	35	9.0	22 × 22	Sc
eg 6822	19	42.1	−14	53	11.0	20 × 10	I
eg 7793	23	55.3	−32	51	9.7	6.0 × 4.0	S

TABLE 17

Harvard College Observatory Plates Used for Photographic Atlas Charts

(see page 138)

No	RA	Dec	Plate No	Date	Exp
1		+90	39209	Sept 13, 1943	90
2	0	+60	38571	Aug 20, 1944	87
3	3	+60	39974	Nov 20, 1946	102
4	6	+60	32373	Oct 4, 1934	80
5	9	+60	38865	Feb 11, 1945	85
6	12	+60	39480	Feb 2, 1946	91
7	15	+60	38419	May 18, 1944	90
8	18	+60	38416	May 18, 1944	92
9	21	+60	38570	Aug 20, 1944	89
10	0	+30	40874	Sept 12, 1948	120
11	2	+30	32372	Oct 4, 1934	87
12	4	+30	39477	Feb 2, 1946	96
13	6	+30	33902	Nov 9, 1937	87

TABLE 17 (CONT'D)

No	RA	Dec	Plate No	Date	Exp
14	8	+30	38955	Apr 5, 1945	115
15	10	+30	40034	Dec 30, 1946	111
16	12	+30	32615	Apr 2, 1935	101
17	14	+30	42221	Mar 21, 1952	90
18	16	+30	40156	Mar 26, 1947	113
19	18	+30	39682	June 22, 1946	86
20	20	+30	42066	Aug 2, 1951	84
21	22	+30	39961	Nov 18, 1946	88
22	0	0	38634	Sept 23, 1944	106
23	2	0	39390	Dec 22, 1945	103
24	4	0	40455	Oct 12, 1947	98
25	6	0	34608	Dec 1, 1938	90
26	8	0	38918	Mar 9, 1945	92
27	10	0	39464	Jan 28, 1946	110
28	12	0	39488	Feb 4, 1946	116
29	14	0	40243	June 10, 1947	91
30	16	0	39681	June 22, 1946	91
31	18	0	25745	Aug 1, 1946	90
32	20	0	38632	Sept 23, 1944	94
33	22	0	40497	Nov 2, 1947	92
34	0	−30	22000	Aug 25, 1941	90
35	2	−30	27256	Dec 21, 1948	90
36	4	−30	25102	Oct 5, 1945	90
37	6	−30	21512	Apr 19, 1941	90
38	8	−30	27422	Apr 27, 1949	90
39	10	−30	25182	Dec 5, 1945	90
40	12	−30	25568	June 21, 1946	90
41	14	−30	25515	June 2, 1946	90
42	16	−30	28356	Aug 18, 1952	90
43	18	−30	16828	June 4, 1935	90

continued on page 382

TABLE 18
Planetary Data

(*see page 292*)

	Diameter		Rotation Period	Distance from Sun		Sidereal Period	Synodic Period (days)	Number of Satellites
	miles	km		miles	km			
Mercury	3007	4840	59 d	36,000,000	57,900,000	87.969 d	115.88	0
Venus	7705	12400	243 d	67,200,000	108,100,000	224.643 d	583.92	0
Earth	7917	12742	$23^h\ 56^m$	92,900,000	149,500,000	365.2564 d		1
Mars	4225	6800	$24^h\ 37^m$	141,500,000	227,800,000	686.792 d	779.94	2
Jupiter	88730	142800	$9^h\ 50^m$	483,300,000	777,800,000	11.86223 y	398.88	12
Saturn	75060	120800	$10^h\ 14^m$	886,100,000	1,426,100,000	29.45772 y	378.09	9
Uranus	29600	47600	$10^h\ 49^m$	1,782,800,000	2,869,100,000	84.01342 y	369.66	5
Neptune	27700	44600	$15^h\ 40^m$	2,793,000,000	4,495,600,000	164.79395 y	367.49	2
Pluto	8900?	14400?	16^h	3,665,000,000	5,898,900,000	248.4302 y	366.74	0

TABLE 17 (CONT'D)

No	RA	Dec	Plate No	Date	Exp
44	20	−30	24127	Aug 17, 1944	90
45	22	−30	19905	June 26, 1939	90
46	0	−60	28017	Aug 14, 1950	60
47	3	−60	20924	Aug 8, 1940	90
48	6	−60	27251	Dec 6, 1948	90
49	9	−60	19628	Apr 20, 1939	90
50	12	−60	26873	June 8, 1948	90
51	15	−60	28001	Aug 2, 1950	90
52	18	−60	19617	Apr 19, 1939	90
53	21	−60	26286	June 17, 1947	90
54		−90	19031	June 3, 1938	90

For Table 18, see preceding page

TABLE 19

Latitudes of Mercury, Venus, and Mars at Unit Distance from the Earth

(see page 301)

MERCURY		VENUS		MARS	
Day	Lat	Day	Lat	Day	Lat
0	+2.2	0	+0.3	0	+0.2
4	2.3	10	1.2	30	−0.6
8	2.3	20	2.1	60	1.2
12	2.0	30	2.7	90	1.9
16	1.5	40	3.2	120	2.3
20	0.9	50	3.4	150	2.5
24	+0.3	60	3.3	180	2.5
28	−0.4	70	3.0	210	2.2
32	1.0	80	2.4	240	1.8
36	1.6	90	1.6	270	0.9
40	2.1	100	+0.8	300	−0.3

TABLE 19 (CONT'D)

MERCURY		VENUS		MARS	
Day	Lat	Day	Lat	Day	Lat
44	2.5	110	−0.2	330	+0.4
48	2.7	120	1.1	360	1.1
52	2.9	130	1.9	390	1.9
56	2.9	140	2.6	420	2.4
60	2.7	150	3.1	450	2.8
64	2.3	160	3.4	480	3.0
68	1.6	170	3.4	510	3.0
72	−0.8	180	3.1	540	2.8
76	+0.1	190	2.6	570	2.6
80	1.0	200	1.7	600	2.1
84	+1.7	210	1.1	630	1.6
		220	−0.1	660	+1.0

TABLE 20

Latitudes of Jupiter and Saturn as Seen from the Sun

(see page 301)

Jupiter				Saturn			
Year	Lat	Year	Lat	Year	Lat	Year	Lat
0	+1.3	7.5	0.6	0	+1.4	15	−1.5
0.5	1.2	8.0	−0.3	1	1.0	16	−0.9
1.0	1.0	8.5	+0.1	2	0.6	17	−0.4
1.5	0.8	9.0	+0.4	3	0.1	18	+0.2
2.0	0.5	9.5	+0.7	4	−0.3	19	+0.8
2.5	+0.2	10.0	+0.9	5	−1.0	20	+1.3
3.0	−0.1	10.5	+1.1	6	−1.5	21	+1.7
3.5	−0.4	11.0	+1.3	7	−1.8	22	+2.1
4.0	−0.8	11.5	+1.3	8	−2.1	23	+2.3
4.5	−1.0			9	−2.4	24	+2.5
5.0	−1.2			10	−2.5	25	+2.5
5.5	−1.3			11	−2.5	26	+2.4
6.0	−1.3			12	−2.4	27	+2.2
6.5	−1.1			13	−2.2	28	+2.0
7.0	−0.9			14	−1.6	29	+1.6

TABLE 21

Meteor Showers

(*see page 307*)

Shower	Mean Evening Date of Maximum	Radiant RA °	Radiant Dec °	Associated Comet if Known	Hourly Number of Shower Meteors for Single Observer	Shower Duration (days)†	Type of Shower‡
Quadrantid	Jan 3	232	+50		40	0.6	An
Corona Australid	Mar 16	245	−48		5	10	An
Virginid	Mar 26	190	00			(15)	St
α Virginid	Apr 9	210	−10			(20)	St
Lyrid	Apr 21	274	+34	1861 I	12	2.3	An
Aquarid	May 4	336	00	Halley?	20	18	An
Ophiuchid	June 15	260	−20			(25)	St
Draconid	June 28	215	+55	Pons-Winnecke	50 (1916)		Ir
Sagittariid	July 6	300	−30			(20)	St
Capricornid	July 22	318	−15			(25)	St
Aquarid	July 29	339	−17		20	20	An
Pisces Australid	July 30	340	−30			(20)	St
α Capricornid	Aug 1	308	−10			(30)	St
N—δ Aquarid	Aug 3	337	00		10	(30)	St
N—Aquarid	Aug 5	331	−06			(40)	St

TABLE 21 (CONT'D)

Shower	Mean Evening Date of Maximum	Radiant RA °	Radiant Dec °	Associated Comet if Known	Hourly Number of Shower Meteors for Single Observer	Shower Duration (days)	Type of Shower
S–Aquarid	Aug 5	335	−15			(40)	St
Perseid	Aug 11	046	+58	1862 III	50	5.0	An
Cygnid	Aug 18	290	+55			(15)	St
Giacobinid	Oct 9	262	+54	Giacobini-Zinner	20,000 (1933) 1,000 (1946)	0.05	Ir
Orionid	Oct 20	095	+15	Halley	25	8	An
S. Taurid-Arietid	Nov 5	053	+14	Encke	15	(30)	St
N. Taurid	Nov 10	057	+22			(45)	St
Bielid	Nov 14	024	−44	Biela 1826	5,000–10,000 (1827 and 1885) 15	0.2	Ir
Leonid	Nov 16	152	+22	1886	10,000 (1883) 1,000 (1867)	4	Ir–An
Geminid	Dec 13	113	+32		50	6.0	An
Ursid	Dec 22	217	+76	Tuttle (1939)	15	2.2	An

† Estimated range between days whose rate is ¼ that of maximum; parentheses mean "uncertain."
‡ St, stream; An, annual shower expected; Ir, intense showers irregular.

TABLE 22

Recommended Telescopes

(*see page 313*)

Refractors:	Unitron, 3-inch or larger diameter Balscope, for beginner, by Bausch and Lomb Prism Binocular 7 × 50 Bausch and Lomb
Reflectors:	Cave, Celestron Pacific, Colonial, Criterion, Ealing, Group 128, Optical Craftsmen, Skilcraft
Catadioptric:	Questar
Professional:	Boller and Chivens
Miscellaneous Surplus Optical Parts:	Edmund, Jaegers

For details, other suggestions, and addresses, see *Sky and Telescope*, published at the Harvard College Observatory, Cambridge, Massachusetts 02138.

TABLE 23

Recommended Cameras

(*see page 321*)

Large Cameras	Crown Graphic "45," 4 × 5, f/4.5 lens Speed Graphic "45," 4 × 5, f/4.5 lens Polaroid Land with 3000 speed film Polaroid Land Model 180 (purchase as used camera) Polaroid Land backs for various makes of cameras
Intermediate Cameras	Rolleiflex 2 × 2, f/3.5 Hasselblad 2 × 2, f/3.5, or various telephoto lenses
35-Millimeter Cameras	Leica M-4 or M-5 Leicaflex Nikon, with various telephoto lenses Canon, with f/1.2 lens and also telephoto lenses Konica Pentax

TABLE 24

Almanac Data, 1959

(*see pages 324 and 325*)

Date	Eqn. of Time m s	Sid Time of 0^h GMT h m s	Date	Eqn. of Time m s	Sid Time of 0^h GMT h m s
Jan 0	− 2 39.1	6 35 39.05	May 25	+ 3 17.1	16 07 19.58
5	− 4 59.8	6 55 21.83	30	+ 2 44.2	16 27 02.36
10	− 7 10.6	7 15 04.61			
15	− 9 07.7	7 34 47.38	Jun 4	+ 1 59.7	16 46 45.13
20	−10 48.2	7 54 30.16	9	+ 1 06.0	17 06 27.91
25	−12 09.8	8 14 12.94	14	+ 0 05.7	17 26 10.69
30	−13 11.8	8 33 55.71	19	− 0 58.0	17 45 53.47
			24	− 2 02.6	18 05 36.24
Feb 4	−13 53.9	8 53 38.49	29	− 3 05.7	18 25 19.02
9	−14 16.0	9 13 21.27			
14	−14 18.3	9 33 04.05	Jul 4	− 4 04.5	18 45 01.80
19	−14 01.6	9 52 46.82	9	− 4 56.0	19 04 44.57
24	−13 27.5	10 12 29.60	14	− 5 37.5	19 24 27.35
			19	− 6 06.6	19 44 10.13
Mar 1	−12 38.0	10 32 12.38	24	− 6 22.2	20 03 52.91
6	−11 35.9	10 51 55.15	29	− 6 23.7	20 23 35.68
11	−10 23.3	11 11 37.93			
16	− 9 02.3	11 31 20.71	Aug 3	− 6 10.6	20 43 18.46
21	− 7 35.0	11 51 03.48	8	− 5 42.5	21 03 01.23
26	− 6 04.2	12 10 46.26	13	− 4 59.3	21 22 44.01
31	− 4 32.7	12 30 29.04	18	− 4 01.8	21 42 26.79
			23	− 2 51.6	22 02 09.55
Apr 5	− 3 03.4	12 50 11.81	28	− 1 30.4	22 21 52.34
10	− 1 38.5	13 09 54.59			
15	− 0 19.9	13 29 37.37	Sep 2	− 0 00.2	22 41 35.11
20	+ 0 50.8	13 49 20.14	7	+ 1 37.4	23 01 17.89
25	+ 1 51.5	14 08 02.92	12	+ 3 20.5	23 21 00.67
30	+ 2 40.3	14 28 45.60	17	+ 5 06.9	23 40 43.45
			22	+ 6 53.7	0 00 26.23
May 5	+ 3 15.5	14 48 28.47	27	+ 8 38.0	0 20 09.00
10	+ 3 36.7	15 08 11.25			
15	+ 3 43.8	15 27 54.03	Oct 2	+10 17.4	0 39 51.78
20	+ 3 37.2	15 47 36.80	7	+11 49.4	0 59 34.56

TABLE 24 (CONT'D)

Date	Eqn. of Time m s	Sid Time of 0^h GMT h m s	Date	Eqn. of Time m s	Sid Time of 0^h GMT h m s
Oct 12	+13 12.1	1 17 17.33	Nov 26	+13 00.3	4 16 42.32
17	+14 23.0	1 39 00.11			
22	+15 19.5	1 58 42.89	Dec 1	+11 19.6	4 36 25.10
27	+15 59.1	2 18 25.66	6	+ 9 22.3	4 56 07.88
			11	+ 7 11.8	5 15 50.65
Nov 1	+16 20.1	2 38 08.44	16	+ 4 51.4	5 35 33.43
6	+16 21.4	2 57 51.22	21	+ 2 24.5	5 55 16.21
11	+16 02.5	3 17 33.99	26	− 0 05.0	6 14 58.99
16	+15 22.7	3 37 16.77	31	− 2 32.9	6 34 41.76
21	+14 21.8	3 59 59.55			

TABLE 25

Code for Use with Table 24

(*see page 325*)

Year	Corr ST m	Corr ST s	Year	Corr ST m	Corr ST s
1959	0	0.00	1977	+2	31.52
1969	+2	16.75	1978	+1	34.23
1970	+1	19.46	1979	+0	36.94
1971	+0	22.16	1980 Jan–Feb 29	−0	20.35
1972 Jan–Feb 29	−0	35.13	1980 Mar–Dec	+3	36.20
1972 Mar–Dec	+3	21.43	1981	+2	38.91
1973	+2	24.13	1982	+1	41.62
1974	+1	26.84	1983	+0	44.33
1975	+0	29.55	1984 Jan–Feb 29	−0	12.96
1976 Jan–Feb 29	−0	27.74	1984 Mar–Dec	+3	43.59
1976 Mar–Dec	+3	28.81	1985	+2	46.30

Index

Illustrations are not generally included because of their complete listing on pages xi–xiii. The 25 Tables are also omitted, because of their coverage on page xiv. Irrespective of their specific form of mention in the text, the constellations are listed under both Latin and English-equivalent names for rapid and easy location of the text references. The principal text reference for a constellation or planet appears in boldface figures if more than one reference occurs. Table 5 serves as an index to constellations alphabetically arranged by Latin names.

Abenezra, 282
aberration, chromatic, 313, 315
 of light, 331, 332
Abulfeda, 282
Acamar, 116
Achernar, 105, 113, 232
achromatic telescope, 310
Air Pump (Antlia), **114,** 217
Aitken, Robert G., 125
Albategnius, 268
Alcor, 150–51
Aldebaran, 105, 109, 163
Algol, 119, 163
Aliacensis, 282
Alioth, 107, 108, 110
Alkaid, 107
Almagest, 117
Almanon, 282
Al Na'ir, 244
Alnilam, 112
Alphard, 105, 111
Alpheratz, 104
Alphonsus, 255, 270, 283
Alpine Valley, 284
Alps, 284
Altair, 104
Altar (Ara), 105–6, **112**
altazimuth mounting, 315–16
altitude, 316
American Association of Variable Star Observers (AAVSO), 120, 289
Ames, Adelaide, 131
Andromeda, 104, **110,** 158–59, 329
Andromeda Nebula, 158, 161, 206
annular eclipse, 291
Antares, 105, 109, 116, 249
Antlia, **114,** 217
Apennines (mountains), 268, 271, 283, 284
Apianus, 282
apparent solar time, 323
apparent sun, 323
Apus, 113
Aquarius, 104, 105, **109,** 112, 113, 205, 228
Aquila, 105, **111,** 112, 115, 200, 202, 203
Ara, 105–6, **112**
Archer (Sagittarius), 105, **109,** 111, 112, 114, 115, 200, 224, 225, 226, 249
Archimedes, 279, 284
Arcturus, 105
Argo, 7, 113
Argonautic expedition, 109, 113
Ariadaeus, 269
Aries, 104, **109,** 113, 161
Aristarchus, 280
Aristillus, 284
Aristoteles, 284
Arrow (Sagitta), 105, **111,** 202
artificial satellites, 307, 319–20
Arzachel, 270, 283
asterisms, definition of, 7
asteroids, 304
astigmatism, 313
astronomical unit, 301

Atlas, 275
atmosphere: of Mars, 294
of Venus, 293
atmospheric tremor, 312
Augean stables, 111
Auriga, 104, 108, **110**, 115, 147, 164, 165
Autolycus, 284
azimuth, 316
Azophi, 282

barred spiral, 131
Bayer, Johann, 7, 113, 114, 139
Bayer Group, 113–14
Bear Driver (Boötes), **108**, 152–53, 173, 175, 196–97
Beehive, 110, 166–67
beetle, 250
Bellatrix, 105
Belt of Orion, 105, 112
Berenice's Hair (Coma Berenices), **108**, 170, 194
Bessel, F. W., 213
Beta Lyrae, 119
Betelgeuse, 105, 112, 116
Big Dipper, 6, 7, 104, 105, 107, 108, 151, 327
Bird of Paradise (Apus), 113
bolides, 306
Bonpland, 270
Boötes, **108**, 152–53, 173, 175, 196–97
Bull (Taurus), 105, **109**, 112, 163, 165, 186, 187
Bullialdus, 282

Caelum, 114
calendar, 322
Camelopardalis, 104, **108**, 109, 141, 146
camera(s), 311, 317–21
Campanus, 265
Cancer, **110**, 111, 166–67
Canes Venatici, **108**, 171, 172, 194
Canis Major, **112**, 212, 214
Canis Minor, 104, **112**
Canopus, 105, 234
Cape Banat, 273
Capella, 104, 105, 146–47
Caph, 104
Capricornus, **109**, 112
Carina, 105, **113**, 115, 234, 236, 237, 238
Carpathian Mountains, 273
Cassini, 279
Cassiopeia, 104, **110**, 115, 141, 142–43, 144, 164
Castor, 105, 110, 166
catadioptric telescope, 311
Catharina, 282
Caucasus Mountains, 277
Cederblad, S., 126
celestial: coordinates, 117
equator, 330
latitude, 297
longitude, 297
Centaur (Centaurus), 106, 111, **112**, 115, 198, 218, 219, 221, 238, 239, 240, 249
Centaur (Sagittarius), **109**, 112
Cepheids, 120
Cepheus, 104, **110**, 111, 115, 141, 156–57
Ceres, 304
Cetus, **110**, 113, 183
Head of, 104
Neck of, 104, 185
Chamaeleon (Chameleon), **113**, 247
Charioteer (Auriga), 104, 108, **110**, 115, 147, 164, 165
chromatic aberration, 313, 315
Circinus, **114**, 241
Circlet, 104, 182
Clark, Alvan, 213
Clavius, 262
Claws. *See* Libra
clefts, 252
Cleomedes, 274
cliffs, 252
Clock (Horologium), 114
clockwork drive, 316
cluster variables, 120
clusters, 115, 125
globular, 115, 125, 174, 198, 222, 224–25, 230, 238, 242–43
open, 108, 115, 125, 142, 162, 170–71, 190, 214, 222, 238
Coalsack, 112, 115, 239
Northern, 115, 248
colors, star, 118
Columba, 113
Coma Berenices, **108**, 170, 194
comes, 125
Comet, Halley's, 306
comet seekers, 306
comets, 304–6, 320
Compasses (Circinus), **114**, 241
conjunction, inferior and superior, 292

Copernicus, 252, 270, 273
corona, 290
Corona Australis, **112,** 113, 227
Corona Borealis, 105, **108,** 175
Corvus, 106, **111,** 218, 219
Crab (Cancer), **110,** 111, 166–67
crab (in moon), 250
Crab Nebula, 126, 165
Crane (Grus), 105, **113,** 207, 229, 244
Crater, **111,** 218
crater(s), 251, 252, 254
 cones, 252
 pits, 251, 252
crepe ring, 297
Cross (Crux), 106, **112,** 115, 238, 249
Crow (Corvus), 106, **111,** 218, 219
Crux, 106, **112,** 115, 238, 249
Cup (Crater), **111,** 218
Cygnus, 104, 105, **111,** 115, 156–57, 178–79, 181, 202, 248
Cyrillus, 282

dark glass(es), 286, 321
Daylight Time, 324
declination, 117, 329, 331
Delambre, 268–69
Delphinus, **112,** 113, 203
Deneb, 104
Denebkaitos, 117
Denebola, 105, 117, 170
depressions, 252
difficult object, 115
diffuse galactic nebulae, 115, 126
Dolphin (Delphinus), **112,** 113, 203
Dorado, **113,** 234, 235
double stars, 120–25
Dove (Columba), 113
Draco, 105, **108,** 111, 141, 152–53, 154
Draconis, 331
Dragon (Draco), 105, **108,** 111, 141, 152–53, 154
Dreyer, J. L. E., 125
Dubhe, 107
Dumbbell Nebula, 178

Eagle (Aquila), 105, **111,** 112, 115, 200, 202, 203
"earth shine" on the moon, 251
Easel (Pictor), **114,** 235
eastern: elongation, 292
 quadrature, 292
eclipses, solar and lunar, 290–91
eclipsing stars, 119
ecliptic, 117, 297, 329, 330
Ecliptic Star Maps, 297–300
elliptical galaxy, 131
elongation, eastern, 292
 western, 292, 298
El Rischa, 104, 184
Endymion, 275
Equation of Time, 324
equator, 4
equatorial mounting, 315, 316
Equuleus, 112
Eratosthenes, 270–71
Eridanus, 105, 110, 111, **113,** 186, 189, 209, 210–11, 232, 233
Erymanthian boar, 112
Eudoxus, 284
Euler, 273
external galaxies, 115, 130
eyepiece, 309

Fabricius, 258–59
faculae, 286
False Cross, 106
field lens, 309
filter(s), 286, 321
fireballs, 306
Fish (Pisces), 104, **109,** 112, 113, 182, 184. *See also* Flying Fish *and* Southern Fish
fixed stars, 292
Flaming Star, 164
Flamsteed, John, 113
 crater, 272–73
 numbers, 7, 139
Flare stars, 119
Fly (Musca), **114,** 249
Flying Fish (Volans), **113,** 235
Fomalhaut, 229
Fornax, 114
Foucault test, 313–15
Fox (Vulpecula), 111
Fracastorius, 259, 282
Franklin, 274–75
Frederik's Glory, 159
full moon, 251
Furnace (Fornax), 114

galactic center, 115
galaxies, 115, 130
Galaxy. *See* Milky Way
Galileo, 296
Garnet Star, 157
gaseous nebulae, 115, 130, 142, 156, 164–65, 200

gases, frozen, 304
Gassendi, 283
gauze ring, 297
Gemini, 104, **110**, 115, 164, 166
Geminus, 274
gibbous phase, 292
Giraffe (Camelopardalis), 104, **108**, 109, 141, 146
globular (star) clusters, 115, 125, 174, 198, 222, 224, 230, 238, 242–43
Goldfish (Dorado), **113**, 234, 235
granular structure (solar surface), 286
Graving Tool (Caelum), 114
Great Bear (Ursa Major), 7, 104, **107**, 108, 148–49, 150, 153, 168, 170
Great Nebula, 248
Greenwich, meridian of, 323
  Civil Time, 323
  Mean Time, 323
  Sidereal Time, 325
  Standard Time, 323
Gregorian calendar, 322
Grimaldi, 272
Grus, 105, **113**, 207, 229, 244
Guards (Guardians of the Pole), 105, 108, 153
Guericke, 270

Haemus Mountains, 268
Halley, Edmund, 114
Halley's Comet, 306
Hare (Lepus), 105, **112**
Heavenly G, 105
Heavenly Waters, 109, 110, 112–13
Helical Nebula, 228
Hell, 282
Hercules, 105, 108, **111**, 153, 154, 175, 198, 199, 248
  crater, 275
Hercules cluster, 174
Hercules Family, 110, 111–12
Herdsman (Boötes), **108**, 152–53, 173, 175, 196–97
Hevelius, John, 108
Hipparchus, Greek astronomer, 117
  crater, 268
Hippocrene, 113
Hippolyta, 113
Horologium, 114
"Horse and Rider," 151
Horsehead Nebula, 248
Horseshoe Nebula, 200–201, 249
hour angle, 332
hour circle, 117, 329
Hubble, Edwin P., 130
Hunting Dogs (Canes Venatici), **108**, 171, 172, 194
Hyades, 7, 109, 162, 163, 186
Hydra, **111**, 190, 193, 218–19, 221
Hydrus, 113
Hyginus, 269

Indus (Indian), 113
inferior: conjunction, 292
  planet, 292
Innes, R. T. A., 125
irregular variables, 120

Jenkins, Louise F., 116
Job's Coffin, 113
Julius Caesar, 283, 322
Juno, 304
Jupiter, 118, 292, **296**, 301, 304
  "red spot of," 296
  satellites of, 296

Keel (Carina), 7, 105, **113**, 115, 234, 236, 237, 238
Kepler (crater), 273
Keystone, 105
Kies, 282
"knife edge" test, 315
Knot of Pisces, 104, 184
Kozyrev, N. A., 255
Kukarkin, B. V., 118

La Caille, 114
La Caille Family, 114
Lacerta, 110
Lacus, 251
Lady of the Chair (Cassiopeia), 104, **110**, 115, 141, 142–43, 144, 164
lady reading a book, 250
Lagoon Nebula, 225
Lalande, 270
Lambert, 284
Langrenus, 266–67
Large Magellanic Cloud, 105, 113, 115, 234
Larger Dog (Canis Major), **112**, 212, 214
latitude, 117, 301, 329
Leighton, R. B., 294
Leo, 104, **109**, 111, 168, 170, 192–93, 194
Leo Minor, **109**, 168–69

Lepus, 105, **112**
Lernean Hydra, 111
Level (Norma), **114**, 115, 240, 249
Libra, 105, **109**
light-year, 116
limb, 285
Linné (crater), 255, 276
Lion (Leo), 104, **109**, 111, 168, 170, 192–93, 194
Little Bear (Ursa Minor), 105, **108**, 140, 141, 152, 153
Little Dipper, 105, 108
Little Horse (Equuleus), 112
Lizard (Lacerta), 110
Local: Apparent Time, 324
  Longitude Difference, 324, 326
  Mean (solar) Time, 7, 324
  Sidereal Time, 324–26, 332
  time, 323
Lohmann, W., 125
longitude, 117, 301, 329
Longomontanus, 262–63
long-period variables, 120
Loop Nebula, 178–79
Lozenge, 105, 154
Lubiniezky, 282
lunar eclipse, 291
Lupus, 106, **112**, 115, 221, 223, 241
Lynx, **108**, 109
Lyra (Lyre), 105, **111**, 153, 176–77, 331

Macrobius, 283
Madonna and Child, 250
Mädler, 282
Magellanic Clouds. *See* Large *and* Small
Maginus, 262
magnetic fields, 288
magnifier, 310
magnifying power, 139
magnitude, 4, 117–18
Mairan, 281
man in the moon, 250
Mare, 251
  Australe, 258
  Crisium, 266, 283
  Foecunditatis, 266
  Frigoris, 284
  Humboldtianum, 275
  Humorum, 265, 283
  Imbrium, 270, 278, 279, 283, 284
  Marginis, 266
Mare (*cont.*)
  Nectaris, 259, 282
  Nubium, 265, 270, 272, 282
  Serenitatis, 252, 268, 276–77
  Smythii, 266
  Spumans, 266
  Tranquillitatis, 266, 268
  Undarum, 266
  Vaporum, 268
Mariner's Compass (Pyxis), 7, **113**
Mars, 116, 250, 292, **294**, 297, 301, 304
  "canals," 294
  polar caps of, 294
Maurolycus, 260–61
mean positions, 323
mean solar time, 323, 324
Megrez, 107
Melotte catalog, 125
Menkalinen, 147
Mensa, 114
Merak, 107
Mercator, 265
Mercury, **292–93**, 297–301
meridian, 322
Messier numbers, 125
  crater, 267
meteors, 306–7
Microscopium (Microscope), 114
Milk Dipper, 105, 109
Milky Way, 109, 110, 111, 112, 114–15, 126, 130, 139, 142–43, 144–45, 148, 156, 164, 188, 192, 202, 214, 222, 224, 236, 240, 242, 248
Mira, 120, 185
Mizar, 107, 108, 150–51, 153
Monoceros, **112**, 115, 188, 248
moon, 117, 250–84, 292
Moonwatch, 307
Mount Hadley, 283
mountain peaks, 251, 252
mountain ranges, 251, 252
Musca, **114**, 249

Neander, 259
Nebula, Andromeda, 158, 161, 206
  Crab, 126, 165
  Dumbbell, 178
  Helical, 228
  Horsehead, 248
  Horseshoe, 200–201, 249
  Lagoon, 225

Nebula (*cont.*)
Loop, 178–79
Network, 178–79
North American, 156
Omega, 200–201, 249
Orion, 188, 248
Owl, 150
Pelican, 156
Ring, 176
Saturn, 204
Swan, 200–201, 249
Trifid, 225
Whirlpool, 172
Witch Head, 189
Wreath, 178–79
nebulae, diffuse galactic, 115, 126
planetary, 126
Nemean Lion, 111
Neptune, 292, **297**
Net (Reticulum), 114
Network Nebula, 178–79
Newton, Sir Isaac, 311
Nonius, 261
noon, 322
Norma (et Regula), **114**, 115, 240, 249
North American Nebula, 156
North Star. *See* Polaris
Northern Coalsack, 115, 248
Northern Crown (Corona Borealis), 105, **108**, 175
Northern Fly, 109
nova(e), 119, 238
nova-like variables, 119
nutation, 325, 331, 332

objective lens, 309
Oceanus Procellarum, 251, 272, 280
Octans (Octant), **114**, 246, 247
Old Bag, 115
"old moon in the new moon's arms," 251
Omega Centauri, 198
Omega Nebula, 200–201, 249
open (star) clusters, 108, 115, 125, 142, 162, 170–71, 190, 214, 222, 238
Ophiuchus, 105, **111**, 115, 198, 200, 201, 202, 223, 249
opposition, 292
Orion, 105, 107, 109, **112**, 186, 188, 189, 248
Belt of, 105, 112
Orion Family, 112
Orion Nebula, 188, 248
Orús, 282
Owl Nebula, 150

Pallas, 304
Palus, 251
Palus Epidemiarum, 265
Parenago, P. P., 118
Parry, 270
partial eclipse, 291
Pavo (Peacock), 105, **113**, 243, 244
Pegasus, 104, **110**, 111, 112, 113, 158
Square of, 104, 159, 182, 329
Pelican Nebula, 156
pentaprism, 287–88
penumbrae, 286
periodic comets, 306
periods of revolution, 125
Perseus, 104, **110**, 115, 144–45, 160, 163, 164
Perseus Family, 110–11
Phecda, 107
Phoenix, 105, **113**, 231
Photographic Atlas Charts, 138–249
Photography, 156, 317–21
Piccolomini, 282
Pickering, W. H., 267
Pico, 279, 284
Pictor, **114**, 235
Pisces, 104, **109**, 112, 113, 182, 184
Knot of, 104, 184
Piscis Austrinus, 105, **113**, 229
Pitatus, 282
Piton, 279, 284
planet(s), 116, 292–301
inferior and superior, 292
planetary nebulae, 115, 126
Plato, 254, 278, 284
Pleiades, 7, 109, 110, 162
Plinius, 283
Plow, 107
Pluto, 292, **297**, 304
Pointers, 104, 108, 327
polar axis, 316
polar caps of Mars, 294
Polaris, 6, 104, 105, 108, 110, 140, 153, 317, 331
Pole Star. *See* Polaris
Pollux, 105, 110
Pope Gregory, 322
Posidonius, 277
position angles, 125

Praesepe, 110, 166
precession of the equinoxes, 325, 329, 331
prime vertical, 3–4
Prinz, 281
Procyon, 104, 105, 111
prominences, solar, 289–90
Proxima Centauri, 116, 241
Ptolemaeus, 270, 283
Ptolemy, 7, 117
Puppis, 7, **113**, 115, 212, 214–15, 235
Pyxis, 7, **113**

quadratures, eastern and western, 292

rabbit, 250
radar meteors, 307
Ram (Aries), 104, **109**, 113, 161
Ramsden, 265
Rasalhague, 117
rays, 252
reflection, 311
refractors, 310–11
Regula. *See* Norma
Regulus, 111
repeating novae, 119
resolving power, 312
Reticulum, 114
revolution of the earth, 322
Rhacticus, 269
Rheita, crater, 258–59
  Valley, 258
Riccioli, 272
Rift, 115, 202
Rigel, 105, 112, 113
right ascension, 117, 329, 331
Rigil Kent, 105, 116, 240
rills, 252
ring formations, 252
Ring Nebula, 176
Rings of Saturn, 296–97
River (Eridanus), 105, 110, 111, **113**, 186, 189, 209, 210–11, 232, 233
rolling plains on moon, 251
Rosenberger, 260
rotation of the earth, 322
Ruler (Norma), 114

Sagitta, 105, **111**, 202
Sagittarius, 105, **109**, 111, 112, 114, 115, 200, 224, 225, 226, 249
Sails (Vela), 7, **113**, 115, 214, 217, 236–37
Santbech, 259
satellites, man-made, 307, 319–20
  of Jupiter, 296
  of Mars, 294
Saturn, 292, **296–97**, 301
  Nebula, 204
  rings of, 296–97
Scales (Libra), 105, **109**
Schickard, 264
Schlesinger, Frank, 116
Schmidt, 276
Schmidt camera, 311
Schröter's Valley, 281
Scorpius (Scorpion), 105, **109**, 111, 112, 115, 222–23, 248, 249
Sculptor (Sculptor's Apparatus), 114
Scutum, **111**, 115, 200, 202
Sea Goat (Capricornus), **109**, 112
sea horse. *See* Equuleus
Sea Monster or Sea Serpent (Cetus), 104, **110**, 113, 183
Sea Serpent (Hydra), **111**, 190, 193, 218–19, 221
semiregular variables, 120
separation, 125, 312
Serpens (Serpent), **111**, 198–99, 201
Serpent Holder (Ophiuchus), 105, **111**, 115, 198, 200, 201, 202, 223, 249
Sextans (Sextant), **111**, 192
Shapley, Harlow, 125, 131
Shield (Scutum), **111**, 115, 200, 202
shooting star, 306
Sickle, 104, 109
sidereal time, 324–29
Sinus, 252
  Iridium, 284
Sirius, 105, 112, 117, 212–13
Small Magellanic Cloud, 105, 113, 115, 230
Smaller Dog (Canis Minor), 104, **112**
Smaller Lion (Leo Minor), **109**, 168–69
smoked glass, 286
solar: diagonal, 287
  eclipse, 290, 291
  prominences, 289–90
  surface, 286
  time, 322–23

solar (*cont.*)
wind, 304
Sosigenes, 322
crater, 269
south celestial pole, 113, 114, 246
Southern Cross (Crux), 106, **112**, 115, 238, 249
Southern Crown (Corona Australis), **112**, 113, 227
Southern Fish (Piscis Austrinus), 105, **113**, 229
Southern Triangle (Triangulum Australe), 105, 106, **112**
spectral type, 118
spectrum, 118
Spica, 105, 196
spiral, 131
Square of Pegasus, 104, 159, 182, 329
Stadius, 271
Standard: Longitude Difference, 323
Mean Time, 7, 323
Time, 323
time zones, 323
star clock, 326, 327–29
star clusters, globular, 115, 125, 174, 198, 222, 224–25, 230, 238, 242–43
open, 108, 115, 125, 142, 162, 170–71, 190, 214, 222, 238
star time, 324–29
stars, 116–31
eclipsing, 119
fixed, 292
flare, 119
novae (new), 119, 238–39
variable, 6, 118–19, 120, 185
wandering, 116, 292
Steavenson, 259
Stern (Puppis), 7, **113**, 115, 212, 214–15, 235
Stevinus, 259
Stöfler, 261
Straight Range, 279
Straight Wall, 283
String of Pearls, 112
Stymphalian Birds, 111
sun, 117, 285–91, 292
"sundial" time, 324
sunspot(s), 285–86, 288
cycle, 289–90
superior: conjunction, 292
planet, 292
surface temperature, 118
Swan (Cygnus), 104, 105, **111**, 115, 156–57, 178–79, 181, 202, 248
Swan Nebula, 200–201, 249
synodic period, 292
Syntaxis, The, 117

Table Mountain (Mensa), 114
Taruntius, 283
Taurus, 105, **109**, 112, 163, 165, 186, 187
Telescope(s), 309–16, 317, 332. *See also* Telescopium
Telescopium, 114
Teneriffe Mountains, 279, 284
terminator, 252
Theophilus, 282
Thuban, 153
Time, 322–32
apparent solar, 324
Daylight, 324
rules for computing, 323–24
sidereal, 324–29
Standard, 323
Standard Mean, 7, 323
Timocharis, 284
tortoise. *See* Lyra
total solar eclipse, 290, 291
Toucan (Tucana), 105, **113**, 198, 230–31
Trapezium, 189
Triangulum (Triangle), 104, **111**, 160, 161
Triangulum Australe, 105, 106, **112**
Triesnecker, 269
Trifid Nebula, 225
Tucana, 105, **113**, 198, 230–31
"twinkling," 292, 312
Twins (Gemini), 104, **110**, 115, 164
Tycho, 252, 263, 271, 282

umbrae, 286
Unicorn (Monoceros), **112**, 115, 188, 248
uniform sidereal time, 325
Universal Time, 323
Uranus, 292, **297**
Ursa Major, 7, 104, **107**, 108, 148–49, 150, 153, 168, 170
Ursa Major Family, 107–9
Ursa Minor, 105, **108**, 140, 141, 152, 153

valleys, 252

variable stars, 6, 118–19, 120, 185
Vaucouleurs, Gérard de, 294
Vega, 104, 111, 154, 177, 331
Vela, 7, **113**, 115, 214, 217, 236, 237
Venus, 117, 250, 292, **293**, 297, 301
Venus' Mirror, 112
vernal equinox, 117, 325
Vesta, 304
Virgo (Virgin), **109**, 111, 194, 195, 196, 197
Vlacq, 260
Volans, **113**, 235
Vorontsov-Velyaminov, Boris A., 130
Vulpecula, 111
Vultur (Vulture), 105, 111

Wagon, 107
Wain, 107
Wallace, 271
walled plains, 252
Walter, 282
wandering stars, 116, 292
Wargentin, 254, 264
"washbowl," 279
water, 304
Water Carrier (Aquarius), 104, 105, **109**, 112, 113, 205, 228
Water Snake (Hydrus), 113
Wazn, 240
Werner, 282
western: elongation, 292, 298
 quadrature, 292
Whale (Cetus), 104, **110**, 113, 183, 185
Whirlpool Nebula, 172
Witch Head Nebula, 189
Wolf (Lupus), 106, **112**, 115, 221, 223, 241
Wreath Nebula, 178–79

year, 322
Yerkes, 283

zero meridian, 323
Zodiac, 109
Zodiacal Family, 109–10

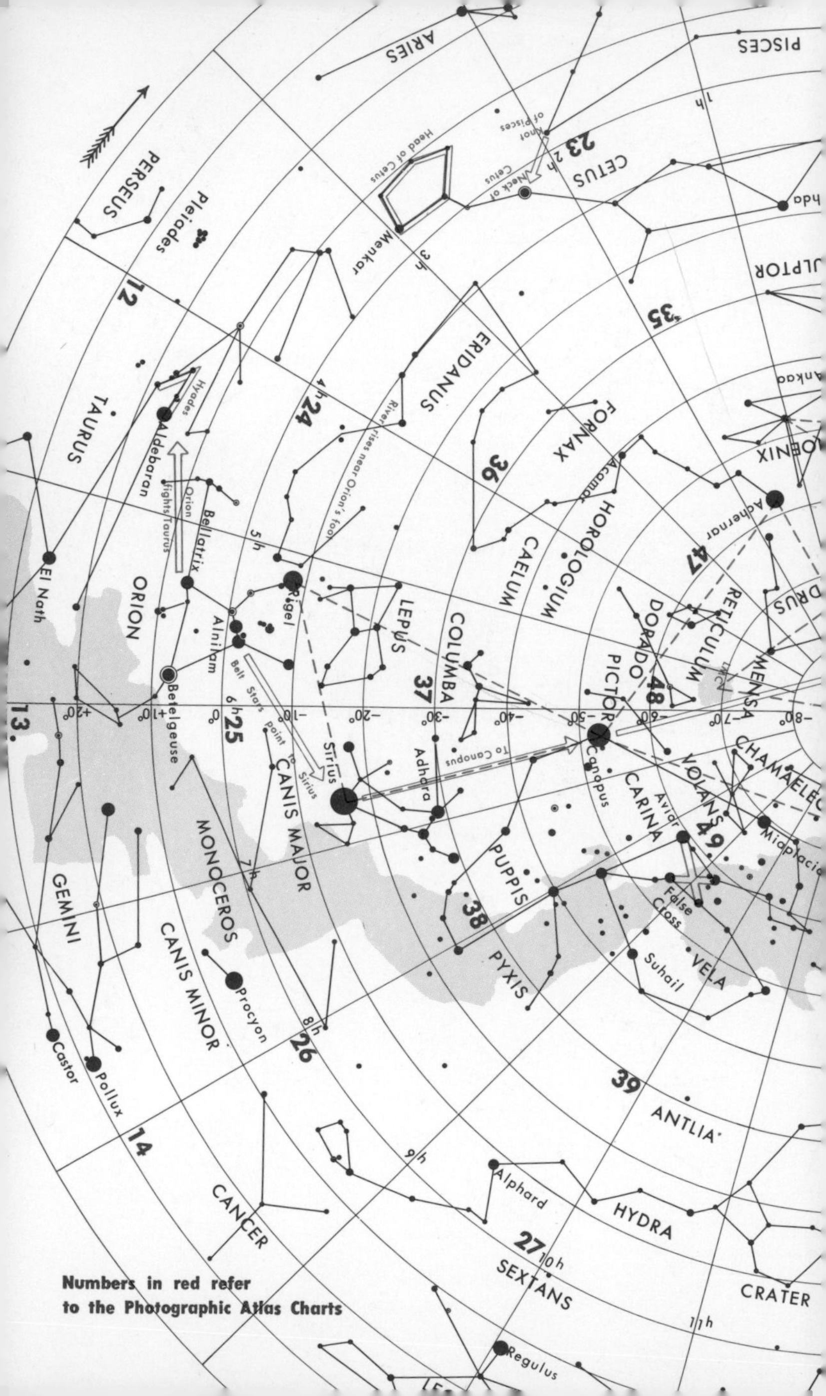

Numbers in red refer to the Photographic Atlas Charts

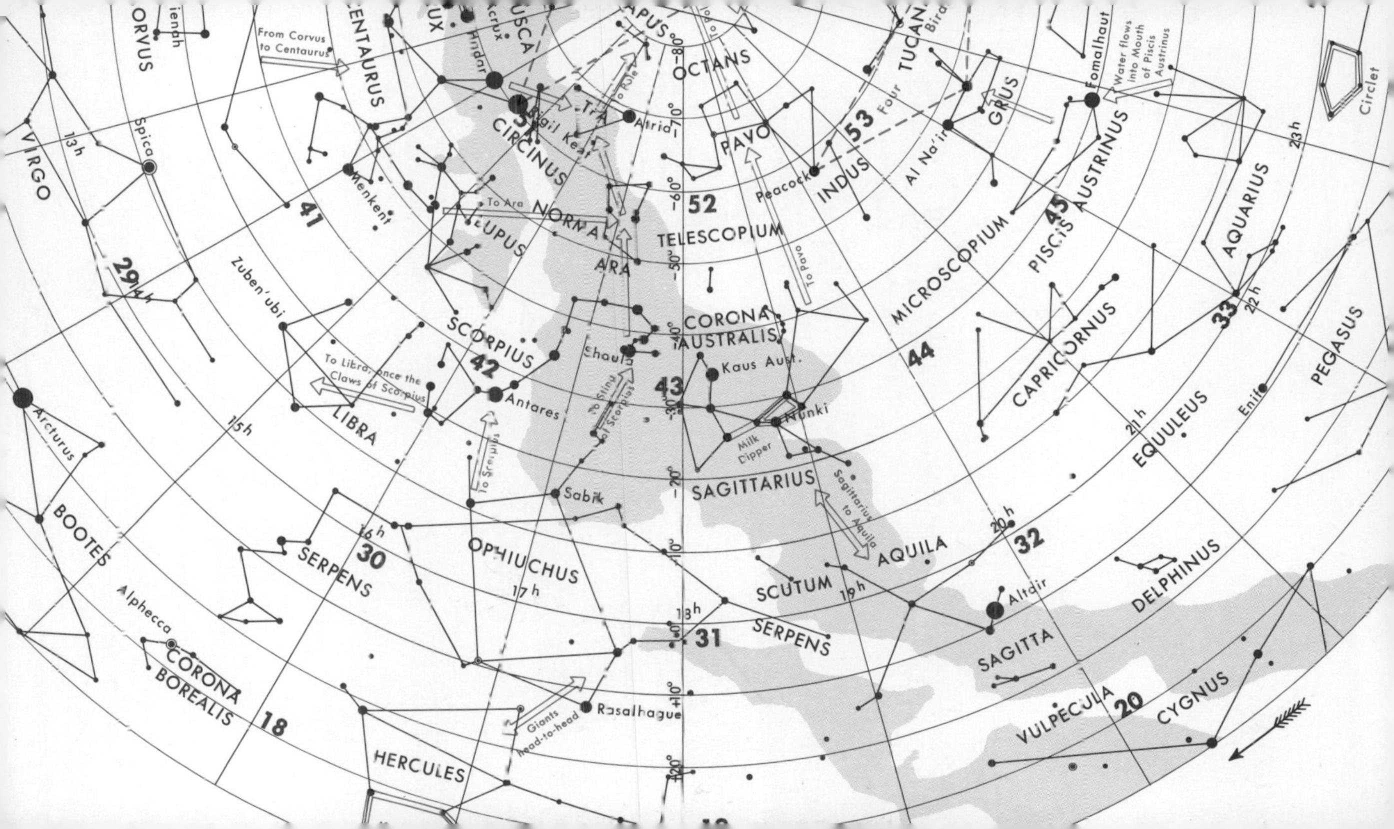

VIRGO
Spica
CORVUS
Gienah
CENTAURUS
From Corvus to Centaurus
Menkent
41
29
13h
14h
Zuben'ubi
LIBRA
To Libra, once the Claws of Scorpius
15h
Arcturus
BOOTES
Alphecca
CORONA BOREALIS
18
SERPENS
30
16h
OPHIUCHUS
17h
HERCULES
Giants head-to-head
Rasalhague
Sabik
To Scorpius
SCORPIUS
42
Antares
Shaula
To Sting of Scorpius
LUPUS
To Ara
NORMA
CIRCINUS
Rigil Kent
TrA
Atria
To Pole
Hadar
Acrux
MUSCA
APUS
OCTANS
To Pole
PAVO
Peacock
To Pavo
INDUS
53
Four
TUCANA
Bird
GRUS
Al Na'ir
52
TELESCOPIUM
ARA
CORONA AUSTRALIS
Kaus Aust.
43
Nunki
Milk Dipper
SAGITTARIUS
Sagittarius to Aquila
AQUILA
SCUTUM
19h
18h
31
SERPENS
44
MICROSCOPIUM
PISCIS AUSTRINUS
45
Fomalhaut
Water flows into Mouth of Piscis Austrinus
AQUARIUS
23h
Circlet
PEGASUS
Enif
33
22h
CAPRICORNUS
21h
EQUULEUS
32
20h
Altair
DELPHINUS
SAGITTA
VULPECULA
20
CYGNUS
-80°
-70°
-60°
-50°
-40°
-30°
-20°
-10°
0°
+10°
+20°